Molecular Mechanisms in Plant Stress Tolerance

Molecular Mechanisms in Plant Stress Tolerance

Guest Editor

Sajid Ali

MDPI

Basel • Beijing • Wuhan • Barcelona • Belgrade • Novi Sad • Cluj • Manchester

Guest Editor
Sajid Ali
Department of Horticulture
and Life Science
Yeungnam University
Gyeongsan
Republic of Korea

Editorial Office
MDPI AG
Grosspeteranlage 5
4052 Basel, Switzerland

This is a reprint of the Special Issue, published open access by the journal *Current Issues in Molecular Biology* (ISSN 1467-3045), freely accessible at: https://www.mdpi.com/journal/cimb/special_issues/2STDI2DF64.

For citation purposes, cite each article independently as indicated on the article page online and as indicated below:

Lastname, A.A.; Lastname, B.B. Article Title. *Journal Name* **Year**, *Volume Number*, Page Range.

ISBN 978-3-7258-7565-8 (Hbk)
ISBN 978-3-7258-7566-5 (PDF)
https://doi.org/10.3390/books978-3-7258-7566-5

Contents

About the Editor

Sajid Ali

Sajid Ali is affiliated with the Department of Horticulture and Life Science at Yeungnam University, Republic of Korea. His research interests include microbial biotechnology, plant–microbe interactions, endophytic fungi and bacteria, microbial plant biostimulants, abiotic stress mitigation, stress tolerance in plants, and sustainable agriculture. His recent scholarly work has focused on how beneficial microbes improve plant resilience under salinity and other abiotic stresses, as well as on the molecular communication of microbial biostimulants in the rhizosphere and their role in crop productivity and postharvest quality. Through his publications and editorial leadership in plant stress biology and crop resilience, he has contributed to advancing both molecular understanding and practical strategies for sustainable agriculture.

current issues in molecular biology

MDPI

Editorial

Molecular Basis of Plant Stress Tolerance: Current Status and Future Perspectives

Sajid Ali

Department of Horticulture and Life Science, Yeungnam University, Gyeongsan 38541, Republic of Korea; drsajid@yu.ac.kr

1. Introduction

I write this interim editorial to record my heartfelt appreciation to all participants and to express sincere gratitude to the authors, reviewers, and editorial staff of *Current Issues in Molecular Biology*, whose intellectual engagement and professional dedication have been pivotal to the success of my Special Issue "*Molecular Mechanisms in Plant Stress Tolerance*".

Plants are continuously exposed to a spectrum of biotic and abiotic stressors, including drought, salinity, extreme temperatures, heavy metals, and pathogen invasion [1,2]. As sessile organisms, they have evolved highly coordinated molecular systems to perceive environmental cues and orchestrate adaptive responses [3,4]. Plant stress biology is transitioning from lists of parts to a clear picture of how signals work together. When plants face drought, heat, or pathogens, they integrate redox cues from reactive oxygen species (ROS) and reactive nitrogen species (RNS), carefully timed calcium ion (Ca^{2+}) pulses, and phosphorylation cascades driven by mitogen-activated protein kinases (MAPKs), calcium-dependent protein kinases (CDPKs), and the calcineurin B-like-CIPK system (CBL-CIPK). These layers interact with hormone pathways, abscisic acid (ABA), salicylic acid (SA), jasmonic acid (JA), and ethylene (ET), while transcription factors translate the combined signals into gene expression programs that enhance antioxidant defenses, maintain protein quality, produce osmolytes, and regulate ion transport. The outcome is a finely calibrated program that strengthens cellular defenses and restores homeostasis while sustaining growth and development [5–7].

Despite significant progress, key questions remain—particularly regarding how concurrent pathways are coordinated and remembered. Many pathways are activated simultaneously, share key components, and can "remember" prior stress, which challenges simple, linear explanations. To close these gaps, we need single-cell and subcellular measurements of Ca^{2+} and ROS/RNS dynamics, quantitative phosphoproteomics linked to causal perturbations, and models that generate testable predictions—all reported with findable, accessible, interoperable, and reusable (FAIR) metadata [8]. With this network-aware, quantitative approach, breeders and biotechnologists can more rationally combine traits—modulating hormone sensitivity, antioxidant capacity, and transport processes—to produce climate-resilient crops without compromising yield or quality [9–11]. Additionally, epigenetic modifications—including DNA methylation, histone acetylation, and non-coding RNAs—contribute to the establishment of "stress memory," enabling plants to mount faster and more effective responses to recurrent stresses. Mechanisms that sustain protein homeostasis, such as chaperone-assisted folding, ubiquitin–proteasome degradation, and autophagy, further reinforce cellular resilience needs to be considered [12,13].

Advancements in omics technologies have revolutionized plant stress biology. Transcriptomic and proteomic analyses now enable high-resolution mapping of stress-induced gene networks, while metabolomics provides functional evidence of altered biochemical fluxes [14]. Integrating transcriptomic, proteomic, and metabolomic datasets through systems biology and computational modeling offers a holistic view of the molecular networks governing stress tolerance [15,16]. Multi-omics analyses reveal how genes, metabolites, and proteins interact under combined stresses. Metabolomic and proteomic studies in crops identify flavonoids, and antioxidative enzymes that are essential for stabilization and detoxification. Gene-editing tools, including CRISPR/Cas, have confirmed key regulators that enhance water-use efficiency, photosynthetic stability, and yield. Functional genomics and genome editing approaches have validated several regulators critical under stress conditions [17,18].

Despite these advances, several challenges persist. Translating molecular insights from model systems to crop improvement requires bridging genotype–environment interactions under field conditions. Many molecular discoveries arise from single-stress experiments, whereas real-world environments impose multifactorial stresses that act simultaneously. Priorities include rigorously designed combined- and sequential-stress studies with defined dose, rate, and order, coupled with continuous in situ monitoring via high-throughput phenotyping and remote sensing to connect cellular events (ROS/RNS, Ca^{2+}, MAPKs, CDPKs, CBL–CIPK) to organ- and canopy-level performance. Another frontier lies in systems-level integration—linking gene expression, metabolic shifts, and physiological performance through machine learning and network modeling. Such integrative frameworks can identify key regulatory nodes and predict adaptive outcomes. In addition, epigenome engineering and synthetic biology offer new tools to design stress-responsive circuits that dynamically fine-tune gene expression. Finally, sustainable crop improvement demands a translational pipeline that couples molecular breeding with agronomic validation. The integration of omics-based biomarkers, precision editing, and microbiome-based strategies holds promise for achieving resilience in the face of global climate change.

Conflicts of Interest: The author declares no conflicts of interest.

References

1. Mircea, D.-M.; Boscaiu, M.; Sestras, R.E.; Sestras, A.F.; Vicente, O. Abiotic stress tolerance and invasive potential of ornamental plants in the Mediterranean area: Implications for sustainable landscaping. *Agronomy* **2024**, *15*, 52. [CrossRef]
2. Du, B.; Haensch, R.; Alfarraj, S.; Rennenberg, H. Strategies of plants to overcome abiotic and biotic stresses. *Biol. Rev.* **2024**, *99*, 1524–1536. [CrossRef] [PubMed]
3. Novoplansky, A.; Souza, G.M.; Brenner, E.D.; Bhatla, S.C.; Van Volkenburgh, E. Exploring the complex information processes underlying plant behavior. *Plant Signal. Behav.* **2024**, *19*, 2411913. [CrossRef] [PubMed]
4. Han, R.; Ma, L.; Terzaghi, W.; Guo, Y.; Li, J. Molecular mechanisms underlying coordinated responses of plants to shade and environmental stresses. *Plant J.* **2024**, *117*, 1893–1913. [CrossRef] [PubMed]
5. Nykiel, M.; Gietler, M.; Fidler, J.; Prabucka, B.; Rybarczyk-Płońska, A.; Graska, J.; Boguszewska-Mańkowska, D.; Muszyńska, E.; Morkunas, I.; Labudda, M. Signal transduction in cereal plants struggling with environmental stresses: From perception to response. *Plants J.* **2022**, *11*, 1009. [CrossRef] [PubMed]
6. Devireddy, A.R.; Zandalinas, S.I.; Fichman, Y.; Mittler, R. Integration of reactive oxygen species and hormone signaling during abiotic stress. *Plant J.* **2021**, *105*, 459–476. [CrossRef] [PubMed]
7. Tarkowski, Ł.P.; Signorelli, S.; Considine, M.J.; Montrichard, F. Integration of reactive oxygen species and nutrient signalling to shape root system architecture. *Plant Cell Environ.* **2023**, *46*, 379–390. [CrossRef] [PubMed]
8. Dumschott, K.; Dörpholz, H.; Laporte, M.-A.; Brilhaus, D.; Schrader, A.; Usadel, B.; Neumann, S.; Arnaud, E.; Kranz, A. Ontologies for increasing the FAIRness of plant research data. *Front. Plant Sci.* **2023**, *14*, 1279694. [CrossRef] [PubMed]
9. Kurepa, J.; Smalle, J. The Evolution of Plant Hormones: From Metabolic Byproducts to Regulatory Hubs. *Int. J. Mol. Sci.* **2025**, *26*, 7190. [CrossRef] [PubMed]
10. Feng, Y.; Xia, P. Heat Shock Transcription Factors as Integrative Hubs for Plant Stress Adaptation: Decoding Regulatory Networks Toward Climate-Resilient Crop Design. *Plant Cell Environ.* **2025**. [CrossRef] [PubMed]

11. Thakur, G.; Singh, P.; Sharma, V.; Sharma, A.; Singh, J.; Kumar, S. Phytohormonal Crosstalk with Flowering Genes Regulating Drought Stress Response in Citrus: A Systematic Review. *J. Plant Growth Regul.* **2025**, 1–20. [CrossRef]
12. Zhou, W.; Wang, M.; Wang, L.; Liu, Y.; Tian, Z.; Xie, L.; Wang, Y. Epigenetics in Plant Response to Climate Change. *J. Biol.* **2025**, *14*, 631. [CrossRef] [PubMed]
13. Kumar, S.; Mohapatra, T. Epigenetic modifications in genome help remembering the stress tolerance strategy adopted by the plant. *Front. Biosci. (Landmark Ed.)* **2024**, *29*, 126. [CrossRef] [PubMed]
14. Tu, M.; Du, C.; Yu, B.; Wang, G.; Deng, Y.; Wang, Y.; Chen, M.; Chang, J.; Yang, G.; He, G. Current advances in the molecular regulation of abiotic stress tolerance in sorghum via transcriptomic, proteomic, and metabolomic approaches. *Front. Plant Sci.* **2023**, *14*, 1147328. [CrossRef] [PubMed]
15. Varadharajan, V.; Rajendran, R.; Muthuramalingam, P.; Runthala, A.; Madhesh, V.; Swaminathan, G.; Murugan, P.; Srinivasan, H.; Park, Y.; Shin, H. Multi-Omics Approaches Against Abiotic and Biotic Stress—A Review. *Plants* **2025**, *14*, 865. [CrossRef] [PubMed]
16. Saleem, M.H.; Noreen, S.; Ishaq, I.; Saleem, A.; Khan, K.A.; Ercisli, S.; Anas, M.; Khalid, A.; Ahmed, T.; Hassan, A. Omics technologies: Unraveling abiotic stress tolerance mechanisms for sustainable crop improvement. *J. Plant Growth Regul.* **2025**, 1–23. [CrossRef]
17. Chen, F.; Chen, L.; Yan, Z.; Xu, J.; Feng, L.; He, N.; Guo, M.; Zhao, J.; Chen, Z.; Chen, H. Recent advances of CRISPR-based genome editing for enhancing staple crops. *Front. Plant Sci.* **2024**, *15*, 1478398. [CrossRef] [PubMed]
18. Mascarenhas, M.S.; Nascimento, F.d.S.; Schittino, L.M.P.; Galinari, L.B.; Lino, L.S.M.; Ramos, A.P.d.S.; Diniz, L.E.C.; Mendes, T.A.d.O.; Ferreira, C.F.; Santos-Serejo, J.A.d. Construction and Validation of CRISPR/Cas Vectors for Editing the PDS Gene in Banana (*Musa* spp.). *Curr. Issues Mol. Biol.* **2024**, *46*, 14422–14437. [CrossRef] [PubMed]

current issues in molecular biology

MDPI

Article

Overexpression of the CAM-Derived NAC Transcription Factor KfNAC83 Enhances Photosynthesis, Water-Deficit Tolerance, and Yield in *Arabidopsis*

Kumudu N. Rathnayake [1], Beate Wone [1], Madhavi A. Ariyarathne [1], Won C. Yim [2] and Bernard W. M. Wone [1,*]

[1] Department of Biology, University of South Dakota, Vermillion, SD 57069, USA
[2] Department of Biochemistry and Molecular Biology, University of Nevada Reno, Reno, NV 89577, USA
* Correspondence: bernie.wone@usd.edu

Abstract

Drought stress is a major constraint on plant photosynthesis, growth, and yield, particularly in the context of increasingly frequent and severe extreme weather events driven by global climate change. Enhancing photosynthetic efficiency and abiotic stress tolerance is therefore essential for sustaining crop productivity. In this study, we functionally characterized *Kalanchoë fedtschenkoi* NAC83 (*KfNAC83*), a transcription factor derived from a heat-tolerant obligate crassulacean acid metabolism (CAM) species, by constitutively overexpressing it in the C_3 model plant *Arabidopsis thaliana*. Transgenic *Arabidopsis* lines overexpressing *KfNAC83* exhibited significantly enhanced tolerance to water-deficit and NaCl stress, along with improved photosynthetic performance, biomass accumulation, and overall productivity. Transcriptomic analysis revealed that *KfNAC83* overexpression increased key components of the jasmonate (JA) signaling pathway in both roots and shoots, suggesting a mechanistic link between *KfNAC83* activity and enhanced abiotic stress responses. Additionally, the transgenic lines displayed increased nighttime decarboxylation activity, indicative of partial CAM-like metabolic traits. These findings demonstrate that *KfNAC83* functions as a positive regulator of abiotic stress tolerance and growth, likely through modulation of jasmonate-mediated signaling and photosynthetic metabolism. This work highlights the potential of CAM-derived transcription factors for bioengineering abiotic stress-resilient crops in the face of climate change.

Keywords: NAC transcription factor; crassulacean acid metabolism (CAM); abiotic stress; carbon assimilation; jasmonate signaling pathway

1. Introduction

Two pressing challenges confronting modern society are the growing global human population [1,2] and the increasing frequency and severity of drought events associated with global climate change [3]. Drought stress poses a significant threat to global agricultural productivity, jeopardizing the supply of food, fiber, and feed [4]. Addressing this challenge requires innovative strategies to enhance crop resilience and sustainability under environmental stress.

One promising approach involves bioengineering the crassulacean acid metabolism (CAM) pathway into C_3 crops to improve water-use efficiency (WUE), particularly for cultivation on marginal lands [5–7]. However, due to the complexity of the CAM pathway, alternative strategies such as manipulating transcription factor (TF) expression offer a more tractable solution. This approach targets a small number of regulatory genes with broad downstream effects on abiotic stress-responsive networks [8,9].

Numerous studies have demonstrated the effectiveness of TF overexpression in enhancing abiotic stress tolerance. For example, overexpression of *EcbHLH57* from *Eleusine coracana* in tobacco improved tolerance to salt, oxidative, and water-deficit stress [10]; *AREB1*, an ABA-dependent TF from *Arabidopsis thaliana*, enhanced water-deficit tolerance in soybean [11]; *AtDREB1A* conferred improved water-deficit tolerance in transgenic *Indica* rice [12]; and *MYB37* increased ABA sensitivity and improved both water-deficit tolerance and seed productivity in *Arabidopsis* [13]. Additionally, members of the MYB, AP2/ERF, WRKY, NAC, and bZIP TF families have been implicated in adaptive responses to water-deficit stress [4,9,14,15].

An underexplored yet promising strategy is the identification of novel regulatory genes from extremophytes, plants adapted to extreme environments such as desiccation-tolerant species, resurrection plants, and CAM plants for use in bioengineering drought tolerance in crops [16]. CAM plants exhibit the highest WUE among all plant groups and possess exceptional tolerance to multiple abiotic stressors [17], making them ideal candidates for the discovery of abiotic stress-adaptive regulatory genes with translational potential [16]. To date, only a handful of studies have characterized such regulators from extremophytes for potential application in crop improvement [18–21].

NAC transcription factors (TFs) are plant-specific regulators known to play pivotal roles in abiotic stress responses [22,23]. However, NAC TFs from CAM species remain largely uncharacterized, despite the high abiotic stress resilience exhibited by these plants [17]. In this study, we investigated the function of *Kalanchoë fedtschenkoi* NAC83 (*KfNAC83*), a transcription factor from a heat-tolerant obligate CAM species [24,25], by constitutively overexpressing it in the C_3 model plant *Arabidopsis thaliana*. Notably, *KfNAC83* expression was previously shown to increase sevenfold during CAM induction in older leaf pairs of *K. fedtschenkoi*, relative to the C_3 state [16]. Our earlier work also provided preliminary evidence that *KfNAC83* enhances abiotic stress tolerance, WUE, and vegetative growth in *Arabidopsis*. These promising findings prompted a deeper investigation into the molecular mechanisms underlying these phenotypes, with the goal of identifying regulatory pathways that could be leveraged for crop improvement. Moreover, elucidating the function of *KfNAC83* in a C_3 background might offer insights into the regulatory transitions associated with the evolution of CAM photosynthesis.

2. Materials and Methods

2.1. Sequence Alignment and Phylogenetic Analysis

Protein sequences homologous to KfNAC83 were retrieved from the Plant Transcription Factor Database (PlantTFDB v4.0; http://planttfdb.cbi.pku.edu.cn/ accessed on 12 August 2025). A total of 62 NAC transcription factor protein sequences from various plant species were used to construct a phylogenetic tree. Initially, a Neighbor-Joining (NJ) tree was generated to provide a starting topology. This was followed by a heuristic search using the Nearest-Neighbor Interchange (NNI) algorithm to explore alternative topologies and identify the one with the highest likelihood. Maximum likelihood (ML) branch lengths were computed for each candidate topology, and the tree with the greatest overall likelihood was retained. Statistical support for the resulting tree was assessed using 1000 bootstrap replicates. Evolutionary distances were estimated using the Whelan and Goldman (WAG) model in MEGA12 software [26]. Multiple sequence alignment was performed using the MUSCLE algorithm, and conserved motifs were identified and analyzed within MEGA12. Nuclear localization signal (NLS) was predicted using cNLS mapper (https://nls-mapper.iab.keio.ac.jp/cgi-bin/NLS_Mapper_form.cgi accessed on 20 August 2025).

2.2. Plasmid Construction

The *KfNAC83* transcription factor was cloned into the pENTR/D-TOPO entry vector via topoisomerase I mediated ligation. The resulting entry clone containing *KfNAC83* with a stop codon was recombined into the binary vector pGWB415 (CaMV35S::3xHA-attR1-attR2-NOS terminator) using Gateway® LR Clonase™ II Enzyme Mix (Invitrogen, Carlsbad, CA, USA) [27]. For subcellular localization, *KfNAC83* without a stop codon was cloned into the binary vector pGWB405 (CaMV35S::attR1-attR2-sGFP-NOS terminator), which contains a C-terminal synthetic green fluorescent protein (sGFP) tag, using the same recombination system. Recombinant plasmids (35S::3xHA-*KfNAC83* and 35S::*KfNAC83*-sGFP) were transformed into *Escherichia coli* (NEB 10-beta competent cells, New England BioLabs, Ipswich, MA, USA), and plasmids were extracted using the QIAprep Spin Miniprep Kit (Qiagen, Hilden, Germany) according to the manufacturer's instructions. All constructs were verified by Sanger sequencing (GENEWIZ, South Plainfield, NJ, USA).

2.3. Floral Dipping and Generation of Homozygous Transgenic Lines

Recombinant plasmids were introduced into *Agrobacterium tumefaciens* strain GV3101 using the freeze–thaw method [28]. *Arabidopsis thaliana* (Col-0) plants were transformed with 35S::3xHA-*KfNAC83* and 35S::*KfNAC83*-sGFP constructs via the floral dip method [29]. T_0 and T_1 seeds were screened on full-strength Murashige and Skoog (MS) medium supplemented with 50 μg/mL kanamycin in a Percival LED-30HL1 growth chamber (Percival, Perry, IA, USA) at 22 °C under a 16/8 h light/dark cycle with 120–150 $\mu mol\ m^{-2}\ s^{-1}$ light intensity. T_2 lines were selected based on a 3:1 segregation ratio for kanamycin resistance. Four independent homozygous T_3 lines overexpressing *KfNAC83* (#6, #11, #23, and #24) were used for further experiments.

2.4. Subcellular Localization of 35S::KfNAC83-sGFP

Homozygous T_3 seedlings expressing 35S::*KfNAC83*-sGFP and control seedlings expressing 35S::sGFP were used to determine subcellular localization. Ten-day-old roots and leaves were stained with Fluoroshield™ containing DAPI (Sigma-Aldrich, St. Louis, MO, USA) for 10 min (roots) or 30–45 min (leaves) at room temperature. Samples were imaged using a Nikon A1 confocal laser-scanning microscope (Nikon, Tokyo, Japan) with 60× and 40× oil immersion objectives for leaves and roots, respectively. GFP and DAPI were excited at 488 nm and 405 nm, respectively.

2.5. Quantitative Real-Time PCR (qRT-PCR) Analysis of OxKfNAC83 Lines

Total RNA was extracted from 100 mg of leaf, root, and stem tissues of *OxKfNAC83* lines and WT using the RNeasy Plant Mini Kit (Qiagen, Hilden, Germany). qRT-PCR was performed using the Luna® Universal One-Step RT-qPCR Kit (New England Biolabs, Ipswich, MA, USA) following the manufacturer's protocol on a QuantStudio 3 Real-Time PCR System (Applied Biosystems, Foster City, CA, USA). Expression levels of *KfNAC83* were quantified using the primers 5′-CGGCATAGACCGCAAGATTA-3′ and 5′-CAGTCCTGGTTCCCTTGTTAG-3′. Expression was normalized to the reference gene *ACT2* (*At3G18780*) using primers 5′-CTACGAGCAGGAGATGGAAAC-3′ and 5′-TCTGAATCTCTCAGCACCAATC-3′ [30], yielding amplicon sizes of 104 bp and 107 bp, respectively. Each reaction was performed with three biological replicates and three technical replicates. Relative expression was calculated using the $2^{-\Delta\Delta CT}$ method [31] with QuantStudio™ Design and Analysis Software v1.5.0. Selected *OxKfNAC83* lines were used for downstream analyses.

2.6. Morphological Characterization of OxKfNAC83 Lines

OxKfNAC83 lines and WT were grown in soilless–perlite potting mixture (Scott, Miracle-Gro, Marysville, OH, USA) in 89 mm square pots (0.3 L rooting volume; Kord, Inc., Toronto, ON, Canada) under a 16/8 h light/dark photoperiod for four weeks. Rosette diameter was measured using a Vernier caliper. Detached rosettes were photographed with an Olympus TG-5 camera, and rosette area was quantified using ImageJ v1.51. Leaf number per rosette was also recorded.

For biomass measurements, sterilized seeds were sown on full-strength MS medium. After stratification, plates were transferred to a Percival LED-30HL1 growth chamber at 22 °C under a 16/8 h light/dark cycle with 120–150 μmol m^{-2} s^{-1} light intensity. Fresh weight was recorded for 3-week-old seedlings; dry weight was measured after drying at 60 °C for 24 h.

2.7. Water-Deficit Stress Assay

To assess water-deficit tolerance, *OxKfNAC83* lines and WT were grown in soilless medium in 89 mm square pots under a 16/8 h light/dark photoperiod. After 14 days of well-watered growth, water was withheld for 15 and 20 days. Plants were then re-watered for 7 days, and survival rates were calculated based on the number of green, healthy plants. After two months of recovery, seed yield per surviving plant was measured. Experiment was repeated three times with 30 plants per genotype.

2.8. In Vitro Water-Deficit Assay Using PEG

Water-deficit tolerance was also evaluated in vitro using polyethylene glycol (PEG 8000) to simulate osmotic stress. Seeds of *OxKfNAC83* lines (#6, #11, #23, and #24) and WT were surface sterilized using chlorine gas [32]. Seeds were sown on half-strength MS medium supplemented with 0%, 25%, 40%, or 55% PEG, corresponding to water potentials of −0.25 MPa (control), −0.5 MPa, −0.7 MPa, and −1.2 MPa, respectively [33]. Seeds were stratified in darkness at 4 °C for 3 days, then transferred to a Percival LED-30HL1 growth chamber at 22 °C under a 16/8 h light/dark cycle with 120–150 μmol m^{-2} s^{-1} light intensity. The experiment was repeated three times with 36 seedlings per genotype. Seed germination and green cotyledon emergence percentages were recorded.

2.9. Seed Yield Measurement of OxKfNAC83 Lines

For seed yield analysis, *OxKfNAC83* lines and WT plants were grown under well-watered conditions in soilless medium in 89 mm square plastic pots (0.3 L rooting volume; Kord, Inc., Toronto, ON, Canada) for two months in a growth chamber under a 16/8 h light/dark photoperiod. After seed maturation, total seed weight per plant was measured for 30 plants per line. The experiment was repeated three times. Seed yield was also compared with that from plants subjected to acute water-deficit treatments.

2.10. Water-Use Efficiency Measurement

Water-use efficiency (WUE) was assessed using a closed system as described by Wituszynska and Karpiński [34] and Wituszynska et al. [35]. Briefly, 50 mL conical tubes were filled with a 1:1 soil-perlite mixture and 35 mL of water. Each tube was capped with a lid containing a central hole, into which 5–10 seeds were placed. Tubes were kept at 4 °C in a humid environment for 2 days, then transferred to a growth chamber under a 16/8 h light/dark photoperiod. After 5–7 days of germination, excess seedlings were removed, leaving one seedling per tube. Initial tube weight (W_0) was recorded. After four weeks, rosettes were harvested and dried at 105 °C for 3 h to determine dry weight (DW). Final tube weight (W) was recorded, and water loss was calculated as $W_0 - W$ (g), assuming 1 g = 1 mL of water. WUE was calculated as DW (mg) per mL of water used.

2.11. In Vitro NaCl Stress Response

To assess salt stress tolerance, sterilized seeds of *OxKfNAC83* lines (#6, #11, #23, and #24) and WT were sown on full-strength MS medium supplemented with 0 (control), 100, 150, 200, or 250 mM NaCl. After 14 days, seed germination and green cotyledon emergence rates were recorded. The experiment was repeated three times with 30 seeds per line per replicate.

For post-germination salt stress assays, seedlings were germinated on full-strength MS medium for 5 days, then transferred to MS medium supplemented with 100, 150, or 200 mM NaCl for 14 days. Survival rate, fresh biomass, and dry biomass were measured for 24 seedlings per line across three replicates.

For root growth assays, five-day-old seedlings grown vertically on MS medium were transferred to MS medium supplemented with 100, 150, or 200 mM NaCl. Root elongation was measured at 3-day intervals for 12 days. After 12 days, seedlings were photographed using an Olympus TG-5 camera, and root length and lateral root number were quantified using ImageJ. Each treatment included 20 seedlings per line across three replicates. All experiments were conducted under the same growth chamber conditions described above.

2.12. Titratable Acidity

OxKfNAC83 lines and WT were grown for four weeks in a Percival LED-30HL1 growth chamber (Perry, IA, USA) under a 12 h photoperiod at 120–150 μmol m^{-2} s^{-1} light intensity and 22 °C. Approximately 0.5 g of leaf tissue was collected at dawn and dusk from each line (n = 3 biological replicates). Samples were flash-frozen in liquid nitrogen and stored at −80 °C. Frozen tissue was ground in liquid nitrogen and extracted with 10 mL of 50% (v/v) methanol at 80 °C for 10 min. Extracts were brought to original volume with water and centrifuged at 4000× g for 20 min (Eppendorf 5430 R, Hamburg, Germany). Supernatants were titrated to pH 7.0 with 100 mM KOH using a calibrated pH meter. Titratable acidity was calculated as (mL KOH × 0.1 M)/g fresh weight and expressed as μmol H^+ g^{-1} FW. Two-way ANOVA followed by Tukey's HSE test was used to assess statistical significance between lines and time points.

2.13. Carbohydrate Analysis

Total soluble sugar content was quantified in *OxKfNAC83* lines and WT grown in soilless medium for four weeks under a 12 h photoperiod at 120–150 μmol m^{-2} s^{-1} light intensity and 22 °C in a Percival LED-30HL1 growth chamber. The assay was performed as described by Fox and Robyt [36] and Lim et al. [37], with modifications. Briefly, 500 mg of leaf tissue was ground in liquid nitrogen and extracted with 10 mL of 50% (v/v) methanol at 80 °C for 30 min. After centrifugation at 3000× g for 10 min, 50 μL of 5% (v/v) phenol was added to an equal volume of supernatant. Samples were mixed gently and placed on ice. Then, 250 μL of concentrated sulfuric acid was added, and samples were incubated at 80 °C for 30 min. Absorbance was measured at 490 nm using a Multiskan™ GO microplate spectrophotometer (Thermo Scientific™, Waltham, MA, USA).

2.14. Photosynthetic and Carboxylation Efficiencies

OxKfNAC83-24 (Line #24) was selected for photosynthetic measurements due to its highest *KfNAC83* expression in leaves and stems and its use in time-series RNA-Seq analysis. Light response (A/I) curves were measured using a LI-6800 portable gas-exchange system with a fluorometer head (LI-COR Biosciences, Lincoln, NE, USA). Plants were dark-adapted for 30 min prior to measurement. Measurements were conducted at a constant CO_2 concentration of 400 µmol mol^{-1} using red-blue actinic light (90%/10%). Assimilation rate (A) and effective quantum yield of Photosystem II (ΦPSII) were recorded simultaneously using a rectangular saturating flash (8000 µmol m^{-2} s^{-1}) under stepwise irradiance (I) increments from 0 to 2000 µmol m^{-2} s^{-1} at 2 min intervals. For each plant, three leaves were measured, and the average was used per replicate. Data were fitted to nine mathematical models, and the model with the lowest sum of squared errors (Equation (2) from Kaipiainen [38] was selected as the best fit [39].

The response of net photosynthesis (A) to intercellular CO_2 concentration (Ci) was also measured using the LI-6800 system to determine carboxylation efficiency. Measurements were conducted at a saturating photosynthetic photon flux density (PPFD) of 200 µmol m^{-2} s^{-1}. CO_2 concentration (Ca) was initially set at 400 µmol mol^{-1}, then decreased stepwise to 300, 200, 100, and 50 µmol mol^{-1}, followed by increases to 100, 200, 300, 400, 500, 700, 900, 1000, and 1500 µmol mol^{-1}. Three leaves per plant were measured, and the average was used per replicate. CO_2 assimilation rate and Ci were calculated using the equations of von Caemmerer and Farquhar [40]. Maximum rates of Rubisco carboxylation (Vc_{max}), electron transport for RuBP regeneration (J_{max}), and triose phosphate utilization (TPU) were estimated at 25 °C using equations from Sharkey et al. [41].

2.15. Statistical Analysis

All phenotypic and physiological data were analyzed using one-way or two-way analysis of variance (ANOVA), followed by Tukey's honestly significant difference (HSE) test for multiple comparisons in RStudio (version 1.1383). Results are presented as mean ± standard deviation (SE) from at least three independent biological replicates. Statistical significance is indicated as *** $p < 0.001$, ** $p < 0.01$.

2.16. RNA Extraction for Water-Deficit and Time-Course Analyses

For water-deficit transcriptomic analysis, total RNA was extracted from 16 leaf samples (two genotypes: *OxKfNAC83-24* and WT × two treatments: well-watered and water-deficit × four biological replicates) using the RNeasy Plant Mini Kit (Qiagen, Hilden, Germany).

For time-course analysis, total RNA was extracted from 32 leaf samples (two genotypes × four time points: 09:00, 15:00, 21:00, and 03:00 h × four biological replicates) using the same kit. RNA quantity and quality were assessed using a Nanophotometer (Implen Inc., Westlake Village, CA, USA) and an Agilent 2100 Bioanalyzer (Agilent Technologies, Santa Clara, CA, USA).

2.17. Illumina Sequencing and Data Quality Control

All RNA samples were processed using the Illumina TruSeq RNA Sample Prep Kit for cDNA library construction, including poly(A) RNA purification, fragmentation, cDNA synthesis, and adaptor/barcode ligation. Sixteen libraries were sequenced on the Illumina NovaSeq 6000 platform (Novogene Corporation Inc., Sacramento, CA, USA) to generate >20 million 150 bp paired-end reads per sample. Quality control included trimming, adapter removal, and filtering of low-quality reads. High-quality reads were mapped to the *Arabidopsis thaliana* reference genome using HISAT2 with default parameters. Mapping results were visualized using the Integrative Genomics Viewer (IGV).

For the time-course experiment, 32 libraries from *OxKfNAC83-24* and WT across four time points were sequenced on the Illumina NovaSeq 6000 platform using 150-cycle paired-end sequencing. After filtering, >40 million paired-end reads were obtained per sample, with an average GC content of 45.91% and a sequencing error rate of 0.02%. Mapping rates ranged from 94.18% to 96.11%, with uniquely mapped reads ranging from 75.19% to 82.4%.

2.18. Identification of Differentially Expressed Genes (DEGs), Gene Ontology (GO), and Pathway Analysis

Normalized gene expression was calculated in fragments per kilobase of exon per million mapped reads (FPKM). Differential expression analysis was performed using DESeq2 v2_1.6.3. Genes with $\geq$2-fold change and FPKM $\geq$ 3 in at least one condition were considered differentially expressed. Volcano plots were generated to visualize DEG distributions.

Gene Ontology (GO) and Kyoto Encyclopedia of Genes and Genomes (KEGG) pathway enrichment analyses were conducted using the ClusterProfiler package to identify significantly enriched biological functions and pathways. Protein–protein interaction (PPI) enrichment analysis was performed using BioGRID (https://thebiogrid.org/ accessed on 30 July 2020). Networks containing 3–500 proteins were further analyzed using the Molecular Complex Detection (MCODE) algorithm via Metascape (https://metascape.org accessed on 30 July 2020). Resulting clusters were visualized in Cytoscape (v3.8.0).

3. Results

3.1. Phylogenetic Analysis of the KfNAC83 Gene

Phylogenetic analysis was performed using 62 NAC transcription factors (TFs) from various plant species to determine the evolutionary relationship of the KfNAC83 protein. The analysis revealed that KfNAC83 clusters closely with several known abiotic stress-responsive NAC proteins, including AtNAC35 which regulates cold stress response, OsNAC45 which is involved in drought and salt stress responses, and SlNAC2, which enhances abiotic stress tolerance through modulation of glutathione metabolism (Figure S1A). Multiple sequence alignment with characterized abiotic stress responsive NAC proteins, such as AaNAC1, AhNAC2, OsIGSNAC1, AmNAC1, BgNAC1, BnNAC1-1, CaNAC5, DgNAC1, GmNAC11 and HvNAC showed that these proteins share conserved motifs and contain a nuclear localization signal (NLS) within the A subdomain (Figure S1B).

3.2. KfNAC83 Localizes to the Nucleus in Root and Leaf Cells

To determine the subcellular localization of *KfNAC83*, *Arabidopsis* plants were transformed with a 35S::*KfNAC83*-sGFP construct and analyzed using confocal laser-scanning microscopy. The control 35S::sGFP protein localized to both the cytoplasm and nuclei of root and leaf cells, whereas the 35S::*KfNAC83*-sGFP fusion protein was exclusively localized to the nuclei (Figure S2).

3.3. KfNAC83 Is Highly Expressed in Leaf and Root Tissues

Four independent homozygous T_3 transgenic lines overexpressing *KfNAC83* (#6, #11, #23, and #24) under the CaMV 35S promoter were generated in *Arabidopsis thaliana*. Transcript levels of *KfNAC83* were quantified in leaf, root, and stem tissues using qRT-PCR. Expression was significantly higher in both leaves and roots, with no significant difference between these two tissues (Figure S3). In contrast, expression in stems was significantly lower.

3.4. Overexpression of KfNAC83 Enhances Growth, Water-Deficit Tolerance, WUE, and Seed Yield

To assess the impact of *KfNAC83* overexpression on growth, fresh and dry biomass were measured in *OxKfNAC83* lines and WT plants grown on MS medium for three weeks. *OxKfNAC83* lines exhibited a 2.5–3.0-fold increase in fresh biomass and a 3.3–3.9-fold increase in dry biomass compared to WT (Figure 1A–C). In four-week-old plants grown in soilless medium, rosette diameter and area were increased by 1.5–1.7-fold and 1.0–1.4-fold, respectively, in *OxKfNAC83* lines relative to WT (Figure 1D–F). Additionally, the number of leaves per rosette was 1.4–1.7-fold higher in *OxKfNAC83* lines (Figure 1G–H), indicating enhanced vegetative growth.

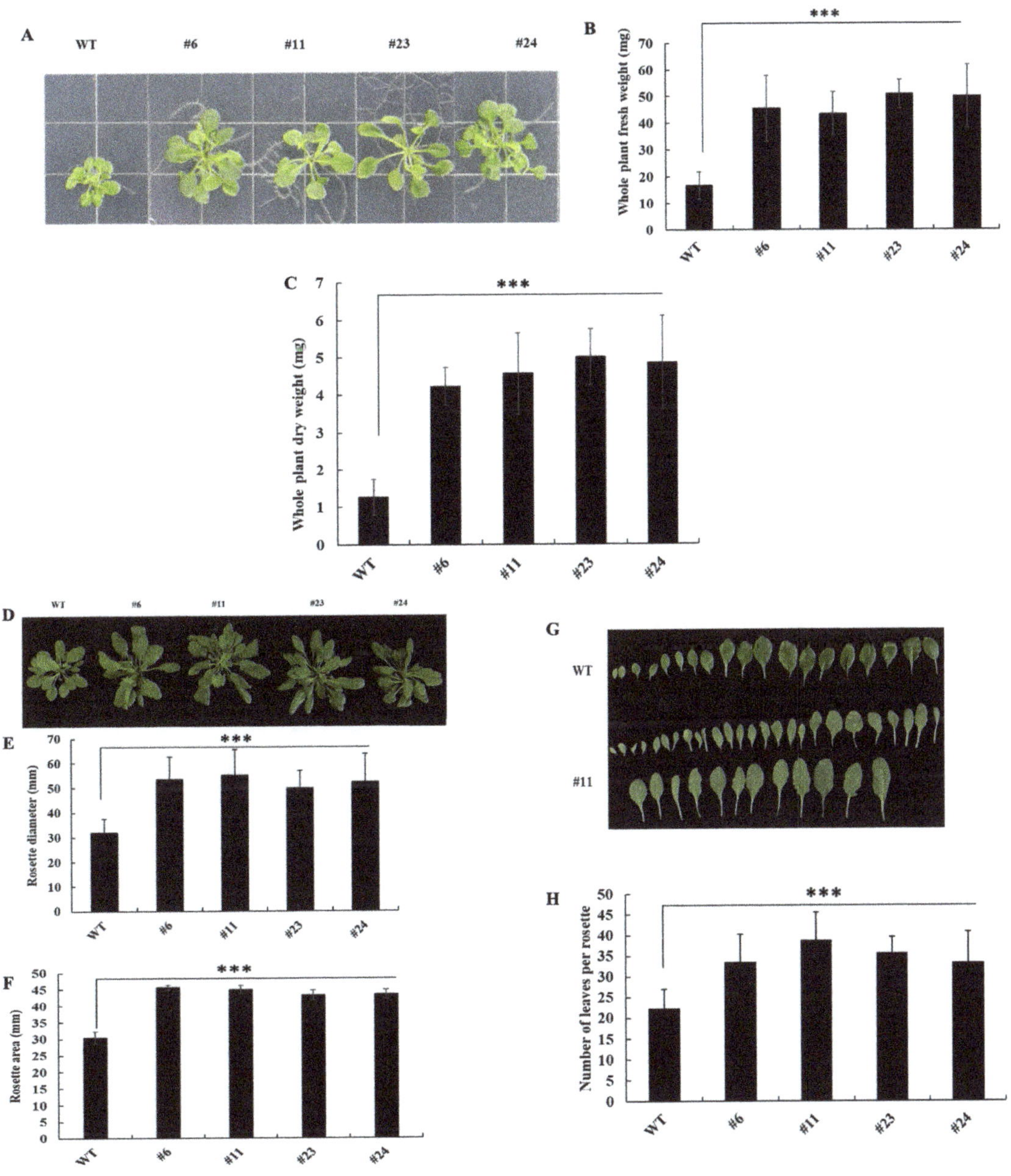

Figure 1. Overexpression of *KfNAC83* in *Arabidopsis* increases plant biomass. (**A**) T_2 homozygous seeds of transgenic lines and wild type (WT) were grown on MS basal medium for three weeks. (**B**) Whole-plant fresh weight and (**C**) dry weight (n = 20) were measured after 21 days of growth. (**D**) Representative images of four-week-old, detached rosettes from transgenic lines and WT. (**E**) Rosette diameter and (**F**) rosette area were quantified using ImageJ (n = 20). (**G**) Representative image of detached leaves from a four-week-old rosette of *KfNAC83*-overexpressing line (#11) and WT. (**H**) Number of leaves per rosette in transgenic lines and WT (n = 20). Values represent means ± SE. *** $p < 0.001$. WT = wild type; #6, #11, #23, and #24 = independent transgenic lines.

To evaluate water-deficit tolerance, *OxKfNAC83* transgenic lines and WT plants were subjected to acute water-deficit treatments lasting 15 and 20 days, following an initial 14 day period of well-watered growth. After a 7 day recovery period with re-watering, *OxKfNAC83* lines exhibited over 90% survival, whereas only 4–6% of WT plants recovered under either the 15 day (Figure 2A) or 20 day (Figure 2B,C) treatment conditions. These results demonstrate that overexpression of *KfNAC83* significantly enhances water-deficit tolerance in transgenic plants.

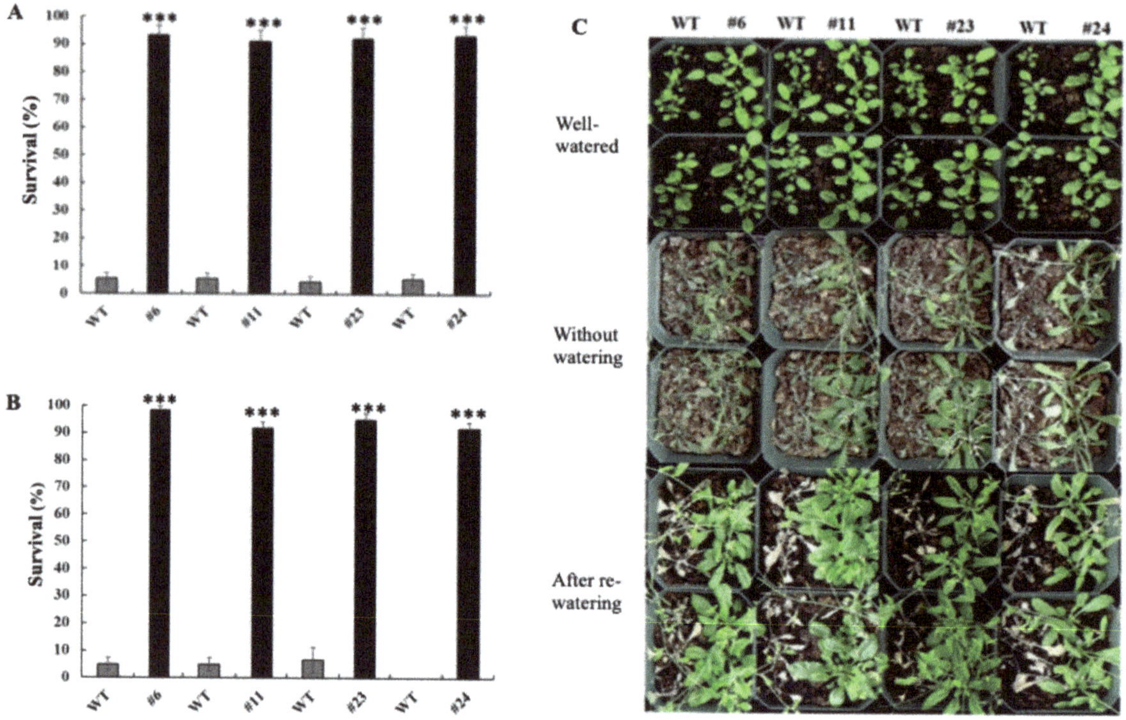

Figure 2. *KfNAC83*-overexpressing lines exhibit enhanced water-deficit tolerance. (**A**) Survival percentages following the 15 day water-deficit treatment. (**B**) Survival percentages following the 20 day water-deficit treatment. (**C**) Representative image from the acute water-deficit treatment, illustrating phenotypic differences between genotypes under well-watered, water-deficit (no watering), and re-watered conditions. Error bars represent ± SE from three independent experiments (n = 90). *** $p < 0.001$. WT = wild type; #6, #11, #23, and #24 = independent transgenic lines.

In vitro water-deficit tolerance was further assessed using PEG8000 to simulate low water potential. Under control conditions (0% PEG), both *OxKfNAC83* and WT seeds germinated at similar rates (Figure S4A). However, at 25% (−0.5 MPa) and 40% (−0.7 MPa) PEG, WT germination and green cotyledon emergence were significantly reduced compared to *OxKfNAC83* lines (Figure S4A,B). No germination occurred at 55% PEG. These results indicate that *OxKfNAC83* seeds exhibit enhanced tolerance to water-deficit stress during germination.

To determine whether water-deficit tolerance affected reproductive output, seed yield per plant was measured after two months of recovery from water-deficit stress. *OxKfNAC83* lines produced significantly more seeds than WT under both water-deficit and well-watered conditions (Figure 3A), indicating that *KfNAC83* enhances water-deficit resilience without compromising productivity.

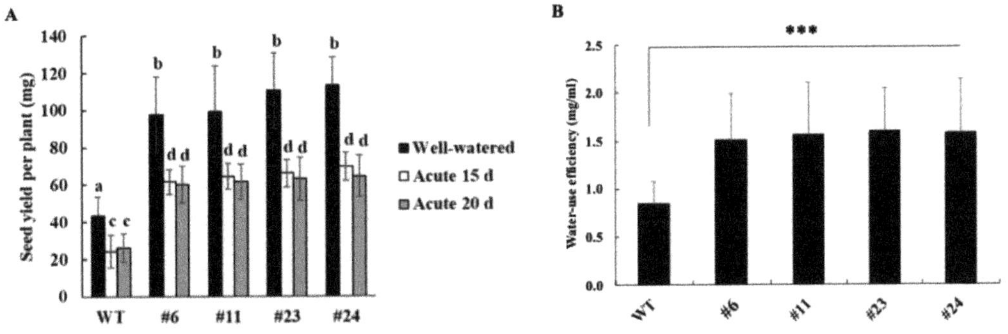

Figure 3. *KfNAC83* overexpression improves seed yield and water-use efficiency (WUE). (**A**) Seed yield per plant was measured under normal conditions and after 15- and 20-day water-deficit stress treatments. (**B**) WUE was calculated from water loss and shoot dry weight in four-week-old plants ($n = 25$). Error bars represent ± SE from three independent experiments. Bars with the same letters are not significantly different ($p < 0.001$); different letters indicate significant differences. *** $p < 0.001$. WT = wild type; #6, #11, #23, and #24 = independent transgenic lines.

Water-use efficiency (WUE), defined as biomass produced per unit of water consumed, was also significantly higher in *OxKfNAC83* lines—by 1.7–1.8-fold compared to WT (Figure 3B).

3.5. Overexpression of KfNAC83 Enhances NaCl Tolerance

To evaluate whether *KfNAC83* overexpression confers salt stress tolerance, seed germination and green cotyledon emergence were assessed in *OxKfNAC83* lines and WT under increasing NaCl concentrations. Under control conditions and 100 mM NaCl, no significant differences were observed between *OxKfNAC83* lines and WT (Figure S5A–C). However, at 150, 200, and 250 mM NaCl, WT germination was significantly reduced (19–40%), while *OxKfNAC83* lines maintained high germination rates (81–93%) (Figure S5B). Similarly, green cotyledon emergence in WT was severely impaired at 150 mM NaCl and completely absent at 200 and 250 mM, whereas *OxKfNAC83* lines produced green cotyledons at 200 mM NaCl, though most became chlorotic by day 14 (<4% remained green) (Figure S5A,C). At 250 mM NaCl, none of the *OxKfNAC83* seedlings remained green after 14 days. These results indicate that *OxKfNAC83* lines exhibit enhanced salt tolerance during seed germination and early seedling development.

To assess salt tolerance at later developmental stages, five-day-old seedlings were transferred to MS medium supplemented with 100, 150, or 200 mM NaCl and grown for 14 days. On control medium, all lines showed 100% survival (Figure S6A,B). At 100 mM NaCl, survival rates were similar between *OxKfNAC83* and WT. However, at 150 mM NaCl, *OxKfNAC83* lines exhibited 93–97% survival, while WT survival dropped to 31%. At 200 mM NaCl, WT survival was <7%, whereas *OxKfNAC83* lines maintained 19–26% survival (Figure S6A,B). Fresh and dry biomass were significantly higher in *OxKfNAC83* lines than WT under all NaCl treatments (100–200 mM), while no differences were observed under control conditions (Figure S6C,D).

To further investigate root growth under salt stress, five-day-old seedlings were transferred to MS medium containing 0, 100, 150, 200, or 250 mM NaCl and grown vertically for 12 days. Root elongation was measured every 3 days. No significant differences were observed under control conditions, but root elongation was significantly less inhibited in *OxKfNAC83* lines than in WT at all NaCl concentrations (Figure S7A–E). Additionally, *OxKfNAC83* lines developed significantly more lateral roots than WT under control,

100 mM, and 150 mM NaCl conditions. At 200 mM NaCl, lateral root numbers were similar between genotypes (Figure S7F).

3.6. *KfNAC83* Overexpression Increases Organic Acid and Carbohydrate Accumulation

To determine whether *KfNAC83* overexpression induces CAM-like metabolic traits, titratable acidity (TA) and carbohydrate content were measured in *OxKfNAC83* lines and WT. All *OxKfNAC83* lines exhibited a significant 1.8–2.2-fold increase in nighttime organic acid accumulation compared to WT (Figure 4A). Conversely, daytime organic acid levels were significantly reduced (1.4–1.7-fold) in *OxKfNAC83* lines relative to nighttime levels (Figure 4A). In WT, no significant diel fluctuation in organic acid content was observed. Total carbohydrate content was significantly higher in *OxKfNAC83* lines than in WT at both daytime and nighttime (Figure 4B). Moreover, *OxKfNAC83* lines showed a significant diel difference in carbohydrate levels, which was not observed in WT.

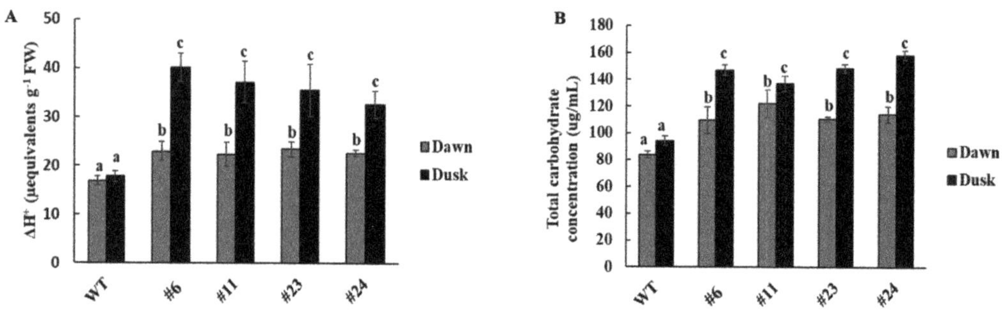

Figure 4. *OxKfNAC83* lines accumulate more organic acids and carbohydrates. T_2 homozygous seeds of *OxKfNAC83* lines and WT were grown on soilless medium for four weeks under a 12 h photoperiod. (**A**) Titratable acidity was measured to quantify organic acid content ($n = 3$). (**B**) Total soluble sugar concentration was measured at dawn and dusk ($n = 3$). Values represent means ± SE. Bars with the same letters are not significantly different ($p < 0.001$); different letters indicate significant differences. WT = wild type; #6, #11, #23, and #24 = independent transgenic lines.

3.7. *KfNAC83* Enhances Photosynthetic and Carboxylation Efficiencies

To assess real-time photosynthetic performance, light response (A/I) curves were measured at a constant CO_2 concentration of 400 μmol mol^{-1}. *OxKfNAC83-24* plants exhibited significantly higher CO_2 assimilation rates than WT at irradiances above 50 μmol m^{-2} s^{-1} (Figure 5A). The initial slope of the A/I curve (quantum efficiency) was significantly steeper in *OxKfNAC83* plants (0.019 ± 0.001) compared to WT (0.009 ± 0.003; $p = 0.006$). Maximum photosynthetic rate (A_{max}) was also significantly higher in *OxKfNAC83* plants (3.7 ± 0.20 μmol CO_2 m^{-2} s^{-1}) than in WT (2.1 ± 0.41 μmol CO_2 m^{-2} s^{-1}; $p = 0.005$).

Quantum yield of PSII (ΦPSII) was significantly higher in *OxKfNAC83* plants across irradiances from 25 to 800 μmol m^{-2} s^{-1} (Figure 5B), indicating more efficient light utilization. Estimated quantum yield of CO_2 at 200 μmol m^{-2} s^{-1} was also significantly higher in *OxKfNAC83* plants (0.010 ± 0.00072) compared to WT (0.006 ± 0.00105; $p = 0.004$) (Figure 5C).

Carboxylation efficiency was assessed by measuring the response of net photosynthesis (*A*) to intercellular CO_2 concentration (Ci) under saturating light (200 μmol m^{-2} s^{-1}). *OxKfNAC83* plants exhibited significantly higher values for Vc_{max} (26 vs. 19 μmol m^{-2} s^{-1}; $p = 0.03$), J_{max} (50 vs. 42 μmol m^{-2} s^{-1}; $p = 0.03$), and TPU (3.6 vs. 3.1 μmol m^{-2} s^{-1}; $p = 0.04$) compared to WT (Figure 6A,B).

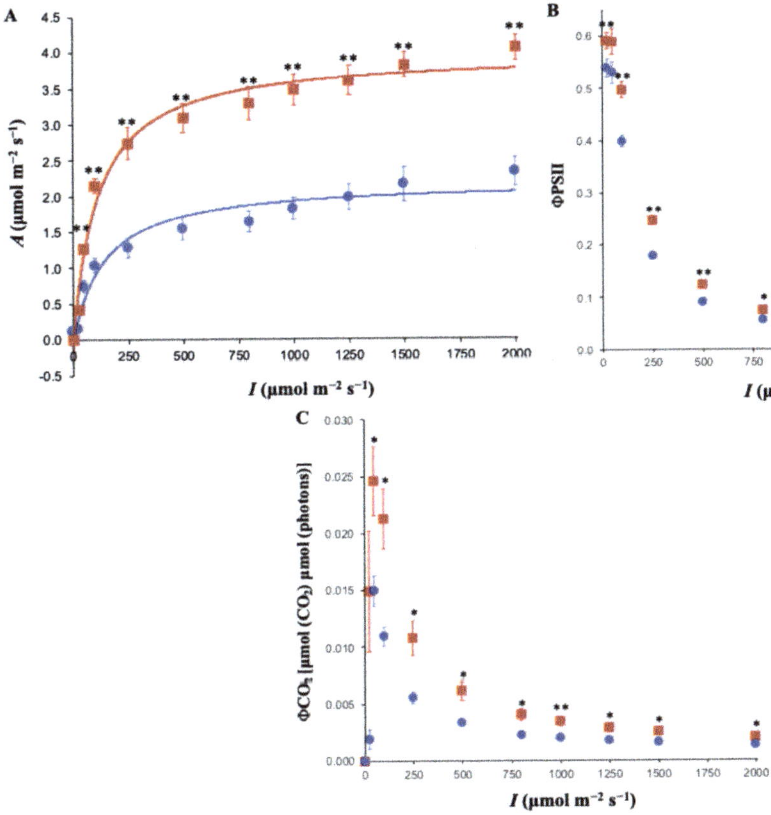

Figure 5. Net photosynthetic light-response (A/I) analysis of *OxKfNAC83* and WT plants. (**A**) Net CO_2 assimilation rate (*A*), (**B**) Quantum yield of Photosystem II (ΦPSII), and (**C**) Quantum yield of CO_2 (ΦCO_2) were measured across a range of irradiance (*I*). Data are from T_2 progeny of line #24 and WT (n = 3 per treatment). Solid lines: red = *OxKfNAC83*, blue = WT. Symbols: red squares = OxKfNAC83, blue circles = WT. Error bars represent standard error of the mean. * $p < 0.05$, ** $p < 0.01$.

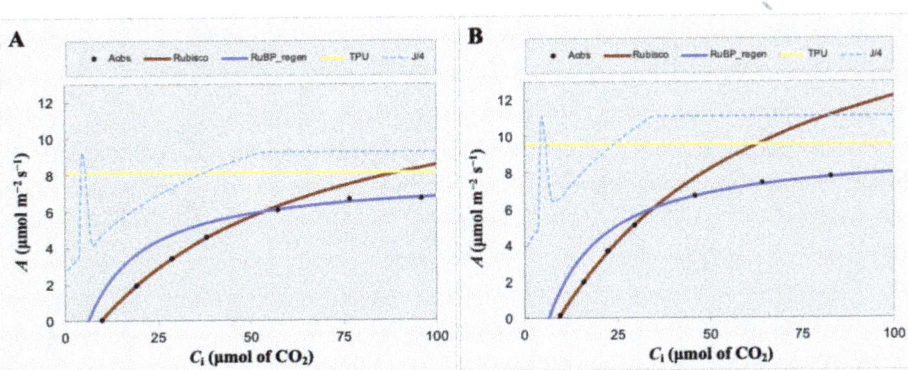

Figure 6. CO_2-response (A/Ci) analysis of *OxKfNAC83* and WT plants. (**A**) WT and (**B**) *OxKfNAC83* plants were analyzed for net CO_2 assimilation rate (**A**) in response to intercellular CO_2 concentration (Ci). Data are from T_2 progeny of line #24 and WT (n = 3 per treatment). Black circles represent observed *A* values.

3.8. Enhanced Water-Deficit Tolerance Involves Extensive Transcriptional Reprogramming

To investigate the transcriptional basis underlying the enhanced water-deficit tolerance in *OxKfNAC83* plants, Illumina-based RNA-Seq was performed on leaf tissues. A total of 5455 differentially expressed genes (DEGs) were identified in water-deficit-treated samples compared to controls. Of these, 1012 and 1245 DEGs were unique to *OxKfNAC83-24* (L24) and WT, respectively, while 3198 DEGs were shared between both genotypes. In L24 under water-deficit conditions, 606 genes (11.1%) were significantly increased, and 406 genes (7.4%) were decreased (Figure 6). In WT, 502 genes (9.2%) were increased, and 743 genes (13.6%) were decreased (Figure 6). Among the 3198 shared DEGs, 1250 (22.9%) were increased and 1948 (35.7%) were decreased (Figure 7).

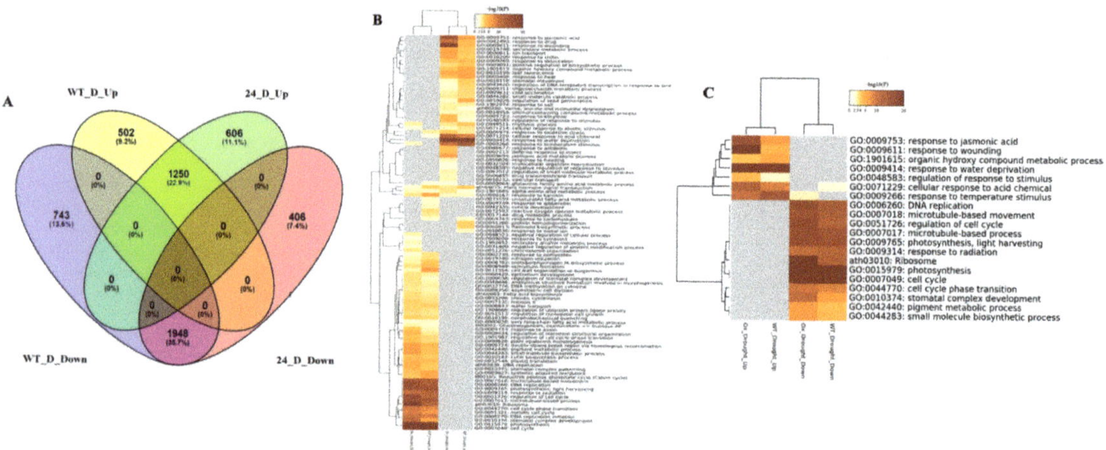

Figure 7. Differentially expressed genes (DEGs) and gene ontology (GO) enrichment in a *OxKfNAC83* line. (**A**) Venn diagram showing increased and decreased DEGs in *OxKfNAC83* line #24 compared to WT. (**B**) Top 100 and (**C**) top 20 enriched GO terms among DEGs, categorized by biological process, presented in hierarchical clusters.

Gene Ontology (GO) enrichment analysis was performed on the DEGs using a false discovery rate (FDR) threshold of <0.005. The top 100 and top 20 enriched GO terms were hierarchically clustered based on Kappa statistics (K = 0.3) (Figure 7B). The top 20 GO terms included "response to jasmonic acid (JA)," "response to water deprivation," "regulation of response to stimulus," and "cellular response to acid chemicals" (Figure 7C).

GO enrichment was also applied to the protein–protein interaction (PPI) network and its MCODE-derived modules to assign biological functions. In L24 under water-deficit stress, five major network hubs were identified: JA-mediated signaling, plant hormone signal tranSEuction, intracellular signal tranSEuction, response to wounding, and response to abscisic acid (ABA) (Figure S8A). One hub included highly expressed jasmonate signaling genes such as *JAZ1*, *JAZ6*, *JAZ8*, *JAZ10*, *TIFY7*, and *TIFY10B*. ABA-responsive genes including *MYC2*, *JAZ5*, *JAZ11*, and *NAC032* were also increased in L24. Notably, Clade A protein phosphatase 2Cs (PP2Cs), including *HAI1*, a key regulator of ABA signaling, were highly expressed in L24 (Figure S8A).

In contrast, WT plants under water-deficit stress showed four major network hubs: JA-mediated signaling, response to water deprivation, response to ABA, and chlorophyll metabolic processes (Figure S8B). Although both genotypes shared JA and ABA-related hubs, the expression levels of associated DEGs were significantly higher in L24. Both L24 and WT also shared decreased hubs related to auxin-activated signaling, cell cycle regulation, and water transport (Figure S8C,D). Additionally, L24 uniquely exhibited decreased expression of seed germination-related genes such as *PDF2*, *ATML1*, and *RGL2* (Figure S8C).

3.9. Diel Shifts in Transcript Abundance Reveal Enhanced Photosynthetic and CAM-like Functional Associations

To explore diel transcriptional dynamics associated with enhanced photosynthesis and growth, RNA-Seq was performed on leaf tissues collected at four time points: 03:00, 09:00, 15:00, and 21:00 h. Volcano plots revealed that L24 samples had 840, 2511, 1713, and 1161 increased DEGs at 03:00, 09:00, 15:00, and 21:00 h, respectively (Figure S9A–D). Corresponding decreased DEGs were 1058, 3061, 1897, and 1102, respectively (Figure S10A–D). Within L24, 9528 DEGs were identified between 03:00 and 15:00 h, with 4717 increased and 4811 decreased genes (Figure S9E). At 09:00 and 21:00 h, 5689 and 5594 DEGs were increased and decreased, respectively (Figure S9F). Among these DEGs, core CAM-related genes such as *PEPCK* and *BCA* were significantly increased in L24 compared to WT (Figure S10A,B). However, other CAM genes including *PPC, PPCK, PPDK,* and *ALMT* did not show significant differences (Figure S10C–F).

KEGG pathway enrichment analysis was performed to further characterize diel DEGs (Figure 8A–H). At 03:00 h, L24 showed enrichment in carbon metabolism and 2-oxocarboxylic acid metabolism, suggesting enhanced nocturnal CO_2 fixation (Figure 8A). Additional enriched pathways included alanine, aspartate, and glutamate metabolism, glucosinolate biosynthesis, and glutathione metabolism. At 09:00 h, increased DEGs were associated with ribosome biogenesis, RNA transport, and DNA replication (Figure 8B). At 15:00 h, DEGs were enriched in abiotic stress-related pathways including plant hormone signal tranSEuction, arginine and proline metabolism, carotenoid and flavonoid biosynthesis, phenylpropanoid biosynthesis, and photosynthesis-antenna proteins (Figure 8C). At 21:00 h, DEGs were enriched in photosynthesis and antenna protein pathways (Figure 8D).

Conversely, carbon metabolism genes were significantly decreased at 15:00 h in L24 (Figure 8G). Plant hormone signaling and plant–pathogen interaction pathways were decreased at 03:00 and 09:00 h (Figure 8E,F). Additional abiotic stress-related pathways, including MAPK signaling, α-linolenic acid metabolism, and lipid metabolism, were also decreased at 09:00 h (Figure 8F).

3.10. Protein–Protein Interaction Enrichment Analysis During Diel Shifts Reveals Plant Defense and Growth Functions

To explore the functional significance of transcriptional reprogramming across diel time points, protein–protein interaction (PPI) networks were constructed for DEGs in *OxKfNAC83-24* (L24) and WT using the BioGRID database [42]. DEGs at each time point were mapped to the global *Arabidopsis* protein association network, and the MCODE algorithm [43] was applied to identify densely connected network modules.

Interestingly, clusters of significantly increased genes at all time points, except L24 09:00 were enriched in regulatory interactions associated with jasmonic acid (JA) signaling, auxin-activated signaling, and plant hormone signal tranSEuction pathways (Figure S11A–D). In the L24 03:00 sample, one cluster included *PAP2, IAA16, IAA18,* and *AXR3,* all associated with the auxin-activated signaling pathway (Figure S11A). A second cluster at the same time point included *JAZ3, JAZ10, TIFY7,* and *TIFY10B,* key components of the JA signaling pathway (Figure S11A).

In contrast, the L24 09:00 sample revealed a distinct network structure. One cluster included *SPA3, SPA4, COP1,* and *DDB1B,* genes associated with cytidine-to-uridine RNA editing. Another cluster included *CKS2, CKS, CDKD1,* and *CYCD2,* which are involved in cell cycle regulation (Figure S11B).

At 15:00, a prominent auxin-related cluster was identified, consisting of *IAA14*, *IAA16*, *IAA5*, *ARF6*, *ARF8*, *PAP2*, and *AXR3* (Figure S11C). Previous studies have shown that *IAA14* regulates lateral root development [44], while *IAA16* influences stem elongation and shoot apical dominance [45]. The auxin response factors *ARF6* and *ARF8* are known to promote JA biosynthesis and floral maturation [46].

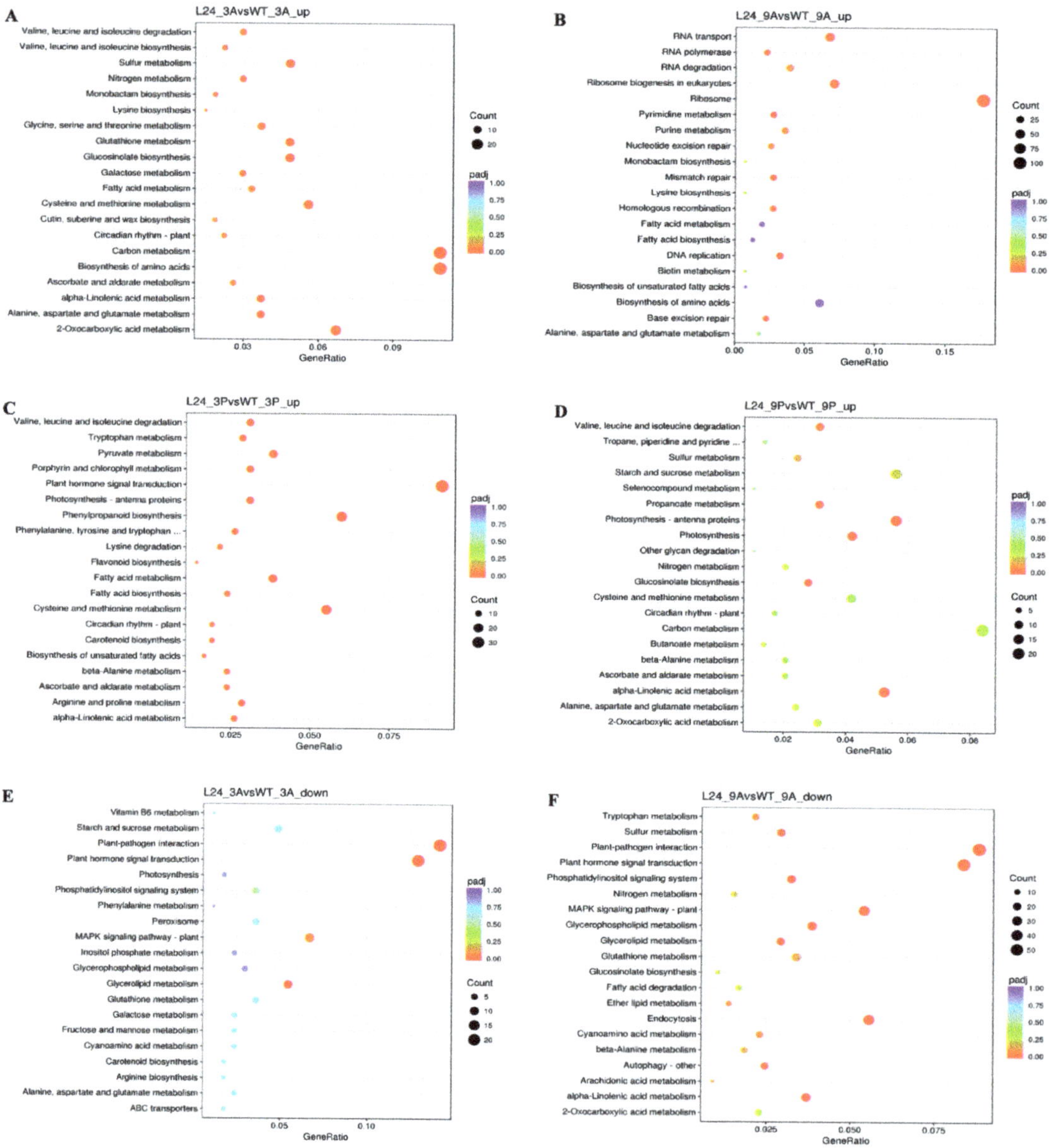

Figure 8. *Cont.*

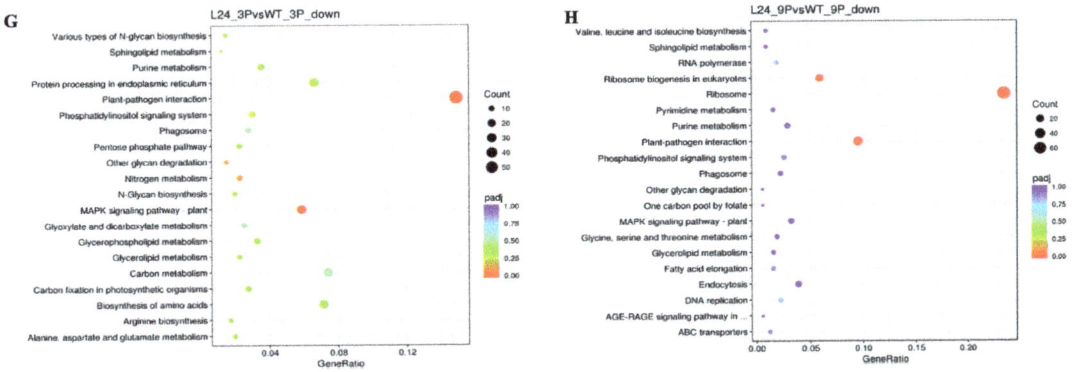

Figure 8. KEGG pathway enrichment analysis of DEGs in a *OxKfNAC83* line. KEGG pathway enrichment scatter plots for significantly increased genes in (**A**) L24_3A vs. WT_3A, (**B**) L24_9A vs. WT_9A, (**C**) L24_3P vs. WT_3P, and (**D**) L24_9P vs. WT_9P; and for decreased genes in (**E**) L24_3A vs. WT_3A, (**F**) L24_9A vs. WT_9A, (**G**) L24_3P vs. WT_3P, and (**H**) L24_9P vs. WT_9P. The x-axis indicates the rich factor; the y-axis lists KEGG pathway names. Dot size represents the number of genes; dot color indicates q-value. Top 20 enriched pathways are shown.

4. Discussion

Collectively, our study demonstrates that overexpression of the CAM-derived transcription factor *KfNAC83* in *Arabidopsis* enhances carbon assimilation, abiotic stress tolerance, growth, and productivity. While NAC proteins are well known to regulate plant development and responses to abiotic stress [47,48], our findings show that ectopic expression of *KfNAC83*, derived from a heat-tolerant obligate CAM species, not only improves water-deficit and NaCl tolerance but also enhances photosynthetic performance and yield in a C_3 model plant.

KfNAC83 is a stress-responsive NAC (SNAC) transcription factor. The NAC superfamily is one of the largest transcription factor families in plants, comprising over 100 members involved in the regulation of both biotic and abiotic stress responses [22,49]. Members specifically associated with abiotic stress responses are classified as SNACs. Phylogenetic analysis revealed that *KfNAC83* is closely related to other well-characterized SNACs from species such as *Arabidopsis thaliana*, *Oryza sativa*, and *Solanum lycopersicum* (Figure S1A). Overexpression of SNAC genes such as *OsNAC2/6* and *OsNAC10* has been shown to enhance tolerance to water-deficit and salt stress [49,50]. Similarly, *ANAC019* and *ANAC055* from *Arabidopsis* are implicated in both abscisic acid (ABA)- and jasmonic acid (JA)-mediated signaling pathways [49,51]. Like other SNACs, KfNAC83 contains a putative nuclear localization signal within the A subdomain (Figure S1B). Consistent with this, *KfNAC83* expression was detected exclusively in the nuclei of root and leaf cells (Figure S2), confirming its nuclear localization. Moreover, *KfNAC83* transcript levels were elevated in both leaves and roots of transgenic lines, although no significant difference was observed between these tissues (Figure S2). Collectively, these findings suggest that *KfNAC83* plays a critical role in enhancing water-deficit stress tolerance in *Arabidopsis*, likely by increasing downstream abiotic stress-responsive genes in both leaves and roots, with a specific influence on the JA signaling pathway.

Functionally, *OxKfNAC83* plants exhibited enhanced water-deficit tolerance, surviving up to 20 days under water-deficit conditions, whereas WT plants did not (Figure 2). Transcriptomic analysis of *OxKfNAC83* plants under water-deficit stress revealed the activation of abiotic stress-responsive genes associated with the JA signaling pathway (Figure S8A). Abiotic stress is known to induce JA biosynthesis and accumulation, which in turn triggers the ubiquitin-mediated degradation of JASMONATE ZIM-DOMAIN (JAZ) proteins. This degradation relieves repression on *MYC2*, leading to the activation of JA-responsive gene expression [52,53]. JA-responsive genes are primarily involved in regulating plant defense mechanisms, including reactive oxygen species (ROS) scavenging and the production of osmoprotectants [53]. JA biosynthesis is regulated by several transcription factors (TFs), such as *MYC2*, *JAZ*, and *WRKY* [49,54]. However, the role of *NAC* TFs in JA biosynthesis and signaling under water-deficit conditions remains relatively unexplored [49,55].

Jasmonates and their derivatives are known to interact synergistically or antagonistically with other phytohormones, including auxin, ethylene, abscisic acid (ABA), salicylic acid, brassinosteroids, and gibberellins, to regulate plant growth, development, and responses to abiotic stresses such as cold, drought, salinity, heavy metals, and light [54,56,57]. Within this complex hormonal network, JA functions as a central signaling molecule [58], and its activity appears to be modulated by abiotic stress-responsive *NAC* TFs [59–63].

Although WT plants showed some induction of *JAZ* and *MYC2* expression under water-deficit stress, only *OxKfNAC83* plants survived prolonged water-deficit conditions. This suggests that *KfNAC83* enhances water-deficit tolerance by regulating the expression of downstream genes involved in abiotic stress adaptation. In addition to JA-related genes, several ABA signaling components, including the ABA-inducible *bHLH*-type transcription factor (*AIB*), *MYC2*, *PP2CA*, and *HAI1*, were differentially expressed in water deficit stressed *OxKfNAC83* plants (Figure S8A). *AIB* is known to play a positive role in ABA signaling in *Arabidopsis*, and *MYC2* also functions as a transcriptional activator in the ABA signaling pathway [64]. These findings suggest that the improved abiotic stress resistance observed in *OxKfNAC83* plants is mediated through transcriptional regulation of both JA and ABA signaling pathways.

Another key observation is the enhanced nighttime carbohydrate availability in *OxKfNAC83* plants, likely due to increased organic acid accumulation and subsequent decarboxylation via the PEP carboxykinase (PEPCK) pathway (Figure 4A and Figure S10A). This metabolic shift supports greater nighttime energy reserves, contributing to improved growth and productivity [65,66]. Diel fluctuations in carbohydrate and organic acid levels are hallmark features of CAM-cycling plants, where carbohydrate accumulation typically occurs during the day [67]. Interestingly, C_4 plants such as *Saccharum* spp. also utilize the PEPCK pathway under water-limited conditions [68], suggesting a convergent mechanism for abiotic stress adaptation. Our study further showed that *KfNAC83* overexpression not only modulated the PEPCK pathway but also enhanced abiotic stress signaling, indicating a dual role in metabolic and hormonal regulation.

These findings are consistent with recent insights from Prusty and Sahoo [69], who reviewed the role of PEPCK in enhancing photosynthetic efficiency and salinity tolerance in rice. Their work highlights PEPCK as a central cytosolic decarboxylase involved in gluconeogenesis, malate metabolism, and the TCA cycle, with additional roles in regulating stomatal conductance, carbon assimilation, and energy metabolism under abiotic stress. They also emphasize PEPCK's contribution to osmotic adjustment, ROS detoxification, and sugar accumulation, mechanisms that align with the physiological and transcriptomic responses observed in *OxKfNAC83* plants. The increased expression of *PEPCK* and associated CAM-like traits in our transgenic lines suggests that *KfNAC83* might regulate or interact with PEPCK-mediated pathways to confer enhanced photosynthetic capacity and abiotic stress resilience, mediated by the JA pathway [70]. The dual role of *KfNAC83* in regulating the PEPCK pathway and abiotic stress-responsive signaling suggests that CAM plants likely have evolved through diel transcriptional rewiring of C_3 regulatory networks [71–73]. These results underscore the potential of CAM-derived transcription factors to reprogram C_3 metabolism toward more efficient, abiotic stress-adaptive states, with implications for bioengineering climate-resilient crops.

Future studies should explore the role of *KfNAC83* and its downstream targets in other C_3 species to assess its broader applicability in improving water-deficit tolerance, nighttime starch metabolism, and agricultural productivity. In particular, it would be valuable to investigate whether the dual role of *KfNAC83* in regulating both the PEPCK pathway and abiotic stress-responsive signaling reflects a broader mechanism by which CAM plants evolved through diel transcriptional rewiring of ancestral C_3 regulatory networks.

5. Conclusions

To our knowledge, this is the first study to functionally characterize an NAC83 transcription factor from the heat-tolerant obligate CAM species *Kalanchoë fedtschenkoi*. Our findings demonstrate that overexpression of KfNAC83 in *Arabidopsis* significantly enhances tolerance to water-deficit and NaCl stress, while also improving carbon assimilation, growth, and seed yield. These phenotypic improvements are underpinned by transcriptional reprogramming of key hormonal pathways, particularly JA and ABA signaling, as well as diel regulation of photosynthetic and metabolic genes.

Importantly, *KfNAC83* overexpression also induced CAM-like traits, including nighttime organic acid accumulation and activation of the PEPCK pathway. These features are consistent with recent findings in rice, where PEPCK has been shown to enhance photosynthetic efficiency and confer salinity tolerance through its roles in gluconeogenesis, malate metabolism, and stomatal regulation [69]. The convergence of these metabolic and regulatory functions suggests that *KfNAC83* might act upstream of or in coordination with PEPCK-mediated processes to promote abiotic stress resilience and metabolic flexibility. Collectively, our results highlight the potential of CAM-derived transcription factors such as *KfNAC83* to reprogram C_3 plant metabolism toward more efficient, abiotic stress-adaptive states, offering a promising strategy for bioengineering climate-resilient crops.

Supplementary Materials: The following supporting information can be downloaded at https://www.mdpi.com/article/10.3390/cimb47090736/s1.

Author Contributions: Conceptualization, B.W.M.W.; methodology, B.W.M.W. and K.N.R.; software, W.C.Y. and M.A.A.; validation, B.W., M.A.A. and B.W.M.W.; formal analysis, K.N.R., M.A.A. and B.W.M.W.; investigation, K.N.R., B.W. and B.W.M.W.; resources, B.W.M.W.; data curation, B.W.M.W.; writing—original draft preparation, K.N.R. and B.W.M.W.; writing—review and editing, B.W.M.W., K.N.R., W.C.Y., M.A.A. and B.W.; visualization, B.W.M.W. and B.W.; supervision, B.W.M.W.; project administration, B.W.M.W.; funding acquisition, B.W.M.W. All authors have read and agreed to the published version of the manuscript.

Funding: The work was supported by South Dakota Board of Regents Competitive Research Grant #A18-0015-001 to B.W.M. Wone.

Institutional Review Board Statement: Not applicable.

Informed Consent Statement: Not applicable.

Data Availability Statement: Data will be made available on request.

Acknowledgments: The authors gratefully acknowledge Annika Lilleberg, and David Melanson for helping with the study. We also thank Kelly Graber for assisting with confocal microscopy imaging. Special thanks to John C. Cushman for providing the *KfNAC83* sequence.

Conflicts of Interest: The authors declare no conflicts of interest. The funders had no role in the design of the study; in the collection, analyses, or interpretation of data; in the writing of the manuscript; or in the decision to publish the results.

References

1. Godfray, H.C.J.; Beddington, J.R.; Crute, I.R.; Haddad, L.; Lawrence, D.; Muir, J.F.; Pretty, J.; Robinson, S.; Thomas, S.M.; Toulmin, C. Food Security: The Challenge of Feeding 9 Billion People. *Science* **2010**, *327*, 812–818. [CrossRef] [PubMed]
2. Gerland, P.; Raftery, A.E.; Ševčíková, H.; Li, N.; Gu, D.; Spoorenberg, T.; Alkema, L.; FoSEick, B.K.; Chunn, J.; Lalic, N.; et al. World population stabilization unlikely this century. *Science* **2014**, *346*, 234–237. [CrossRef]
3. Williams, A.P.; Cook, E.R.; Smerdon, J.E.; Cook, B.I.; Abatzoglou, J.T.; Bolles, K.; Baek, S.H.; Badger, A.M.; Livneh, B. Large contribution from anthropogenic warming to an emerging North American megadrought. *Science* **2020**, *368*, 314–318; Erratum in *Science* **2020**, *370*, eabf3676. https://doi.org/10.1126/science.abf3676. [CrossRef] [PubMed]
4. Wang, H.; Wang, H.; Shao, H.; Tang, X. Recent Advances in Utilizing Transcription Factors to Improve Plant Abiotic Stress Tolerance by Transgenic Technology. *Front. Plant Sci.* **2016**, *7*, 67. [CrossRef]
5. Borland, A.M.; Hartwell, J.; Weston, D.J.; Schlauch, K.A.; Tschaplinski, T.J.; Tuskan, G.A.; Yang, X.; Cushman, J.C. Engineering crassulacean acid metabolism to improve water-use efficiency. *Trends Plant Sci.* **2014**, *19*, 327–338. [CrossRef]
6. Yang, X.; Cushman, J.C.; Borland, A.M.; Edwards, E.J.; Wullschleger, S.D.; Tuskan, G.A.; Owen, N.A.; Griffiths, H.; Smith, J.A.; De Paoli, H.C.; et al. A roadmap for research on crassulacean acid metabolism (CAM) to enhance sustainable food and bioenergy production in a hotter, drier world. *New Phytol.* **2015**, *207*, 491–504. [CrossRef]
7. Lim, S.D.; Lee, S.; Choi, W.G.; Yim, W.C.; Cushman, J.C. Laying the Foundation for Crassulacean Acid Metabolism (CAM) Biodesign: Expression of the C(4) Metabolism Cycle Genes of CAM in Arabidopsis. *Front. Plant Sci.* **2019**, *10*, 101. [CrossRef]
8. Rabara, R.C.; Tripathi, P.; Rushton, P.J. The potential of transcription factor-based genetic engineering in improving crop tolerance to drought. *Omics A J. Integr. Biol.* **2014**, *18*, 601–614. [CrossRef]
9. Joshi, R.; Wani, S.H.; Singh, B.; Bohra, A.; Dar, Z.A.; Lone, A.A.; Pareek, A.; Singla-Pareek, S.L. Transcription Factors and Plants Response to Drought Stress: Current Understanding and Future Directions. *Front. Plant Sci.* **2016**, *7*, 1029. [CrossRef] [PubMed]
10. Babitha, K.C.; Vemanna, R.S.; Nataraja, K.N.; Udayakumar, M. Overexpression of EcbHLH57 Transcription Factor from Eleusine coracana L. in Tobacco Confers Tolerance to Salt, Oxidative and Drought Stress. *PLoS ONE* **2015**, *10*, e0137098. [CrossRef]
11. Gargioni Grisoste Barbosa, E.; Leite, J.; Marin, S.; Marinho, J.; Carvalho, J.; Pagliarini, R.; Farias, J.; Neumaier, N.; Marcelino-Guimaraes, F.; Neves De Oliveira, M.; et al. Overexpression of the ABA-Dependent AREB1 Transcription Factor from Arabidopsis thaliana Improves Soybean Tolerance to Water Deficit. *Plant Mol. Biol. Rep.* **2012**, *31*, 719–730. [CrossRef]
12. Ravikumar, G.; Manimaran, P.; Voleti, S.R.; Subrahmanyam, D.; Sundaram, R.M.; Bansal, K.C.; Viraktamath, B.C.; Balachandran, S.M. Stress-inducible expression of AtDREB1A transcription factor greatly improves drought stress tolerance in transgenic indica rice. *Transgenic Res.* **2014**, *23*, 421–439. [CrossRef]
13. Yu, Y.-T.; Wu, Z.; Lu, K.; Bi, C.; Liang, S.; Wang, X.-F.; Zhang, D.-P. Overexpression of the MYB37 transcription factor enhances abscisic acid sensitivity, and improves both drought tolerance and seed productivity in Arabidopsis thaliana. *Plant Mol. Biol.* **2015**, *90*, 267–279. [CrossRef]
14. Baldoni, E.; Genga, A.; Cominelli, E. Plant MYB Transcription Factors: Their Role in Drought Response Mechanisms. *Int. J. Mol. Sci.* **2015**, *16*, 15811–15851. [CrossRef]
15. Roy, S. Function of MYB domain transcription factors in abiotic stress and epigenetic control of stress response in plant genome. *Plant Signal. Behav.* **2016**, *11*, e1117723. [CrossRef]
16. Amin, A.B.; Rathnayake, K.N.; Yim, W.C.; Garcia, T.M.; Wone, B.; Cushman, J.C.; Wone, B.W.M. Crassulacean Acid Metabolism Abiotic Stress-Responsive Transcription Factors: A Potential Genetic Engineering Approach for Improving Crop Tolerance to Abiotic Stress. *Front. Plant Sci.* **2019**, *10*, 129. [CrossRef]
17. Borland, A.M.; Griffiths, H.; Hartwell, J.; Smith, J.A.C. Exploiting the potential of plants with crassulacean acid metabolism for bioenergy production on marginal lands. *J. Exp. Bot.* **2009**, *60*, 2879–2896. [CrossRef]
18. Villalobos, M.A.; Bartels, D.; Iturriaga, G. Stress Tolerance and Glucose Insensitive Phenotypes in Arabidopsis Overexpressing the *CpMYB10* Transcription Factor Gene. *Plant Physiol.* **2004**, *135*, 309–324. [CrossRef] [PubMed]
19. Ariyarathne, M.A.; Wone, B.W. Overexpression of the Selaginella lepidophylla bHLH transcription factor enhances water-use efficiency, growth, and development in Arabidopsis. *Plant Sci.* **2022**, *315*, 111129. [CrossRef] [PubMed]
20. Liang, J.; Liu, X.; Xu, L.; Mu, R.; Shen, N.; Li, S.; Cheng, C.; Ren, Y.; Ma, L.; Wang, B.; et al. A novel NAC transcription factor from Haloxylon ammodendron promotes reproductive growth in Arabidopsis thaliana under drought stress. *Environ. Exp. Bot.* **2024**, *228*, 106043. [CrossRef]
21. Malwattage, N.R.; Wone, B.; Wone, B.W. A CAM-Related NF-YB Transcription Factor Enhances Multiple Abiotic Stress Tolerance in Arabidopsis. *Int. J. Mol. Sci.* **2024**, *25*, 7107. [CrossRef]
22. Nakashima, K.; Takasaki, H.; Mizoi, J.; Shinozaki, K.; Yamaguchi-Shinozaki, K. NAC transcription factors in plant abiotic stress responses. *Biochim. Biophys. Acta* **2012**, *1819*, 97–103. [CrossRef] [PubMed]
23. Shao, H.; Wang, H.; Tang, X. NAC transcription factors in plant multiple abiotic stress responses: Progress and prospects. *Front. Plant Sci.* **2015**, *6*, 902. [CrossRef]

24. Hu, R.; Zhang, J.; Jawdy, S.; Sreedasyam, A.; Lipzen, A.; Wang, M.; Ng, V.; Daum, C.; Keymanesh, K.; Liu, D.; et al. Comparative genomics analysis of drought response between obligate CAM and C3 photosynthesis plants. *J. Plant Physiol.* **2022**, *277*, 153791. [CrossRef] [PubMed]
25. Hu, R.; Zhang, J.; Jawdy, S.; Sreedasyam, A.; Lipzen, A.; Wang, M.; Ng, V.; Daum, C.; Keymanesh, K.; Liu, D.; et al. Transcriptomic Analysis of the CAM Species Kalanchoë fedtschenkoi Under Low-and High-Temperature Regimes. *Plants* **2024**, *13*, 3444. [CrossRef] [PubMed]
26. Kumar, S.; Stecher, G.; Suleski, M.; Sanderford, M.; Sharma, S.; Tamura, K. MEGA12: Molecular Evolutionary Genetic Analysis Version 12 for Adaptive and Green Computing. *Mol. Biol. Evol.* **2024**, *41*, msae263. [CrossRef]
27. Nakagawa, T.; Suzuki, T.; Murata, S.; Nakamura, S.; Hino, T.; Maeo, K.; Tabata, R.; Kawai, T.; Tanaka, K.; Niwa, Y.; et al. Improved Gateway binary vectors: High-performance vectors for creation of fusion constructs in transgenic analysis of plants. *Biosci. Biotechnol. Biochem.* **2007**, *71*, 2095–2100. [CrossRef]
28. Wise, A.A.; Liu, Z.; Binns, A.N. Three methods for the introduction of foreign DNA into Agrobacterium. *Methods Mol. Biol.* **2006**, *343*, 43–53.
29. Zhang, X.; Henriques, R.; Lin, S.S.; Niu, Q.W.; Chua, N.H. Agrobacterium-mediated transformation of Arabidopsis thaliana using the floral dip method. *Nat. Protoc.* **2006**, *1*, 641–646. [CrossRef]
30. Czechowski, T.; Stitt, M.; Altmann, T.; Udvardi, M.K.; Scheible, W.R. Genome-wide identification and testing of superior reference genes for transcript normalization in Arabidopsis. *Plant Physiol.* **2005**, *139*, 5–17. [CrossRef]
31. Livak, K.J.; Schmittgen, T.D. Analysis of relative gene expression data using real-time quantitative PCR and the 2(-Delta Delta C(T)) Method. *Methods* **2001**, *25*, 402–408. [CrossRef]
32. Lindsey, B.E.; Rivero, L., 3rd; Calhoun, C.S.; Grotewold, E.; Brkljacic, J. Standardized Method for High-throughput Sterilization of Arabidopsis Seeds. *J. Vis. Exp. JoVE* **2017**, 56587.
33. Verslues, P.E. Quantification of water stress-induced osmotic adjustment and proline accumulation for Arabidopsis thaliana molecular genetic studies. *Methods Mol. Biol.* **2010**, *639*, 301–315. [PubMed]
34. Wituszynska, W.; Slesak, I.; Vanderauwera, S.; Szechynska-Hebda, M.; Kornas, A.; Van Der Kelen, K.; Muhlenbock, P.; Karpinska, B.; Mackowski, S.; Van Breusegem, F.; et al. Lesion simulating disease1, enhanced disease susceptibility1, and phytoalexin deficient4 conditionally regulate cellular signaling homeostasis, photosynthesis, water use efficiency, and seed yield in Arabidopsis. *Plant Physiol.* **2013**, *161*, 1795–1805. [CrossRef]
35. Wituszynska, W.; Karpiński, S. Determination of Water Use Efficiency for Arabidopsis thaliana. *Bio-Protoc. J.* **2014**, *4*, e1041. [CrossRef]
36. Fox, J.D.; Robyt, J.F. Miniaturization of three carbohydrate analyses using a microsample plate reader. *Anal. Biochem.* **1991**, *195*, 93–96. [CrossRef] [PubMed]
37. Lim, S.D.; Yim, W.C.; Liu, D.; Hu, R.; Yang, X.; Cushman, J.C. A Vitis vinifera basic helix-loop-helix transcription factor enhances plant cell size, vegetative biomass and reproductive yield. *Plant Biotechnol. J.* **2018**, *16*, 1595–1615. [CrossRef]
38. Kaipiainen, E.L. Parameters of photosynthesis light curve in Salix dasyclados and their changes during the growth season. *Russ. J. Plant Physiol.* **2009**, *56*, 445–453. [CrossRef]
39. Lobo, F.D.A.; De Barros, M.P.; Dalmagro, H.J.; Dalmolin, Â.C.; Pereira, W.E.; De Souza, E.C.; Vourlitis, G.L.; Rodríguez Ortíz, C.E. Fitting net photosynthetic light-response curves with Microsoft Excel—A critical look at the models. *Photosynthetica* **2013**, *51*, 445–456; Erratum in *Photosynthetica* **2014**, *52*, 479–480. https://doi.org/10.1007/s11099-014-0045-6. [CrossRef]
40. Von Caemmerer, S.V.; Farquhar, G.D. Some relationships between the biochemistry of photosynthesis and the gas exchange of leaves. *Planta* **1981**, *153*, 376–387. [CrossRef]
41. Sharkey, T.D. What gas exchange data can tell us about photosynthesis. *Plant Cell Environ.* **2015**, *39*, 1161–1163. [CrossRef] [PubMed]
42. Chatr-Aryamontri, A.; Breitkreutz, B.J.; Oughtred, R.; Boucher, L.; Heinicke, S.; Chen, D.; Stark, C.; Breitkreutz, A.; Kolas, N.; O'Donnell, L.; et al. The BioGRID interaction database: 2015 update. *Nucleic Acids Res.* **2015**, *43*, D470–D478. [CrossRef]
43. Bader, G.D.; Hogue, C.W.V. An automated method for finding molecular complexes in large protein interaction networks. *BMC Bioinform.* **2003**, *4*, 2. [CrossRef]
44. Fukaki, H.; Nakao, Y.; Okushima, Y.; Theologis, A.; Tasaka, M. Tissue-specific expression of stabilized SOLITARY-ROOT/IAA14 alters lateral root development in Arabidopsis. *Plant J. Cell Mol. Biol.* **2005**, *44*, 382–395. [CrossRef] [PubMed]
45. Li, H.; Wang, B.; Zhang, Q.; Wang, J.; King, G.J.; Liu, K. Genome-wide analysis of the auxin/indoleacetic acid (Aux/IAA) gene family in allotetraploid rapeseed (*Brassica napus* L.). *BMC Plant Biol.* **2017**, *17*, 204. [CrossRef]
46. Nagpal, P.; Ellis, C.M.; Weber, H.; Ploense, S.E.; Barkawi, L.S.; Guilfoyle, T.J.; Hagen, G.; Alonso, J.M.; Cohen, J.D.; Farmer, E.E.; et al. Auxin response factors ARF6 and ARF8 promote jasmonic acid production and flower maturation. *Development* **2005**, *132*, 4107–4118. [CrossRef] [PubMed]
47. Liu, C.; Wang, B.; Li, Z.; Peng, Z.; Zhang, J. TsNAC1 Is a Key Transcription Factor in Abiotic Stress Resistance and Growth. *Plant Physiol.* **2018**, *176*, 742–756. [CrossRef]

48. Ali, M.S.; Baek, K.-H. Jasmonic Acid Signaling Pathway in Response to Abiotic Stresses in Plants. *Int. J. Mol. Sci.* **2020**, *21*, 621. [CrossRef]
49. Nuruzzaman, M.; Sharoni, A.M.; Kikuchi, S. Roles of NAC transcription factors in the regulation of biotic and abiotic stress responses in plants. *Front. Microbiol.* **2013**, *4*, 248. [CrossRef]
50. Jeong, J.S.; Kim, Y.S.; Baek, K.H.; Jung, H.; Ha, S.H.; Do Choi, Y.; Kim, M.; Reuzeau, C.; Kim, J.K. Root-specific expression of OsNAC10 improves drought tolerance and grain yield in rice under field drought conditions. *Plant Physiol.* **2010**, *153*, 185–197. [CrossRef]
51. Bu, Q.; Jiang, H.; Li, C.B.; Zhai, Q.; Zhang, J.; Wu, X.; Sun, J.; Xie, Q.; Li, C. Role of the Arabidopsis thaliana NAC transcription factors ANAC019 and ANAC055 in regulating jasmonic acid-signaled defense responses. *Cell Res.* **2008**, *18*, 756–767. [CrossRef] [PubMed]
52. Ruan, J.; Zhou, Y.; Zhou, M.; Yan, J.; Khurshid, M.; Weng, W.; Cheng, J.; Zhang, K. Jasmonic Acid Signaling Pathway in Plants. *Int. J. Mol. Sci.* **2019**, *20*, 2479. [CrossRef] [PubMed]
53. Kim, H.; Seomun, S.; Yoon, Y.; Jang, G. Jasmonic Acid in Plant Abiotic Stress Tolerance and Interaction with Abscisic Acid. *Agronomy* **2021**, *11*, 1886. [CrossRef]
54. Wasternack, C.; Hause, B. Jasmonates: Biosynthesis, perception, signal tranSEuction and action in plant stress response, growth and development. An update to the 2007 review in Annals of Botany. *Ann. Bot.* **2013**, *111*, 1021–1058. [CrossRef]
55. Su, L.; Fang, L.; Zhu, Z.; Zhang, L.; Sun, X.; Wang, Y.; Wang, Q.; Li, S.; Xin, H. The transcription factor VaNAC17 from grapevine (Vitis amurensis) enhances drought tolerance by modulating jasmonic acid biosynthesis in transgenic Arabidopsis. *Plant Cell Rep.* **2020**, *39*, 621–634. [CrossRef]
56. Huang, H.; Liu, B.; Liu, L.; Song, S. Jasmonate action in plant growth and development. *J. Exp. Bot.* **2017**, *68*, 1349–1359. [CrossRef]
57. Wang, J.; Song, L.; Gong, X.; Xu, J.; Li, M. Functions of Jasmonic Acid in Plant Regulation and Response to Abiotic Stress. *Int. J. Mol. Sci.* **2020**, *21*, 1446. [CrossRef]
58. Yang, X.; Kim, M.Y.; Ha, J.; Lee, S.-H. Overexpression of the Soybean NAC Gene GmNAC109 Increases Lateral Root Formation and Abiotic Stress Tolerance in Transgenic Arabidopsis Plants. *Front. Plant Sci.* **2019**, *10*, 1036. [CrossRef]
59. Yoon, Y.; Seo, D.H.; Shin, H.; Kim, H.J.; Kim, C.M.; Jang, G. The Role of Stress-Responsive Transcription Factors in Modulating Abiotic Stress Tolerance in Plants. *Agronomy* **2020**, *10*, 788. [CrossRef]
60. Ren, Y.; Huang, Z.; Jiang, H.; Wang, Z.; Wu, F.; Xiong, Y.; Yao, J. A heat stress responsive NAC transcription factor heterodimer plays key roles in rice grain filling. *J. Exp. Bot.* **2021**, *72*, 2947–2964. [CrossRef]
61. Zhao, X.; Wu, T.; Guo, S.; Hu, J.; Zhan, Y. Ectopic expression of AeNAC83, a NAC transcription factor from Abelmoschus esculentus, inhibits growth and confers tolerance to salt stress in Arabidopsis. *Int. J. Mol. Sci.* **2022**, *23*, 10182. [CrossRef]
62. Xu, Y.; Cheng, J.; Hu, H.; Yan, L.; Jia, J.; Wu, B. Genome-wide identification of NAC family genes in oat and functional characterization of AsNAC109 in abiotic stress tolerance. *Plants* **2024**, *13*, 1017. [CrossRef]
63. Chen, Y.; Xia, P. NAC transcription factors as biological macromolecules responded to abiotic stress: A comprehensive review. *Int. J. Biol. Macromol.* **2025**, *308*, 142400. [CrossRef]
64. Hao, Y.; Zong, X.; Ren, P.; Qian, Y.; Fu, A. Basic Helix-Loop-Helix (bHLH) Transcription Factors Regulate a Wide Range of Functions in Arabidopsis. *Int. J. Mol. Sci.* **2021**, *22*, 7152. [CrossRef]
65. Graf, A.; Schlereth, A.; Stitt, M.; Smith, A.M. Circadian control of carbohydrate availability for growth in Arabidopsis plants at night. *Proc. Natl. Acad. Sci. USA* **2010**, *107*, 9458–9463. [CrossRef] [PubMed]
66. Scialdone, A.; Howard, M. How plants manage food reserves at night: Quantitative models and open questions. *Front. Plant Sci.* **2015**, *6*, 204. [CrossRef] [PubMed]
67. Kerbauy, G.B.; Takahashi, C.A.; Lopez, A.M.; Matsumura, A.T.; Hamachi, L.; Félix, L.M.; Pereira, P.N.; Freschi, L.; Mercier, H. *Crassulacean Acid Metabolism in Epiphytic Orchids: Current Knowledge, Future Perspectives*; InTechOpen: Rijeka, Croacia, 2012.
68. Cacefo, V.; Ribas, A.F.; Zilliani, R.R.; Neris, D.M.; Domingues, D.S.; Moro, A.L.; Vieira, L.G.E. Decarboxylation mechanisms of C4 photosynthesis in *Saccharum* spp.: Increased PEPCK activity under water-limiting conditions. *BMC Plant Biol.* **2019**, *19*, 144. [CrossRef] [PubMed]
69. Prusty, S.; Sahoo, R.K. *PEPCK* Gene for Enhanced Photosynthesis and Salinity Stress Tolerance in Rice: A Review. *Agric. Rev.* **2024**, *45*, 636. [CrossRef]
70. Rahman, M.; Mostofa, M.G.; Keya, S.S.; Ghosh, P.K.; Abdelrahman, M.; Anik, T.R.; Gupta, A.; Tran, L.-S.P. Jasmonic acid priming augments antioxidant defense and photosynthesis in soybean to alleviate combined heat and drought stress effects. *Plant Physiol. Biochem.* **2024**, *206*, 108193. [CrossRef]
71. Heyduk, K.; Ray, J.N.; Ayyampalayam, S.; Leebens-Mack, J. Shifts in gene expression profiles are associated with weak and strong Crassulacean acid metabolism. *Am. J. Bot.* **2018**, *105*, 587–601. [CrossRef] [PubMed]

72. Moseley, R.C.; Mewalal, R.; Motta, F.; Tuskan, G.A.; Haase, S.; Yang, X. Conservation and Diversification of Circadian Rhythmicity Between a Model Crassulacean Acid Metabolism Plant Kalanchoë fedtschenkoi and a Model C3 Photosynthesis Plant Arabidopsis thaliana. *Front. Plant Sci.* **2018**, *9*, 1757. [CrossRef] [PubMed]
73. Yin, H.; Guo, H.B.; Weston, D.J.; Borland, A.M.; Ranjan, P.; Abraham, P.E.; Jawdy, S.S.; Wachira, J.; Tuskan, G.A.; Tschaplinski, T.J.; et al. Diel rewiring and positive selection of ancient plant proteins enabled evolution of CAM photosynthesis in Agave. *BMC Genom.* **2018**, *19*, 588; Erratum in *BMC Genom.* **2019**, *20*, 279. https://doi.org/10.1186/s12864-019-5663-8. [CrossRef] [PubMed]

current issues in molecular biology

MDPI

Article

Transcriptome Analysis Reveals Critical Genes Involved in the Response of *Stropharia rugosoannulata* to High Temperature and Drought Stress

Shengze Yan [1,†], Shengyou Wang [1,†], Meirong Zhan [1], Xianxin Huang [1], Ting Xie [1], Ruijuan Wang [2,3], Huan Lu [2,3], Qingqing Luo [1] and Wei Ye [1,*]

1 Fujian Key Laboratory of Crop Genetic Improvement and Innovative Utilization for Mountain Area, Institute of Edible Fungi, Sanming Academy of Agricultural Sciences, Sanming 365500, China; yanshengze1102025@163.com (S.Y.); smwsy2022@163.com (S.W.); zhmrnhkl@163.com (M.Z.); hxianxin96@163.com (X.H.); 19376776419@163.com (T.X.); luoqingqing6@126.com (Q.L.)

2 National Research Center for Edible Fungi Biotechnology and Engineering, Key Laboratory of Applied Mycological Resources and Utilization, Shanghai Academy of Agricultural Sciences, Shanghai 201403, China; wangruijuan@saas.sh.cn (R.W.); luhuan@saas.sh.cn (H.L.)

3 Ministry of Agriculture, Shanghai Key Laboratory of Agricultural Genetics and Breeding, Institute of Edible Fungi, Shanghai Academy of Agricultural Sciences, Shanghai 201403, China

* Correspondence: yewei922@sina.com

† These authors contributed equally to this work.

Abstract

In this study, the differences in gene expression of *Stropharia rugosoannulata* at different treatment times under high temperature and drought stress were analyzed by transcriptomics. Here, a total of 74,571 transcripts and 16,233 unigenes were identified, with an average assembly length of 3002 bp. A total of 10,248 differentially expressed genes (DEGs) were identified. DEG analysis indicated that the numbers of DEGs under high-temperature stress for 1 d, 2 d, and 3 d were 798, 851, and 1484, respectively. These DEGs were involved in 96 GO functional categories and 69 KEGG metabolic pathways. Meanwhile, the numbers of DEGs under drought stress for 3 d, 6 d, and 9 d were 421, 1072, and 2880, respectively. These DEGs were involved in 108 GO functional categories and 78 KEGG metabolic pathways. Further analysis of the metabolic pathway (ko04011) commonly enriched by DEGs identified 15 candidate genes responding to high-temperature or drought stress. Eight candidate genes were randomly selected for qRT-PCR verification, and the qRT-PCR results were basically consistent with the transcriptome datasets. These findings provide critical candidate genes for understanding the molecular regulation mechanism of *S. rugosoannulata* in response to high temperature and drought stress and have important reference value for its stress resistance breeding.

Keywords: *Stropharia rugosoannulata*; abiotic stress; DEG; candidate genes; transcriptome analysis

1. Introduction

Stropharia rugosoannulata Farl. Ex Murrill is a kind of edible mushroom with high nutritional, low-fat, and medicinal values. It was found that the crude protein content of *S. rugosoannulata* was 25.75%, which was higher than the protein content of most edible fungi [1]. The total amino acid content ranges from 18.89% to 31.01% in recent reports, comprising 18 to 20 kinds of amino acids, with a crude fiber content of 13.3 g/100 g fruiting body [2,3]. The contents of phosphorus and potassium in the mineral elements were higher, at 3.48% and 0.82%, respectively [4–6]. *S. rugosoannulata* is also an extremely precious medicinal fungus, which has the effect of treating or improving many diseases in humans. In addition, *S. rugosoannulata* is one of the cultivated varieties of edible fungi recommended by the FAO, which offers good economic, ecological, and social benefits [1,5,7].

The growth and development of *S. rugosoannulata* require suitable humidity, temperature, light, pH, and air, with humidity and temperature being the most important factors [8–10]. When fungi are exposed to high temperatures and other adverse conditions, reactive oxygen species (ROS) accumulate in large quantities, leading to membrane peroxidation, damage to the cell membrane structure, and destruction of normal metabolism [11,12]. Under high-temperature stress, reactive oxygen species such as intracellular superoxide radical (O^{2-}), hydrogen peroxide (H_2O_2), and hydroxyl (OH-) accumulate rapidly, which leads to membrane lipid peroxidation, which damages membrane lipids, nucleic acids, and membrane proteins, and finally leads to cell death [13]. High temperatures will reduce the quality and yield of fruiting bodies, making them susceptible to infection by various bacteria. Therefore, the tolerance of edible fungi to high temperatures is very important. High temperature will affect the mycelium stage, primordial stage, and fruiting stage of edible fungi to varying degrees [14,15]. In the mycelium stage, high temperature will change its morphology, affect its growth rate, make its biomass change significantly, and form an obvious heat shock circle. The primordial stage is very sensitive to changes in the external environment. High-temperature stress slows primordial differentiation, resulting in a low differentiation rate, and even mushroom growth is hindered, making it very easy to cause fungus bag pollution [16,17]. If the temperature is too high during the fruiting period, the fruiting body will be deformed and not fruit, which will lead to a decline in yield and even no harvest. Understanding the molecular mechanism of edible fungi responding to heat stress can lay a foundation for cultivating high-temperature varieties of edible fungi [10,11]. The effects of high temperature on macrofungi have been reported in *Pleurotus ostreatus* [18], *Pleurotus eryngii* [19], *Agaricus bisporus* [20], and *Ganoderma lucidum* [21]. With the aggravation of land drought, it is necessary to maintain soil water content and air relative humidity by increasing water consumption and water use times to cultivate *S. rugosoannulata* [22]. According to national standards, when the relative water content of soil is lower than 50%, it belongs to severe soil drought. For example, when the soil is subject to severe drought, it will lead to the slow growth of *Morchella esculenta* and the decline of fruiting body quality and yield [11,23].

In recent years, with the rapid development of high-throughput sequencing technology and the substantial reduction in sequencing cost, high-throughput sequencing has become the first choice for genome and transcriptome analysis of plants, animals, and edible fungi. Increasingly, reports have investigated the transcription and expression of genes in edible fungi under abiotic stress from an omics perspective, aiming to elucidate the regulatory mechanisms of physiological processes related to edible fungi [24]. Transcriptomics can study the transcriptional regulation of genes in cells at an overall level and reveal the molecular mechanisms of complex biological pathways and trait regulation networks. Exploring the transcriptomics changes in edible fungi under adversity stress is helpful to deeply understand the complex molecular regulatory network of edible fungi under adversity stress and provide a reference for improving the stress resistance of edible fungi [24]. Kim et al. [25] used RNA-seq to compare the morphological changes and gene expression of *Lentinula edodes* fruiting bodies under blue light and dark conditions. The results showed that blue light mainly induced cap growth, while dark cultivation promoted stalk length development, and a total of 221 DEGs were significantly up-regulated and 541 DEGs were significantly down-regulated, and 8 up-regulated genes under blue light conditions were identified, such as *DDR48* (heat shock protein), Fasciclin-domain-containing protein, carbohydrate esterase family 4 protein, and FAD NAD-binding domain-containing protein that are involved in morphological development of primordium and embryonic muscle development, cell adhesion, and affect the structure of cellulose. Yu et al. [26] studied the response pattern of *L. edodes* to heat stress by transcriptome and found that heat-tolerant strains showed more DEGs at high temperature, and two DEGs were highly expressed at all developmental stages. In addition, several possible candidate genes involved in the Cd accumulation were identified, including the major facilitator superfamily genes, heat shock proteins, and laccase 11, a multicopper oxidase. Hao et al. [27] performed an RNA-seq analysis of *S. rugosoannulata* strain DQ1 (high-sensitivity) and *S. rugosoannulata* strain DQ3 (low-sensitivity) under low-temperature stress, and finally, a total of 9499 DEGs were screened. Additionally, further research showed that carbohydrate enzyme (*GT, CE, GH,* and *AA*) genes were down-regulated more significantly in DQ-1 than DQ-3, and several cellulase activities were also reduced to a greater extent. Moreover, the *GR, POD, CAT2,* and *CAT1* genes and more heat shock protein genes (*HSP78* and *HSP20*) were up-regulated in the two strains after low-temperature stress, and the *GPX* gene and more heat shock protein genes were up-regulated in DQ-3. At present, there is no report on analyzing the gene expression and identifying candidate genes of *S. rugosoannulata* in response to drought and high-temperature stress via transcriptome sequencing.

In this study, *S. rugosoannulata* strain D-3 was selected as the experimental material, and three treatments of normal, high temperature, and drought stress were conducted. Transcriptome sequencing was used, and bioinformatics analysis was carried out. The gene expression of *S. rugosoannulata* in response to stress was detected from the mRNA level, the candidate genes closely related to drought and high-temperature stress were screened, and the expression patterns of randomly selected genes were verified. In conclusion, these findings aim to provide a theoretical basis for stress resistance regulation and breeding of *S. rugosoannulata*.

2. Materials and Methods

2.1. Samples and Stress Treatments

In this study, the *S. rugosoannulata* strain D-3 was derived from the strain preservation center of Sanming Academy of Agricultural Sciences and was used as research material to study its stress resistance mechanism. For the control and drought stress and high-temperature stress treatments, the fermented cultivation materials were placed in the mushroom incubator, the temperature was set to 20–25 °C, and the carbon dioxide concentration was 1000–1500 ppm. Next, 35–45 d after sowing, the temperature was adjusted to 13–15 °C, and after 7 d, the mycelium began to kink and produce mushrooms. Then, the illumination time was set to 6 h/d, the humidity was 80–90% and 60–70%, respectively, and the temperature and humidity detector was used to ensure that the water content of soil and cultivation material was 55–60% and 40–45%, respectively. The drought stress treatment was carried out for 3 d, 6 d, and 9 d. In addition, during the fruiting period, *S. rugosoannulata* (fruiting body diameter of 1–2 mm) with the same growth vigor were placed in the fruiting incubator at 15 °C (control) and 35 °C (stress treatment), respectively, for 1 d, 2 d, and 3 d of high-temperature stress treatment. After harvesting all the treated *S. rugosoannulata*, mushrooms with the same growth vigor (three replicates) were randomly selected and quickly placed in liquid nitrogen for freezing, then stored at −80 °C in a freezer for later use.

2.2. Total RNA Extraction, Construction of the cDNA Library, and RNA-seq

The total RNA of all samples was extracted by Trizol (Invitrogen, Carlsbad, CA, USA), and the genomic DNA residue was removed by DNase I, and the purity and concentration of the extracted total RNA were determined by a NanoDrop One (Thermo Fisher Sci, Waltham, MA, USA) and an Agilent 2100 Bioanalyzer (Agilent Technologies, Waldbronn, Germany). All qualified RNA samples were sent to Biomarker Technologies Co., Ltd. (Beijing, China) for cDNA library construction and transcriptome sequencing. According to the manufacturer's instructions, cDNA libraries were constructed using the TruSeq RNA sample preparation kit (Illumina RS-122-2101, Illumina, CA, USA). Then, all the acceptable cDNA libraries were sequenced on an Illumina HiSeqTM 3500 sequencing platform. The clean reads were processed by data filtering, and low-quality reads were discarded. The RNA-seq raw sequence datasets were deposited in the National Genomics Data Center (NGDC), China National Center for Bioinformation, Chinese Academy of Sciences (CRA028883) (https://ngdc.cncb.ac.cn/gsa/browse/CRA028883, released on 19 August 2025, accessed on 11 August 2025). In the NCBI database, the genome of Stropharia rugosoannulata strain QGU27 (GCA_028532985.1) has been posted (https://www.ncbi.nlm.nih.gov/datasets/genome/GCA_028532985.1/, accessed on 21 September 2025). Considering that large differences between the sequencing data and the reference genome may occur, which will affect the alignment results, we use Trinity v2.15.1 (California, USA) [28] to splice clean reads to obtain transcripts and then carry out subsequent analysis. According to the expression level of genes in different samples, DEG analysis, functional annotation, and functional enrichment of DEG analysis were carried out to explore the critical candidate genes involved in the response to abiotic stress and their regulatory pathways in *S. rugosoannulata*.

2.3. Discovery of New Genes and Functional Annotation

Based on the transcriptome data, the software Trinity v2.15.1 (California, USA) [28] is used to construct reference transcripts and discover new transcripts or unigenes of this species. Based on the basic biological principle that structure determines function, the homologous sequence with the highest similarity to the new gene sequence in each gene function annotation database is found by sequence comparison, and the gene function annotation information of the homologous sequence is used as the functional description of the new gene discovered. Based on NR (ftp://ftp.ncbi.nih.gov/blast/db/, accessed on 5 July 2025), Pfam (http://pfam.xfam.org/, accessed on 5 July 2025), PlantTFDB (http://plntfdb.bio.uni-potsdam.de/v3.0/, accessed on 5 July 2025), GO (http://geneontology.org/, accessed on 5 July 2025), KOG (https://www.ncbi.nlm.nih.gov/research/cog-project/, accessed on 5 July 2025), KEGG (https://www.genome.jp/kegg/kegg1.html, accessed on 5 July 2025), and other databases, we used BLAST 2.9.0+ (NCBI, Bethesda, USA) [29] to annotate the new genes in depth and obtain their annotation information.

2.4. Differential Expression Gene Analysis

After comparing the sequence of clean reads with the assembled reference transcript to obtain the corresponding position information, the expression level of the transcript and the gene can be quantified by the position information of mapped reads on the gene. In order to make the number of fragments truly reflect the transcript expression level, it is necessary to normalize the number of mapped reads and the transcript length in the sample and adopt FPKM (Fragments per kilobase of transcript per million fragments mapped) as an index to measure the transcript or gene expression level. In the process of detecting DEG, the $\log_2$FC (fold change) ≥ 1 and the false discovery rate (FDR) < 0.05 are used as screening criteria. The Benjamini–Hochberg correction method is adopted to correct the significant p-value obtained from the original hypothesis test. Subsequently, EBSeq (Madison, WI, USA) [30] was used for differential analysis to obtain a set of DEGs between the two samples. In addition, hierarchical clustering analysis was performed on the screened DEG, and the genes with the same or similar expression patterns were clustered, and the *K*-means method was used for statistical and clustering analysis of the DEG.

2.5. Enrichment Analysis of GO Function and KEGG Metabolic Pathway

The enrichment analysis of GO and KEGG metabolic pathways of DEGs was performed by the R package clusterProfiler Version 4.1.1 (Nanjing, China) [31], and both were accurately tested by Fisher. After correction, GO entries and KEGG metabolic pathways with $p < 0.05$ were considered to be significantly enriched.

2.6. Screening of Critical Differentially Expressed Genes and qRT-PCR Analysis

According to the results of KEGG enrichment analysis, the DEGs enriched in the same metabolic pathway were further annotated to screen candidate genes in response to high temperature and drought stress. Based on the identified candidate genes, eight genes were randomly selected to analyze their expression patterns in all samples by qRT-PCR. The first strand cDNA synthesis was carried out according to the protocols of the Prime-ScriptTM RT Reagent Kit (TaKaRa, Shimogyo-ku, Kyoto, Japan). The glyceraldehyde-3-phosphate dehydrogenase (*GAPDH*) was chosen as an internal control gene to normalize the expression data. The eight primer pairs were designed by the software Primer Premier 5.0 according to the coding sequences (CDS) of the selected candidate genes, and primer details are provided in Supplementary Table S1. The qRT-PCR amplification and procedure were carried out according to the report of Cheng et al. [32]. Three biological replicates for each sample were performed, and the relative expression level of each selected candidate gene in different stress treatments was calculated from the $2^{-\Delta\Delta Ct}$ value [33]. Histograms and the significance of the difference in expression levels between different samples were analyzed by the software GraphPad Prism 8.

3. Results

3.1. Summary of the Transcriptome Sequencing and Data Processing

In our study, 21 cDNA libraries were established, including six stress treatment groups (gh3d, which means response to drought for three days; gh6d, which means response to drought for six days; gh9d, which means response to drought for nine days; gw1d, which means response to high temperature for a day; gw2d, which means response to high temperature for two days; and gw3d, which means response to high temperature for three days) and a control group (ck) with three replicates in each group. RNA-seq generated 41.96 million to 49.15 million raw reads for each library (Table 1). Subsequently, after removing the low-quality raw reads and adaptors, a total of 142.43 G clean reads were produced, and the average number of clean reads for the gh3d, gh6d, gh9d, gw1d, gw2d, gw3d, and CK treatments was 6.78, 6.58, 7.12, 6.83, 6.36, 6.83, and 6.97 Gb, respectively. The percentage of bases with Q30 was more than 97.00% in each group (Table 1). The above results suggest that the RNA-seq data are reliable and can meet the requirements of subsequent transcriptome bioinformatics analysis.

Table 1. Summary of RNA-seq of all samples.

Items	Raw Reads [a] (M)	Clean Reads [b] (G)	The Average Clean Data (G) [c]	Q30 [d]
gh3d_1	45.06	6.74		97.99
gh3d_2	46.19	6.91	6.78	98.00
gh3d_3	44.78	6.7		97.49
gh6d_1	44.54	6.67		97.96
gh6d_2	45.42	6.8	6.58	98.02
gh6d_3	41.96	6.28		98.25

Table 1. *Cont.*

Items	Raw Reads [a] (M)	Clean Reads [b] (G)	The Average Clean Data (G) [c]	Q30 [d]
gh9d_1	46.12	6.9		97.90
gh9d_2	48.04	7.19	7.12	97.8
gh9d_3	48.54	7.26		97.84
gw1d_1	45.24	6.77		98.05
gw1d_2	47.52	7.11	6.83	98.02
gw1d_3	44.12	6.6		97.77
gw2d_1	43.54	6.52		97.88
gw2d_2	40.78	6.09	6.36	98.02
gw2d_3	43.32	6.48		98.12
gw3d_1	43.35	6.49		98.07
gw3d_2	49.15	7.36	6.83	97.80
gw3d_3	44.45	6.65		98.00
ck_1	45.29	6.77		97.74
ck_2	47.95	7.18	6.97	97.77
ck_3	46.50	6.96		98.03
Total	951.86	142.43	-	-

a~d represent the raw reads number of sequencing data; the number of clean reads; the average of clean reads per stress treatment sample; and the percentage of bases with a quality value greater than or equal to 30 to the total number of bases, respectively.

3.2. Transcriptome Sequence Assembly and Gene Function Annotation

The sequences with more than 99% similarity were merged and assembled by Trinity software for transcriptome assembly, and the parameters were set to the default parameters. Finally, 74,571 transcripts were generated, with an average length of 3002 bp. The smallest transcript sequence length was 190 bp, and the number of transcripts longer than 2000 bp was the largest, accounting for 56.41% of the total. The N50 was 4550 bp, and the GC content was 49.74% (Supplementary Table S2). Subsequently, the filtered sequences were aligned and annotated in databases such as KEGG, KOG, NR, GO, and Pfam. The results showed that the number of unigenes successfully annotated in NR, KEGG, KOG, GO, and Pfam databases accounted for 55.11% (8591), 19.49% (3039), 30.21% (4709), 23.91% (3728), and 54.85% (8550) of the total number of genes, respectively. 65.74% (10,248) of the transcript sequences were successfully annotated in at least one database (Supplementary Table S3).

3.3. Sample Correlation Analysis and Screening of Differentially Expressed Genes

To reflect the correlation of gene expression among all samples, Pearson correlation coefficients of all gene expression between every two samples were calculated, and these coefficients were reflected in the form of heat maps. As shown in Supplementary Figure S1, the repeatability and correlation of all the samples were high, and the subsequent analysis was highly reliable. The screening results of DEGs between ck vs. gw1d, ck vs. gw2d, and ck vs. gw3d showed that *S. rugosoannulata* differentially expressed 798 genes after 1 d of 35 °C high-temperature stress (gw1d), including 494 up-regulated and 304 down-regulated genes (Figure 1A). After 2 d of high-temperature stress at 35 °C (gw2d), a total of 851 genes were differentially expressed, and 548 and 303 genes were up-regulated and down-regulated, respectively (Figure 1B). After 3 d of high-temperature stress at 35 °C (gw3d), the total number of DEGs in *S. rugosoannulata* was 1484, including 963 up-regulated and 521 down-regulated genes (Figure 1C). With the prolongation of high-temperature stress time, the total number of DEGs in *S. rugosoannulata* gradually increased. In addition, the screening results of DEGs between ck vs. gh3d, ck vs. gh6d, and ck vs. gh9d showed that after 3 d of drought stress (gh3d), *S. rugosoanulata* differentially expressed a total of 421 genes, including 269 up-regulated and 152 down-regulated genes (Figure 2A). After

6 d of drought stress (gh6d), a total of 1072 genes were differentially expressed, with 711 up-regulated and 361 down-regulated genes, respectively (Figure 2B). After 9 d of drought stress (gh9d), the total number of DEGs in *S. rugosoanulata* was 2880, including 1501 up-regulated and 1379 down-regulated genes (Figure 2C). Clearly, as the duration of high temperature and drought stress prolongs, the total number of DEGs in *S. rugosoanulata* gradually increases. These findings suggest that as the duration of high temperature and drought stress increases, *S. rugosoanulata* will resist external adversity by regulating more genes.

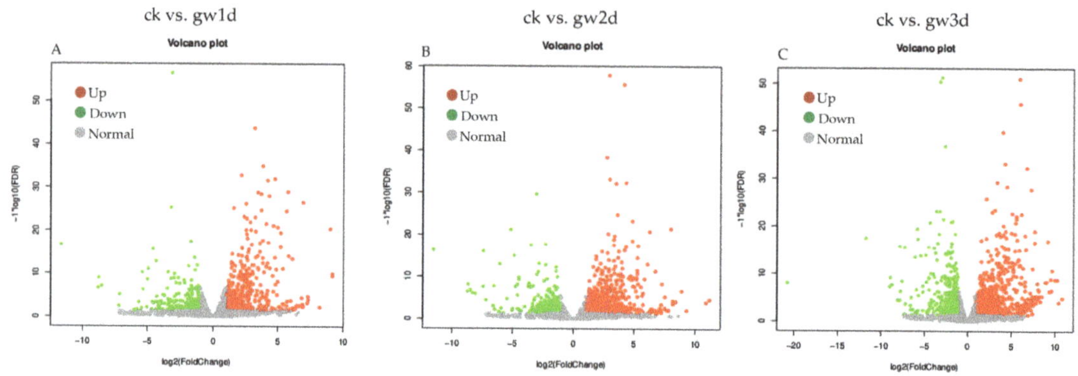

Figure 1. DEG analysis between the high-temperature stress treatment and the control group. Volcano plot analysis of DEGs for the pairwise comparisons ck vs. gw1d (**A**), ck vs. gw2d (**B**), and ck vs. gw3d (**C**).

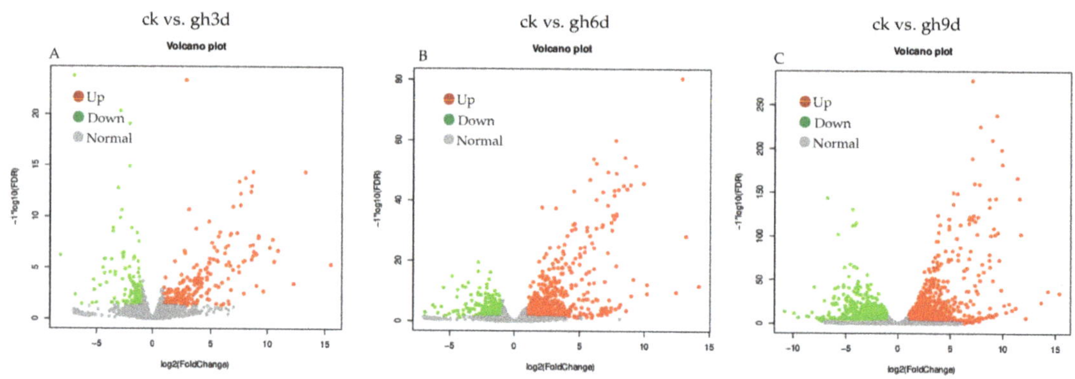

Figure 2. DEG analysis between the drought stress treatment and the control group. Volcano plot analysis of DEGs for the pairwise comparisons ck vs. gh3d (**A**), ck vs. gh6d (**B**), and ck vs. gh9d (**C**).

3.4. GO Enrichment Analysis of Differentially Expressed Genes

GO enrichment analysis was conducted on the DEGs screened between the group of high-temperature and drought-stress treatment versus the control group. Under high-temperature stress, 212 DEGs were enriched in the ck vs. gw1d group (Figure 3A). In terms of cellular components (CCs), membrane, nucleus, fungal-type cell wall, and other related DEGs were significantly enriched. In terms of molecular function (MF), ATP binding, protein binding, oxidoreductase activity, flavin adenine dinucleotide binding, catalytic activity, and other related DEGs were significantly enriched. After high-temperature stress, obsolete oxidation-reduction processes, transmembrane transport, biosynthetic processes, and other processes play an important role in biological processes (BPs). A total of 214 DEGs were enriched by GO in the ck vs. gw2d group, and the DEGs were significantly enriched in transmembrane transport, obsolete oxidation-reduction process, response to stress, and other items in BPs. In CCs, differential genes were significantly enriched in the membrane, nucleus, and other items. In MF, differential genes were significantly enriched in oxidoreductase activity, ATP binding, heme binding, and catalytic activity (Figure 3B). A total of 390 DEGs were enriched by GO in the ck vs. gw3d group, among which the obsolete oxidation-reduction process, transmembrane transport, oxidoreductase activity, metabolic process, and other BPs were significantly enriched. MF, such as ATP binding, heme binding, flavin adenine dinucleotide binding, and response to stress, were enriched, as were CCs, such as the membrane, nucleus, and fungal-type cell wall (Figure 3C).

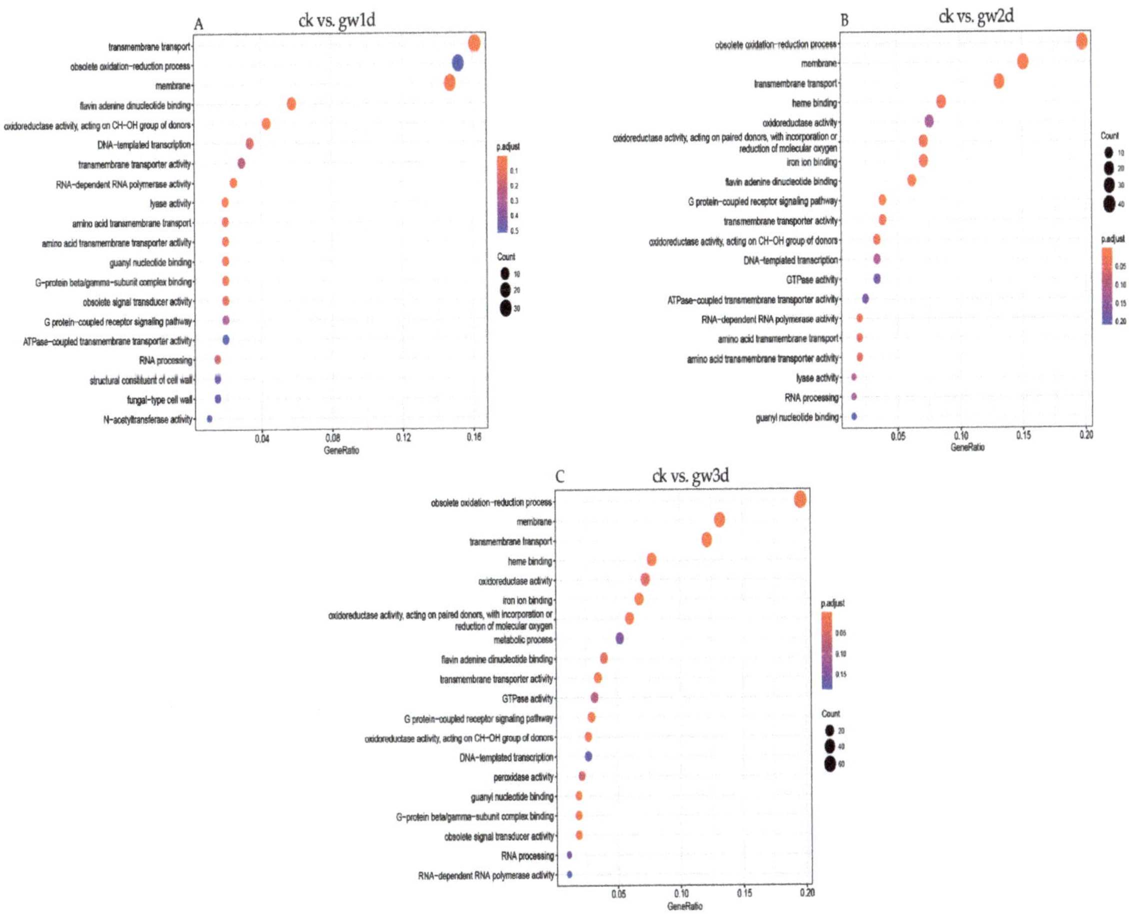

Figure 3. GO enrichment analysis of DEGs. Enrichment analysis of DEGs for the pairwise comparisons ck vs. gw1d (**A**), ck vs. gw2d (**B**), and ck vs. gw3d (**C**).

GO enrichment analysis of DEGs in *S. rugosoannulata* under drought stress indicated that 107 DEGs in ck vs. gh3d group were significantly enriched in CCs (including the membrane, nucleus, and fungal-type cell wall), MF (including ATP binding, protein binding, oxidoreductase activity, flavin adenine dinucleotide binding, and catalytic activity), and BPs (including obsolete oxidation-reduction processes, transmembrane transport, and metabolic processes) (Figure 4A). A total of 294 DEGs were enriched in the ck vs. gh6d group, and the DEGs were significantly enriched in obsolete oxidation-reduction processes, transmembrane transport, metabolic processes, protein kinase activity, response to stress, and other items in BPs. In CCs, the differential genes were significantly enriched in the membrane, nucleus, cytoplasm, and other items. In MF, differential genes were significantly enriched in iron ion binding, oxidoreductase activity, flavin adenine dinucleotide binding, protein binding, and transmembrane transporter activity (Figure 4B). In addition, 812 DEGs enriched by GO were found in the ck vs. gh9d. In terms of BPs, obsolete oxidation-reduction processes, transmembrane transport, metabolic processes, responses to stress, and other related DEGs were significantly enriched. As far as CCs are concerned, differential genes were significantly enriched in the membrane, nucleus, fungal-type cell wall, and cytoplasm. In terms of MF, ATP binding, protein binding, oxidoreductase activity, heme binding, iron ion binding, protein kinase activity, and other related DEGs were significantly enriched (Figure 4C).

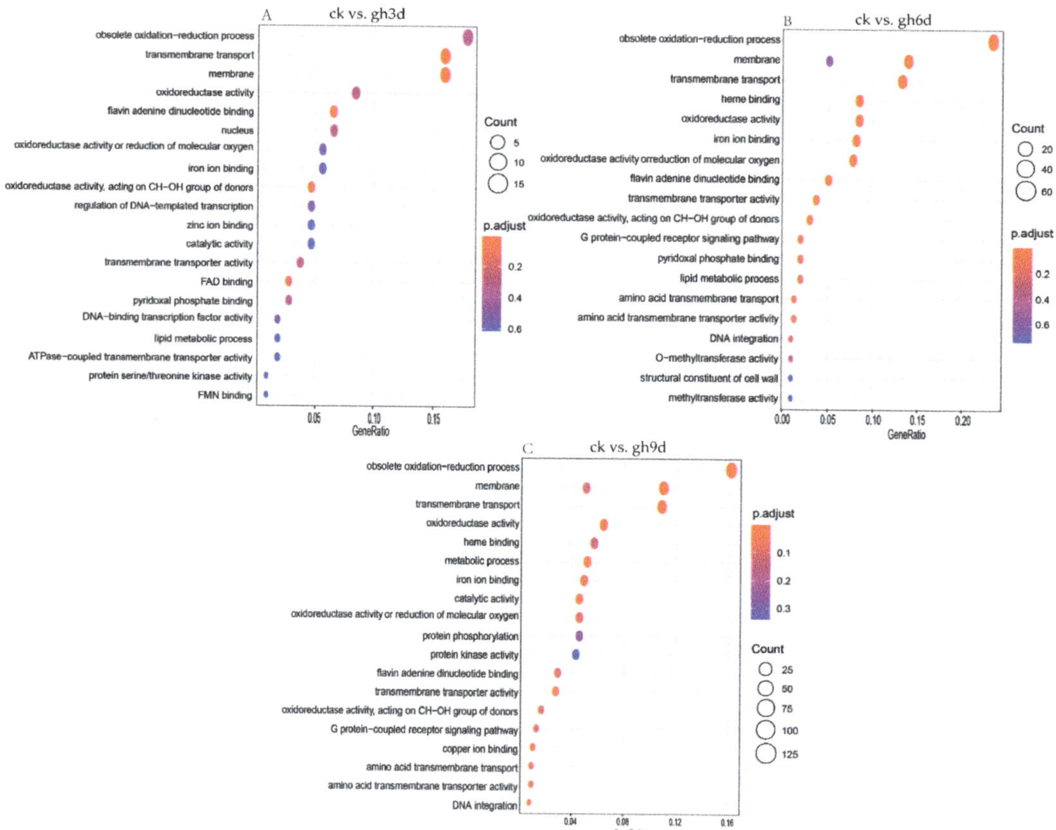

Figure 4. GO enrichment analysis of DEGs. Enrichment analysis of DEGs for the pairwise comparisons ck vs. gh3d (**A**), ck vs. gh6d (**B**), and ck vs. gh9d (**C**).

3.5. KEGG Enrichment Analysis of Differentially Expressed Genes

Similarly, KEGG signaling pathway enrichment analysis was carried out on the DEGs screened between the group of high temperature and drought stress treatment versus the control group, and the top 20 most significantly enriched pathways were listed based on the number of enriched DEGs. The results revealed that the pathways enriched in the ck vs. gw1d group mainly included protein processing in endoplasmic reticulum (ko04141), amino sugar and nucleotide sugar metabolism (ko00520), DNA replication (ko03030), pyruvate metabolism (ko00620), purine metabolism (ko00230), and the MAPK signaling pathway (ko04011) (Figure 5A). The pathways enriched in the ck vs. gw2d group mainly include the MAPK signaling pathway (ko04011), cysteine and methionine metabolism (ko00270), pyruvate metabolism (ko00620), phenylalanine metabolism (ko00360), and cyanoamino acid metabolism (ko00460) (Figure 5B). In addition, the pathways enriched in the ck vs. gw3d group mainly include the MAPK signaling pathway (ko04011), protein processing in the endoplasmic reticulum (ko04141), pyruvate metabolism (ko00620), steroid biosynthesis (ko00100), and cysteine and methionine metabolism (ko00270) (Figure 5C).

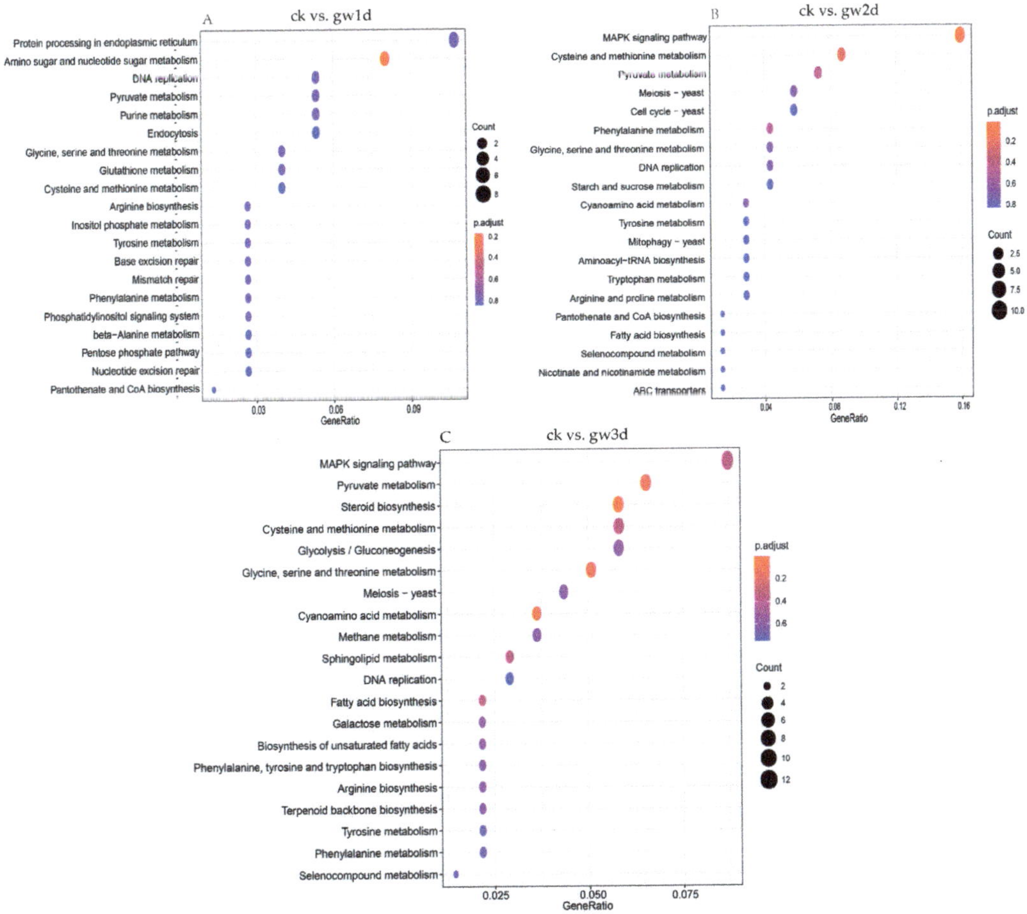

Figure 5. KEGG enrichment analysis of DEGs. Enrichment analysis of DEGs for the pairwise comparisons ck vs. gw1d (**A**), ck vs. gw2d (**B**), and ck vs. gw3d (**C**). The longitudinal axis represents the name of the pathway, and the size of the bubble represents the number of DEGs in this pathway.

KEGG pathway enrichment analysis was conducted on DEGs in *S. rugosoannulata* under different drought stress treatments, and the top 20 most significantly enriched pathways were listed based on the number of enriched DEGs. The results showed that the pathways enriched in the ck vs. gh3d group mainly include pentose and glucuronate interconversions (ko00040), butanoate metabolism (ko00650), alanine, aspartate, and glutamate metabolism (ko00250), spliceosome (ko03040), and the MAPK signaling pathway (ko04011)

(Figure 6A). The pathways enriched in the ck vs. gh6d group mainly include the MAPK signaling pathway (ko04011), steroid biosynthesis (ko00100), pentose and glucuronate interconversions (ko00040), meiosis (ko04113), and cysteine and methionine metabolism (ko00270) (Figure 6B). In addition, the pathways enriched in the ck vs. gh9d group mainly include the MAPK signaling pathway (ko04011), cell cycle (ko04111), meiosis (ko04113), DNA replication (ko03030), and nucleotide excision repair (ko03420). The results revealed that DEGs in *S. rugosoannulata* may adapt to drought stress through dynamic regulation of different pathways (Figure 6C).

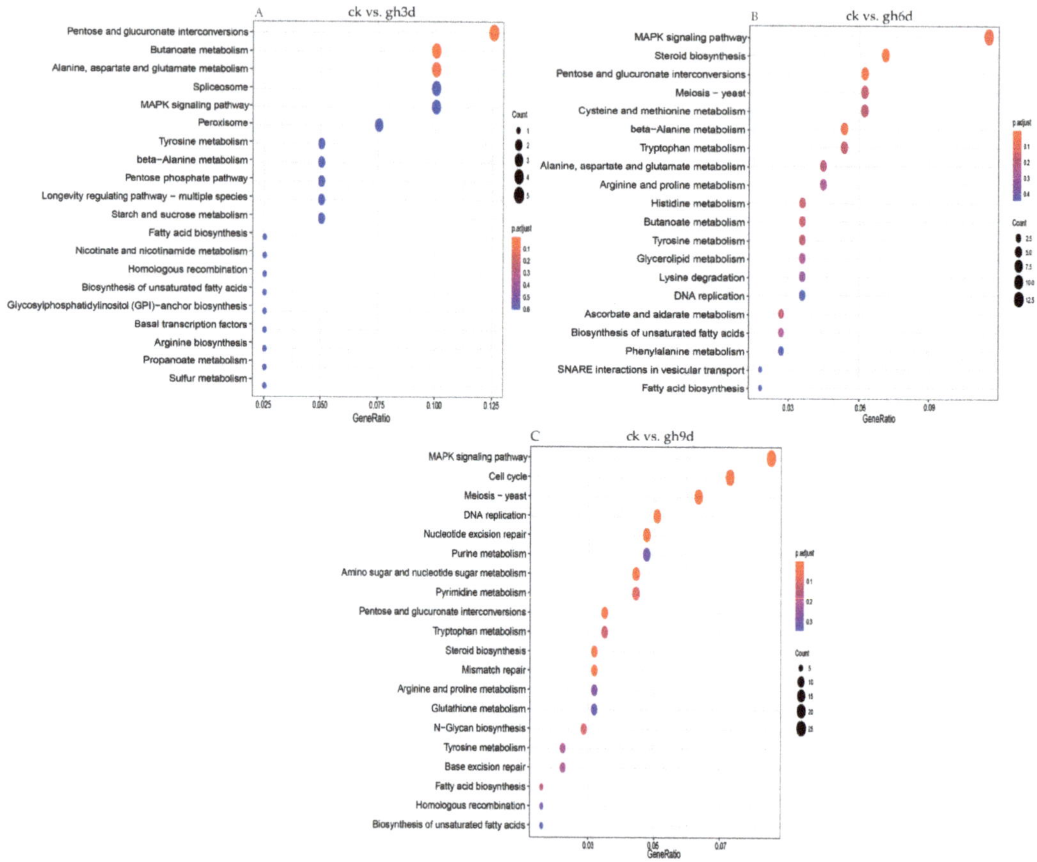

Figure 6. KEGG enrichment analysis of DEGs. Enrichment analysis of DEGs for the pairwise comparisons ck vs. gh3d (**A**), ck vs. gh6d (**B**), and ck vs. gh9d (**C**). The longitudinal axis represents the name of the pathway, and the size of the bubble represents the number of DEGs in this pathway.

In conclusion, the MAPK signaling pathway (ko04011) was significantly enriched in the DEGs between all stress treatment groups and the control group, indicating that this signaling pathway may play an important regulatory role in response to high temperature and drought stress in *S. rugosoanulata*. Therefore, DEGs in this pathway can be used as candidate genes for stress responses in subsequent gene annotation analyses.

3.6. Screening of Critical Candidate Genes' Response to High Temperature and Drought

In order to explore candidate genes that may be involved in stress resistance regulation of *S. rugosoannulata*, differential genes annotated in the KEGG pathway (ko04011) enriched in all treatment and control groups were selected for further analysis. The results showed that four, 13, and 27 genes were annotated in ck vs. gh3d, ck vs. gh6d, and ck vs. gh9d, respectively. Among them, two genes (*TRINITY_DN164_c0_g1*, *ATF2*, up-regulated; and *TRINITY_DN351_c0_g1*, *PTP2_3*, up-regulated) responded simultaneously among all groups, and 10 genes were repeatedly annotated in ck vs. gh6d and ck vs. gh9d (Supplementary Tables S4 and S5). In addition, four, 11, and 12 genes were annotated in ck vs. gw1d, ck vs. gw2d, and ck vs. gw3d under high-temperature stress, respectively. Among them, just one gene (*TRINITY_DN164_c0_g1*, *ATF2*, up-regulated) responded simultaneously among all groups, and eight genes were repeatedly annotated in ck vs. gw2d and ck vs. gw3d (Supplementary Tables S5 and S6). In summary, these 15 annotated candidate genes (Supplementary Tables S7 and S8) play a crucial role in response to high temperature and drought stress in *S. rugosoannulata*, providing genetically supported evidence for further analysis of the regulatory mechanism.

3.7. qRT-PCR Expression Analysis of Candidate Genes

In order to verify the accuracy of RNA-seq data and the expression pattern of candidate genes under high temperature and drought stress, eight candidate genes were randomly selected for qRT-PCR. As shown in Figure 7, the expression patterns of each candidate gene were differentially expressed under different stress treatments. Specifically, candidate genes *PTP2_3*, *BCK1*, *SHO1*, *GPCR*, and *SAC* were significantly expressed at 1 d of high-temperature stress, *ATF2* was significantly expressed at 2 d of high-temperature stress, while *GPA* and *PRMT5* were significantly expressed at 5 d of high-temperature stress. In addition, *BCK1*, *SHO1*, and *GPCR* were significantly expressed at 3 d of drought stress; *PTP2_3* and *PRMT5* were significantly expressed at 6 d of drought stress; while *ATF2*, *SAC*, and *GPA* were significantly expressed at 9 d of drought stress. The qRT-PCR results were basically consistent with the transcriptome sequencing data, which proved the reliability of the transcriptome sequencing data. In conclusion, the differential expression of each candidate gene under different stress treatments may be due to its response to high temperature and drought stress through different defense mechanisms in different regulatory pathways. These findings will provide important gene resources for stress-resistance breeding of *S. rugosoannulata*.

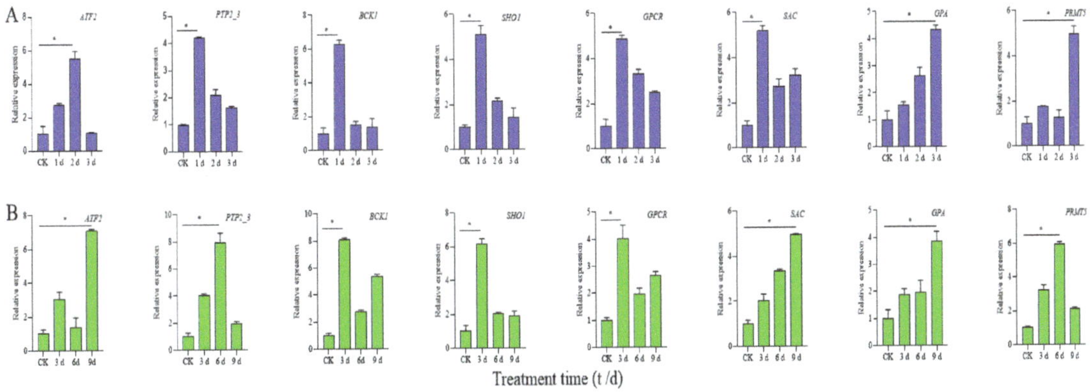

Figure 7. Relative expression levels of eight selected candidate genes were analyzed using qRT-PCR. (**A**) The expression patterns of eight candidate genes under high-temperature stress at different time points in *S. rugosoannulata*; (**B**) the expression patterns of eight candidate genes under drought stress at different time points in *S. rugosoannulata*. The x-axes show different stress treatments, and the y-axes show the relative expression level of each selected gene. Bars represent the standard deviation (SD) (n = 3) of three technical replicates. * Duncan's new multiple range test indicates that candidate genes are significantly differentially expressed at the level of $p < 0.05$.

4. Discussion

S. rugosoannulata is an important edible and medicinal fungus, which is delicious and nutritious and has health care and medicinal values such as enhancing human immunity and lowering blood sugar [2,34–36]. *S. rugosoannulata* belongs to medium-low temperature edible fungi, and its mycelium growth temperature ranges from 5 °C to 35 °C, and the temperature range of fruiting body formation and growth is from 10 °C to 20 °C [10,37]. In cultivation practice, the mycelium is easy to burn when the feed temperature is above 30 °C, and when the temperature rises to 20 °C, the fruiting body is easy to open, the quality becomes worse, and the primordium does not differentiate at 25 °C, which leads to cultivation failure or yield reduction [38–40]. Up to now, there are few reports on screening candidate genes of *S. rugosoannulata* responding to high temperature and drought stress by the high-throughput sequencing technique. In this study, transcriptome sequencing, GO functional classification, KEGG metabolic pathway analysis, and gene annotation were used to explore the critical candidate gene of *S. rugosoannulata* under stress to provide research reference for obtaining stress-tolerant strains in the future.

When organisms are exposed to high temperature, drought, and other adverse conditions for a long time, it will cause cell dehydration, an increase in reactive oxygen species (ROS), and intracellular level disorder [12,41,42]. Whether it is edible fungi or animal and plant breeding, the aim is to obtain new varieties with excellent characteristics. The high temperature and drought tolerance of edible fungi is a concrete manifestation of its extensive adaptation to the environment, which is closely linked with high yield. Hence, the screening of resistant strains is also one of the main characteristics of high-yield traits in *S. rugosoannulata* [11,40]. Therefore, high temperature and drought environments will not only reduce the metabolic activity of mycelium and affect the growth of mycelium but also cause damage to mycelium and reduce economic benefits. In order to cultivate varieties with stronger resistance to high temperature and drought stress, here, we used transcriptome analysis to study the *S. rugosoannulata* strain D-3 to better understand its expression pattern and response mechanism under high temperature and drought stress. Chen et al. [43] compared the mycelium growth of the collected *Auricularia auricula* at different growth temperatures for different treatment times and preliminarily identified three *A. auricula* varieties with high temperature resistance. In the preliminary experiment of this study, we found that long-term high temperature and drought stimulation could easily exceed the tolerance limit of *S. rugosoannulata* mycelia, resulting in degradation or even death of mycelia. Even if the suitable growth environment were restored, the growth rate and growth of mycelia would also deteriorate. In addition, 15 candidate genes of *S. rugosoannulata* in response to high temperature and drought stress were finally identified in this study. It is worth noting that one candidate gene was repeatedly identified in all treatment groups. The excavation of these genes is of great significance to improve the regulation mechanism of *S. rugosoannulata* in response to high temperature and drought stress and provides gene resources for its stress resistance breeding in the future.

Transcriptome sequencing technology, as an important research technique for identifying DEGs and screening stress-resistant genes, can not only help researchers quickly and globally understand all transcriptional changes in plant response to stress but also accurately obtain the target transcript sequence and the information of expression level, thereby quickly and accurately screening stress-related genes. Ma et al. [44] carried out transcriptomic analysis using drought- or high-temperature-resistant seedlings, and the results indicated that 11,731 and 9639 DEGs were identified, including 4444 drought-induced genes and 7287 drought-inhibited genes, and 5493 significantly up-regulated genes and 4146 significantly down-regulated genes under high-temperature stress treatment, respectively. Meanwhile, the critical genes that respond to both drought and high-temperature stress were identified, such as *WRKY*, *HSF*, and *MYB*. A report by Ren et al. [10] indicated that the catalase (*CAT*) and superoxide dismutase (*SOD*) activity of the thermotolerant *S. rugosoannulata* strain L3 were higher than those of the heat-sensitive *S. rugosoannulata* strain SM. The expression patterns of trehalose-6-phosphate phosphatase (*TPP*), *SOD*, and *CAT* in the thermotolerant strain L3 were higher than those in the heat-sensitive strain SM. Furthermore, Minina et al. [45] revealed that the up-regulated expression of the *atg5* and *atg8* mutant was closely related to the antioxidant response. Under drought stress, the significantly up-regulated expression of *atg5* and *atg8* in *Morchella sextelata* may affect the stress response of cells, resulting in increased autophagy, which may enhance the resistance to drought stress, thereby promoting the viability of *M. sextelata* in harsh environments. This study identified 15 candidate genes responding to drought stress through RNA-seq technology in *S. rugosoannulata*. The results in the present study will be useful in expanding our insight into drought and high-temperature stress biology. Notably, two candidate genes were repeatedly identified among all treatment groups. The isolation, functional research, and study on the molecular mechanism of stress resistance of these candidate genes will be the focus of future research.

Under drought stress, *app1* in fungi can regulate the expression of important antioxidant enzymes and reduce oxidative stress caused by drought. In addition, it is also related to the steady-state regulation of copper ions, which further affects the redox balance of cells, thus playing a protective role under drought stress [46]. In addition, *PRO1* can regulate multiple signaling pathways under drought stress, including cell wall integrity, NADPH oxidase, and pheromone signal transduction, and plays an important role in the growth and development of fungi [47]. Moreover, Xin et al. [48] speculated that the expression of the hydrophobin *hyd1* gene could enhance the ability of *L. edodes* to resist high-temperature stress by using transcriptome technology. In summary, in combination with the 15 candidate genes screened in this study, no repeatedly reported candidate genes were found. This may be due to the different degrees of stress treatments of different edible fungi or different strains of *S. rugosoannulata*, resulting in different molecular mechanisms for responding to stress treatment. At present, there are few reports on the molecular mechanisms of *S. rugosoannulata* under stress treatment. However, with the rapid development of sequencing technology and molecular biology technology, elucidating the molecular mechanisms of its response to abiotic and biotic stresses will become a top priority for researchers. At present, the assembled genome of the *S. rugosoannulata* strain QGU27 (GCA_028532985.1) was posted in the NCBI database in 2023 (https://www.ncbi.nlm.nih.gov/datasets/genome/GCA_028532985.1/, accessed on 21 September 2025); however, there is no *S. rugosoannulata* strain D-3 genome information published, and there is no report on the molecular response mechanism of *S. rugosoannulata* to high temperature and drought stress at the transcriptome level, and the genetic information available for reference is insufficient, which limits the mining of *S. rugosoannulata* functional genes to a certain extent. In this study, based on RNA-seq technology, the DEGs of *S. rugosoannulata* in response to high temperature and drought stress were preliminarily screened, and the expression patterns of eight candidate genes were verified. However, the in-depth analysis and functional identification of these genes are worthy of further work.

5. Conclusions

Our study provides the first comprehensive discovery of the transcriptome of *S. rugosoannulata* under drought and high-temperature stress. We identified 74,571 transcripts and 16,233 unigenes, out of which 7506 DEGs were differentially regulated. GO analysis showed that the DEGs were mainly involved in biological processes such as the obsolete oxidation-reduction process, transmembrane transport, and response to stress. In addition, the MAPK signaling pathway (ko04011) was significantly enriched in all the stress treatment groups versus the control group. Further, the present study identifies 15 candidate genes that respond to both high temperature and drought stress through annotation. The expression levels of eight selected candidate genes were carried out via qRT-PCR, and the results were consistent with the transcriptome datasets, indicating the data's reliability. Until now, this is the first study to screen stress-related candidate genes by RNA-seq in *S. rugosoannulata* under high temperature and drought stress conditions. Therefore, our study provides valuable information and gene resources for the molecular breeding and the regulation mechanism for drought and high temperature resistance in *S. rugosoannulata* and other edible fungi species.

Supplementary Materials: The following supporting information can be downloaded at https://www.mdpi.com/article/10.3390/cimb47100835/s1: Figure S1. Correlation heatmap of all samples. Table S1. Primer for qRT-PCR of eight candidate genes. Table S2. The distribution of transcript assembly length. Table S3. Statistics of function annotation results. Table S4. Screening of candidate genes response to drought. Table S5. Annotation of candidate genes. Table S6. Screening of candidate genes response to high temperature. Table S7. Annotation of 15 candidate genes. Table S8. The sequence of 15 candidate genes.

Author Contributions: Conceptualization, W.Y. and R.W.; methodology, W.Y.; software, S.W., S.Y., M.Z., H.L., and X.H.; validation, S.Y., T.X., H.L., and Q.L.; formal analysis, M.Z., Q.L., and T.X.; data curation, S.W.; writing—original draft preparation, S.Y.; writing—review and editing, W.Y.; supervision, W.Y. and S.W.; project administration, S.Y. All authors have read and agreed to the published version of the manuscript.

Funding: This research was funded by Fujian Provincial Department of Science and Technology (2024S0015).

Data Availability Statement: The original contributions presented in this study are included in the article/Supplementary Materials. Further inquiries can be directed to the corresponding author(s).

Acknowledgments: The authors are thankful to all team members in Ye's lab, Institute of Edible Fungi, Sanming Academy of Agricultural Sciences, for helpful discussions and suggestions.

Conflicts of Interest: The authors declare no conflicts of interest.

References

1. Yang, Y.; Meng, G.L.; Ni, S.J.; Zhang, H.F.; Dong, C.H. Genomic analysis of *Stropharia rugosoannulata* reveals its nutritional strategy and application potential in bioremediation. *J. Fungi* **2022**, *8*, 162. [CrossRef] [PubMed]
2. Li, S.R.; Wang, L.; Ni, S.J.; Wang, H.H.; Liu, X.F.; Liu, C.X. The amino acids content of different part of *Stropharia rugosoannulata* and their nutrition evaluation. *Food Res. Dev.* **2017**, *38*, 95–99.
3. Liu, M.M.; Zhang, X.L.; Xu, L.L.; Jiang, P. Analysis of amino acids content and food safety assessment of *Stropharia rugosoannulata* cultivated in the imitated wild environment under forest. *Edible Fungi China* **2021**, *40*, 67–70.
4. Tang, S.J.; Fan, T.T.; Jin, L.; Lei, P.; Shao, C.X.; Wu, S.L.; Yang, Y.; He, Y.L.; Ren, R.; Xu, J. Soil microbial diversity and functional capacity associated with the production of edible mushroom *Stropharia rugosoannulata* in croplands. *Peer J.* **2022**, *10*, e14130. [CrossRef]
5. Wang, H.; Chen, H.; Zhang, J.J.; Chen, M.J. Research progresses on bioactive components in *Stropharia rugosoannulata* and their pharmacological effects. *Acta Edulis Fungi* **2018**, *25*, 115–120.
6. Wang, X.W.; Zhan, W.; Tao, M.X.; Ji, Z.S.; Liu, Y.; Wang, F.; Cheng, G.Y. Analysis of nutritional components and antioxidant active substances of *Stropharia rugosoannulata*. *Edible Fungi* **2007**, *6*, 62–63.
7. Huang, L.; Si, C.; He, C.M.; Liu, X.C.; Duan, J. The rise of Stropharia rugosoannulata industry in China: Current state and prospects. *Appl. Microbiol. Biotechnol.* **2025**, *109*, 188. [CrossRef]
8. Huang, L.; Si, C.; Shi, H.Y.; He, C.M.; Duan, J. Research on the stipe cracking of wine-cap mushroom (*Stropharia rugosoannulata*) in different humidity conditions. *Sci. Rep.* **2023**, *13*, 21122. [CrossRef]
9. Wu, X.; Du, Z.H.; Liu, L.; Chen, Z.L.; Li, Y.R.; Fu, S.B. Integrative analysis of transcriptome and metabolome sheds light on flavonoid biosynthesis in the fruiting body of *Stropharia rugosoannulata*. *J. Fungi* **2024**, *10*, 254. [CrossRef]

10. Ren, J.F.; Wang, Q.J.; Zuo, J.; Jiang, S.X. Study of thermotolerant mechanism of *Stropharia rugosoannulata* under high temperature stress based on the transcriptome sequencing. *Mycoscience* **2021**, *62*, 95–105. [CrossRef]
11. Sakamoto, Y. Influences of environmental factors on fruiting body induction, development and maturation in mushroom-forming fungi. *Fungal Biol. Rev.* **2018**, *32*, 236–248. [CrossRef]
12. Miller, G.; Shulaev, V.; Mittler, R. Reactive oxygen signaling and abiotic stress. *Physiol. Plant* **2008**, *133*, 481–489. [CrossRef] [PubMed]
13. Skyba, M.; Petijova, L.; Kosuth, J.; Koleva, D.; Ganeva, T.; Kapchinatoteva, V.; Cellarova, E. Oxidative stress and antioxidant response in *Hypericum perforatum* L. plants subjected to low temperature treatment. *J. Plant Physiol.* **2012**, *169*, 955–964. [CrossRef] [PubMed]
14. Bellettini, M.B.; Fiorda, F.A.; Maieves, H.A.; Teixeira, G.L.; Ávila, S.; Hornung, P.S.; Júnior, A.M.; Ribani, R.H. Factors affecting mushroom *Pleurotus* spp. *Saudi J. Biol. Sci.* **2019**, *26*, 633–646.
15. Liang, Y.Q.; Luo, K.M.; Wang, B.L.; Huang, B.Q.; Fei, P.; Zhang, G.G. Inhibition of polyphenol oxidase for preventing browning in edible mushrooms: A review. *J. Food Sci.* **2024**, *89*, 6796–6817. [CrossRef]
16. Casu, A.; Camardo Leggieri, M.; Toscano, P.; Battilani, P. Changing climate, shifting mycotoxins: A comprehensive review of climate change impact on mycotoxin contamination. *Compr. Rev. Food Sci. Food Saf.* **2024**, *23*, e13323. [CrossRef]
17. Adhikari, M.; Isaac, E.L.; Paterson, R.R.M.; Maslin, M.A. A review of potential impacts of climate change on coffee cultivation and mycotoxigenic fungi. *Microorganisms* **2020**, *8*, 1625. [CrossRef]
18. Yan, Z.Y.; Zhao, M.R.; Huang, C.Y.; Zhang, L.J.; Zhang, J.X. Trehalose alleviates high-temperature stress in *Pleurotus ostreatus* by affecting central carbon metabolism. *Microb. Cell Tact.* **2021**, *20*, 82. [CrossRef]
19. Kong, W.W.; Huang, C.Y.; Chen, Q.; Zou, Y.J.; Zhang, J.X. Nitric oxide alleviates heat stress-induced oxidative damage in *Pleurotus eryngii* var. *tuoliensis*. *Fungal Genet. Biol.* **2012**, *49*, 15–20. [CrossRef]
20. Hao, H.B.; Huang, J.C.; Wang, Q.; Juan, J.X.; Xiao, T.T.; Song, X.X.; Chen, H.; Zhang, J.J. Effects of heat stress on differential expression of antioxidant enzymes and heat shock protein genes of *Agaricus bisporus*. *Mycosystema* **2021**, *40*, 616–625.
21. Liu, R.; Zhu, T.; Yang, T.; Yang, Z.Y.; Ren, A.; Shi, L.; Zhu, J.; Yu, H.S.; Zhao, M.W. Nitric oxide regulates ganoderic acid biosynthesis by the S-nitrosylation of aconitase under heat stress in *Ganoderma lucidum*. *Environ. Microbiol.* **2021**, *23*, 682–695. [CrossRef]
22. Sato, H.; Mizoi, J.; Shinozaki, K.; Yamaguchi-Shinozaki, K. Complex plant responses to drought and heat stress under climate change. *Plant J.* **2024**, *117*, 1873–1892. [CrossRef] [PubMed]
23. Ma, J.C.; Ma, X.J. Cultivation techniques of Morchella in greenhouse. *Agricul. Technol. Equip.* **2022**, *4*, 136–141.
24. Wijayawardene, N.N.; Boonyuen, N.; Ranaweera, C.B.; de Zoysa, H.K.S.; Padmathilake, R.E.; Nifla, F.; Dai, D.Q.; Liu, Y.; Suwannarach, N.; Kumla, J.; et al. OMICS and other advanced technologies in mycological applications. *J. Fungi* **2023**, *9*, 688. [CrossRef]
25. Kim, J.Y.; Kim, D.Y.; Park, Y.J.; Jang, M.J. Transcriptome analysis of the edible mushroom *Lentinula edodes* in response to blue light. *PLoS ONE* **2020**, *15*, e0230680. [CrossRef]
26. Yu, H.L.; Li, Q.Z.; Shen, X.F.; Zhang, L.J.; Liu, J.Y.; Tan, Q.; Li, Y.; Lv, B.B.; Shang, X.D. Transcriptomic analysis of two *Lentinula edodes* genotypes with different cadmium accumulation ability. *Front. Microbiol.* **2020**, *11*, 558104. [CrossRef]
27. Hao, H.B.; Zhang, J.J.; Wu, S.D.; Bai, J.; Zhuo, X.Y.; Zhang, J.X.; Kuai, B.K.; Chen, H. Transcriptomic analysis of *Stropharia rugosoannulata* reveals carbohydrate metabolism and cold resistance mechanisms under low-temperature stress. *AMB Expr.* **2022**, *12*, 56, Erratum in *AMB Expr.* **2022**, *12*, 72. [CrossRef]
28. Haas, B.J.; Papanicolaou, A.; Yassour, M.; Grabherr, M.; Blood, P.D.; Bowden, J.; Couger, M.B.; Eccles, D.; Li, B.; Lieber, M.; et al. De novo transcript sequence reconstruction from RNA-seq using the Trinity platform for reference generation and analysis. *Nat. Protoc.* **2013**, *8*, 1494–1512. [CrossRef]
29. Conesa, A.; Götz, S. Blast2GO: A comprehensive suite for functional analysis in plant genomics. *Int. J. Plant Genom.* **2008**, *2008*, 619832. [CrossRef]
30. Leng, N.; Dawson, J.A.; Thomson, J.A.; Ruotti, V.; Rissman, A.I.; Smits, B.M.; Haag, J.D.; Gould, M.N.; Stewart, R.M.; Kendziorski, C. EBSeq: An empirical Bayes hierarchical model for inference in RNA-seq experiments. *Bioinformatics* **2013**, *29*, 1035–1043. [CrossRef]
31. Xu, S.B.; Hu, E.Q.; Cai, Y.T.; Xie, Z.J.; Luo, X.; Zhan, L.; Tang, W.L.; Wang, Q.W.; Liu, B.D.; Wang, R.; et al. Using clusterProfiler to characterize multiomics data. *Nat. Protoc.* **2024**, *19*, 3292–3320. [CrossRef]
32. Cheng, W.; Wang, Z.T.; Xu, F.; Lu, G.L.; Su, Y.C.; Wu, Q.B.; Wang, T.; Que, Y.X.; Xu, L.P. Screening of candidate genes associated with brown stripe resistance in sugarcane via BSR-seq analysis. *Int. J. Mol. Sci.* **2022**, *23*, 15500. [CrossRef] [PubMed]
33. Livak, K.J.; Schmittgen, T.D. Analysis of relative gene expression data using real-time quantitative PCR and the 2(-Delta Delta CT) Method. *Methods* **2001**, *25*, 402–408. [CrossRef] [PubMed]
34. Liu, D.; Chen, Y.Q.; Xiao, X.W.; Zhong, R.T.; Yang, C.F.; Liu, B.; Zhao, C. Nutrient properties and nuclear magnetic resonance-based metabonomic analysis of macrofungi. *Foods* **2019**, *8*, 397. [CrossRef] [PubMed]

35. El-Ramady, H.; Abdalla, N.; Badgar, K.; Llanaj, X.; Törős, G.; Hajdú, P.; Eid, Y.; Prokisch, J. Edible mushrooms for sustainable and healthy human food: Nutritional and medicinal attributes. *Sustainability* **2022**, *14*, 4941. [CrossRef]
36. Hamza, A.; Mylarapu, A.; Krishna, K.V.; Kumar, D.S. An insight into the nutritional and medicinal value of edible mushrooms: A natural treasury for human health. *J. Biotechnol.* **2024**, *381*, 86–99. [CrossRef]
37. Yan, P.S.; Li, G.F.; Jiang, J.H.; Zhang, Y.C.; Wang, H.; Xu, H.T. Effects of nutrients and environmental factors on the mycelial growth of *Stropharia rugosoannulata*. *Acta Edulis Fungi* **2001**, *1*, 5–9.
38. Hu, Y.W.; Kakumyan, P.; Bandara, A.R.; Mortimer, P.E. The nutrition, cultivation and biotechnology of *Stropharia rugosoannulata*. *Fungal Biotec.* **2021**, *1*, 13–25.
39. Bruhn, J.N.; Abright, N.; Mihail, J.D. Forest farming of wine-cap *Stropharia* mushrooms. *Agrofor. Syst.* **2010**, *79*, 267–275. [CrossRef]
40. Pekşen, A.; Kibar, B. Determination of optimum culture conditions for mycelial growth of *Macrolepiota procera* mushroom. *Acta Sci. Pol. Hortorum Cultus* **2020**, *19*, 11–20. [CrossRef]
41. Choudhury, F.K.; Rivero, R.M.; Blumwald, E.; Mittler, R. Reactive oxygen species, abiotic stress and stress combination. *Plant J.* **2017**, *90*, 856–867. [CrossRef]
42. Suzuki, N.; Koussevitzky, S.; Mittler, R.; Miller, G. ROS and redox signalling in the response of plants to abiotic stress. *Plant Cell Environ.* **2012**, *35*, 259–270. [CrossRef] [PubMed]
43. Chen, X.F.; Wu, S.J.; Pang, H.; Wang, C.Q.; Wu, X.J.; Wei, S.Y. High temperature resistance strains screening of *Auricularia auricula-judae* in Guangxi. *Southwest China J. Agricul. Sci.* **2018**, *31*, 131–135.
44. Ma, P.P.; Guo, G.L.; Xu, X.Q.; Luo, T.Y.; Sun, Y.; Tang, X.M.; Heng, W.; Jia, B.; Liu, L. Transcriptome analysis reveals key genes involved in the response of *Pyrus betuleafolia* to drought and high-temperature stress. *Plants* **2024**, *13*, 309. [CrossRef] [PubMed]
45. Minina, E.A.; Moschou, P.N.; Vetukuri, R.R.; Sanchez-Vera, V.; Cardoso, C.; Liu, Q.; Elander, P.H.; Dalman, K.; Beganovic, M.; Lindberg Yilmaz, J.; et al. Transcriptional stimulation of rate-limiting components of the autophagic pathway improves plant fitness. *J. Exp. Bot.* **2018**, *69*, 1415–1432. [CrossRef]
46. Zhou, X.M.; Xiang, Y.; Li, C.L.; Yu, G.H. Modulatory role of reactive oxygen species in root development in model plant of *Arabidopsis thaliana*. *Front. Plant Sci.* **2020**, *11*, 485932. [CrossRef]
47. Steffens, E.K.; Becker, K.; Krevet, S.; Teichert, I.; Kück, U. Transcription factor *PRO1* targets genes encoding conserved components of fungal developmental signaling pathways. *Mol. Microbiol.* **2016**, *102*, 792–809. [CrossRef]
48. Xin, M.M.; Zhao, Y.; Huang, J.L.; Song, C.Y.; Chen, M.J. Expression and bioinformatics analysis of hydrophobin protein gene (*hyd1*) in *Lentinula edodes* under high temperature stress. *Mol. Plant Breed.* **2016**, *14*, 2645–2652.

current issues in molecular biology

MDPI

Article

ZmNLR-7-Mediated Synergistic Regulation of ROS, Hormonal Signaling, and Defense Gene Networks Drives Maize Immunity to Southern Corn Leaf Blight

Bo Su [1,2,†], Xiaolan Yang [1,†], Rui Zhang [1], Shijie Dong [1], Ying Liu [1], Hubiao Jiang [1], Guichun Wu [3,*] and Ting Ding [1,*]

1 School of Plant Protection, Key Laboratory of Biology and Sustainable Management of Plant Diseases and Pests of Anhui Higher Education Institutes, Anhui Agricultural University, Anhui Province Key Laboratory of Integrated Pest Management on Crops, Hefei 230036, China; subowsm31415@163.com (B.S.); yxiaolan77@163.com (X.Y.); 19505560818@163.com (R.Z.); dongshijie0823@163.com (S.D.); 15855053647@163.com (Y.L.); 12116088@zju.edu.cn (H.J.)
2 School of Biological and Food Engineering, Su Zhou University, Suzhou 234000, China
3 National Engineering Laboratory of Crop Stress Resistance Breeding, School of Life Sciences, Anhui Agricultural University, Hefei 230036, China
* Correspondence: guichunwu@ahau.edu.cn (G.W.); dtdingting98@ahau.edu.cn (T.D.)
† These authors contributed equally to this work.

Abstract

The rapid evolution of pathogens and the limited genetic diversity of hosts are two major factors contributing to the plant pathogenic phenomenon known as the loss of disease resistance in maize (*Zea mays* L.). It has emerged as a significant biological stressor threatening the global food supplies and security. Based on previous cross-species homologous gene screening assays conducted in the laboratory, this study identified the maize disease-resistance candidate gene *ZmNLR-7* to investigate the maize immune regulation mechanism against *Bipolaris maydis*. Subcellular localization assays confirmed that the ZmNLR-7 protein is localized in the plasma membrane and nucleus, and phylogenetic analysis revealed that it contains a conserved NB-ARC domain. Analysis of tissue expression patterns revealed that *ZmNLR-7* was expressed in all maize tissues, with the highest expression level (5.11 times) exhibited in the leaves, and that its transcription level peaked at 11.92 times 48 h post *Bipolaris maydis* infection. Upon inoculating the *ZmNLR-7* EMS mutants with *Bipolaris maydis*, the disease index was increased to 33.89 and 43.33, respectively, and the lesion expansion rate was higher than that in the wild type, indicating enhanced susceptibility to southern corn leaf blight. Physiological index measurements revealed a disturbance of ROS metabolism in *ZmNLR-7* EMS mutants, with SOD activity decreased by approximately 30% and 55%, and POD activity decreased by 18% and 22%. Moreover, H_2O_2 content decreased, while lipid peroxide MDA accumulation increased. Transcriptomic analysis revealed a significant inhibition of the expression of the key genes *NPR1* and *ACS6* in the SA/ET signaling pathway and a decrease in the expression of disease-related genes *ERF1* and *PR1*. This study established a new paradigm for the study of NLR protein-mediated plant immune mechanisms and provided target genes for molecular breeding of disease resistance in maize. Overall, these findings provide the first evidence that *ZmNLR-7* confers resistance to southern corn leaf blight in maize by synergistically regulating ROS homeostasis, SA/ET signal transduction, and downstream defense gene expression networks.

Keywords: *ZmNLR-7*; maize; *Bipolaris maydis*; plant disease susceptibility

1. Introduction

Plant diseases often cause large-scale reductions in grain production. Breeding disease-resistant varieties is one of the most economical and environmentally friendly means to resist crop diseases. Plants have built a complex and sophisticated immune defense system during the long-term evolution process and precisely regulate their resistance to pathogenic microorganisms through multi-level signaling networks. Its core mechanisms include basic immunity (pattern-triggered immunity, PTI) mediated by cell-surface pattern-recognition receptors (PRRs) and effector-triggered immunity (effector-triggered immunity, ETI) dominated by intracellular nucleotide-binding leucine-rich repeat proteins (nucleotide-binding leucine-rich repeat receptors, NLRs) [1]. PTI activates basic defense responses such as MAPK cascades and callose deposition by recognizing pathogen-associated molecular patterns (PAMPs; such as bacterial flagellar protein flg22 and fungal chitin), while ETI specifically recognizes effector proteins secreted by pathogens through NLR proteins, triggering strong anti-disease phenotypes, such as hypersensitive response (HR) and reactive oxygen species (ROS) bursts [2]. Research breakthroughs in recent years have shown that PTI and ETI have a synergistic mechanism: the two jointly regulate the phosphorylation activation of NADPH oxidase RBOHD, significantly enhancing the ROS synthesis capacity and forming an "immune signal amplifier" effect [3]. For example, in *Arabidopsis*, FLS2 (PTI receptor) and RPS4 (ETI receptor) promote the dual phosphorylation of Ser39/343 sites of RBOHD through interaction, which increases the ROS production by 3–5 times compared with a single pathway, thereby breaking through the antioxidant defense threshold of pathogens [4]. At the level of systemic immune regulation, the three major hormone pathways of salicylic acid (SA), jasmonic acid (JA), and ethylene (ET) constitute a dynamic regulatory network. SA signals regulate the expression of pathogenesis-related genes (*PR* genes) through NPR1 receptor proteins, specifically resisting biotrophic pathogens, while the JA/ET pathway activates the synthesis of defense substances, such as protease inhibitors, through *MYC2/ERF* transcription factors to resist the invasion of necrotrophic pathogens [5]. It is worth noting that there is a fine cross-regulation between hormone pathways: SA can antagonize JA signals by inhibiting the stability of JAZ proteins, while ET can synergistically enhance JA responses by enhancing the activity of *EIN3/EIL1* transcription factors. This "disease resistance signal allocation" mechanism enables plants to optimize defense resource allocation according to the type of pathogen, while maintaining a balance between growth and development and immune response through the SA-JA-ET signal integration node mediated by *EIN2* [6].

As a core member of plant disease-resistance genes (R genes), NLR (nucleotide-binding leucine-rich Repeat) genes play an irreplaceable biological function in effector-triggered immunity (ETI) [7]. Its molecular characteristics are composed of three typical domains: the amino-terminal signal transduction domain (CC or TIR), the central nucleotide binding site (NBS/NB-ARC), and the carboxyl-terminal leucine-rich repeat sequence (LRR). According to the differences in the amino-terminal domain, NLR proteins can be divided into two major categories: CNL (CC-NBS-LRR) and TNL (TIR-NBS-LRR) [8,9]. It is worth noting that dicotyledons have both CNL and TNL types, while monocotyledons (such as wheat) usually only retain the CNL type [10]. Functional studies have shown that the amino-terminal domain (such as CC or TIR) is responsible for downstream immune signal transduction, while the LRR domain specifically recognizes pathogenic effectors through conformational changes [11,12]. Among them, the NBS domain acts as a molecular switch, and its nucleotide binding properties are closely related to the NTPase activity of the STAND (Signal Transduction ATPases with Numerous Domains) superfamily, regulating the protein activation state through the ATP/GTP binding–hydrolysis cycle [13,14]. NLR proteins are not only the core executors of ETI, but their modular structure also provides unique advantages for molecular design. Through strategies such as domain recombination and gene stacking, the precise expansion and persistence of the anti-disease spectrum can be achieved. For example, the wheat stem rust-resistance NLR genes *Sr22*, *Sr35*, *Sr45*, *Sr50*, and *Sr55* were transferred in series into the susceptible variety Fielder, which can give it broad-spectrum resistance to stem rust fungi with different physiological subspecies around the world [15]. Similarly, the wheat *Sr22*, *Sr33*, *Sr35*, and *Sr45* genes were introduced into the barley variety Golden Promise, which successfully obtained specific resistance to stem rust fungus *TTKSK* without affecting agronomic traits [16]. These breakthrough results not only verify the feasibility of NLR genetic engineering but also reveal its great potential in crop disease-resistance breeding [17,18]. The current research focus has shifted to the optimization of multi-gene stacking systems and the analysis of disease-resistance–developmental balance mechanisms, laying a theoretical foundation for the design of the next generation of intelligent disease-resistant varieties.

By examining the disease-resistance phenotype of EMS mutant strains, integrating disease resistance phenotypic transcriptomics, and measuring oxidative stress physiology, this study finished the multi-level functional analysis of the maize disease-resistance candidate gene *ZmNLR-7* (*Zm00001d020117*) in the maize–microbe *Bipolaris maydis* interaction system, which was based on the cross-species homologous gene-screening strategy. This study provided a novel functional module for the molecular design and breeding of maize resistance genes by revealing, for the first time, the molecular mechanism by which *ZmNLR-7* coordinates disease-resistance responses by controlling reactive oxygen homeostasis and the ET/SA signaling pathway.

2. Materials and Methods

2.1. Bioinformatics Analysis of ZmNLR-7

The full-length coding sequence (CDS) and corresponding amino acid sequence of *ZmNLR-7* (Table S6) were retrieved from the maize genome database (MaizeGDB; https://www.maizegdb.org, accessed on 18 March 2024). Physicochemical properties of the encoded protein, including molecular weight, isoelectric point (pI), and amino acid composition, were analyzed using the ProtParam tool on the Expasy online platform (https://web.expasy.org/protparam/, accessed on 8 June 2024). Structural domains of ZmNLR-7 protein sequences were predicted via the SMART web server (http://smart.embl-heidelberg.de/, accessed on 10 June 2024). Homologous sequences of ZmNLR-7 in *Zea mays*, *Oryza sativa*, *Sorghum bicolor*, *Setaria italica*, and *Arabidopsis thaliana* were identified through systematic screening of the NCBI GenBank database. Phylogenetic analysis was performed using MEGA7.0 software with the neighbor-joining (NJ) method, and the resultant tree file was imported into the iTOL v6 platform (https://itol.embl.de/, accessed on 16 July 2024) for multidimensional annotation and visualization optimization.

2.2. Vector Construction and Subcellular Localization

The *ZmNLR-7* coding sequence was amplified from maize-leaf mRNA using RT-PCR, with primers *pst1*-F (5′-TCCTCTAGAGTCGACCTGCAGATGCCGCATGGTCACGCC-3′) and *pst1*-R (5′-TAAAGCAGGGCATGCCTGCAGTTTTGCACAGCCCTTTTGCA-3′). The verified product was directionally inserted into the *pCAMBIA2300* vector via homologous recombination (Figure S5). Recombinant plasmids were electroporated into Agrobacterium tumefaciens GV3101 (containing pSoup helper plasmid) for agroinfiltration-mediated transient expression. For subcellular localization, Nicotiana benthamiana leaves were infiltrated with recombinant Agrobacterium suspensions and analyzed 48 h post-infiltration, following established protocols [19]. Maize Protoplast Source and Transformation B73 seeds were sown in soil at a depth of ~2 cm and dark-incubated at 28 °C, with soil moisture maintained by regular watering. After 10–11 days of growth, when the second leaf of etiolated seedlings became fully expanded (~12–15 cm in length), protoplasts were isolated and subjected to transformation assays. Successfully constructed plasmids were transiently transformed into maize protoplasts, followed by 12–16 h of dark incubation at 22 °C. Transformed protoplasts were then observed under a confocal laser scanning microscope [20]. Tissue sections were imaged under fluorescence microscopy (488 nm excitation), with *pCAMBIA2300*-GFP empty vector infiltrations serving as controls.

2.3. Pathogen Cultivation

To induce sporulation, *Bipolaris maydis* cultures from PDA medium were inoculated onto sterilized sorghum grain medium and incubated for 3–5 days at 26 °C. Surface mycelium was removed, and the colonized grains were evenly spread on trays. UV irradiation (60 min) was applied to trigger conidiation, followed by incubation in a controlled environment (25 °C, 80% RH) for an additional 3–5 days to maximize spore production.

2.4. Plant Materials and Treatments

Maize (B73) seeds were germinated in vermiculite (28 °C, 3–5 days) and transferred to pots under controlled conditions (28 °C, 16/8 h light/dark). Prior to inoculation, the conidia of Bipolaris maydis underwent germination rate assessment, with 95.23% germination observed. These activated conidia were subsequently used for infection assays. Five-leaf-stage plants were sprayed with either *Bipolaris maydis* conidial suspension (1×10^5 CFU/mL) or sterile water (both containing 0.1% Tween-20) for experimental and control groups, respectively. Leaf samples were collected at 0, 24, 48, 60, and 72 h post-inoculation (hpi) for RNA extraction and *ZmNLR-7* expression analysis. Three biological replicates ($n = 3$ plants each) were included per group.

Mature maize plants were flash-frozen in liquid nitrogen and stored at −80 °C for RNA extraction and *ZmNLR-7* expression profiling. For pathogen inoculation, field-grown plants at maturity were sprayed with *Bipolaris maydis* conidial suspension (1×10^5 CFU/mL, 0.1% Tween-20), while controls received sterile water with 0.1% Tween-20. Leaf samples were collected at 0–72 hpi for molecular analysis, with disease severity assessed at 7 dpi using standardized disease indices [19,21]. The completely randomized block design included three biological replicates (10 plants each).

2.5. RNA Extraction and RT-PCR Analysis

Total RNA was extracted from maize leaf tissues using established protocols [22]. qPCR reactions contained 10 μL of AceQ qPCR SYBR Green Master Mix, 25 ng of cDNA, and gene-specific primers (1 μL each; Table S1) per reaction. Cycling parameters included initial denaturation (94 °C, 30 s), followed by 40 cycles of 94 °C for 5 s and 55 °C for 15 s. Data normalization and analysis followed the $2^{-\Delta\Delta Ct}$ method [23].

2.6. EMS Mutant Identification

The EMS mutants were obtained from the Maize EMS Mutant Library (http://maizeems.qlnu.edu.cn/search/geneid.html, accessed on 6 July 2024). EMS-mutant seedlings were genotyped using primers flanking the target site *ZmNLR-7-1*-F: CATGACACATAACTAGAAGCTG; *ZmNLR-7-1*-R: GTACCCTTCAG-CATTTCCTTC; *ZmNLR-7-2*-F: ATTCCCGGTGAGTAGGTTG; *ZmNLR-7-2-R*: CCAGGCAATCAACGGTAGG, amplifying a 700 bp region (±350 bp from mutation locus). Leaf tissue (1 cm^2) was collected for genomic DNA isolation via modified *Arabidopsis* protocols [24]. PCR-amplified fragments were sequenced and aligned against wild-type references to identify homozygous lines. Successful editing was confirmed by detecting insertions/deletions (indels) at target loci, with unedited sequences excluded.

2.7. EMS Mutant Identification Physiological Parameter Analysis of ZmNLR-7

Maize plants at the five-leaf stage were spray-inoculated with conidia of *Bipolaris maydis* (1×10^5 CFU /mL, 0.1% Tween-20), and the control group was inoculated with an equal amount of Tween-20 supplemented with sterile water. Leaf tissues of the wild type and ZmNLR-7 mutant were collected 48 h after inoculation for oxidative stress analysis. Hydrogen peroxide (H_2O_2) accumulation, SOD/POD activity, and malondialdehyde (MDA) content were determined spectrophotometrically, using commercial kits produced by Suzhou Keming Biotechnology Co., Ltd. (Suzhou, China). Data represent the mean $\pm$ SD of three biological replicates.

2.8. RNA Sequencing (RNA-Seq) Analysis

RNA sequencing was conducted on the leaves of wild-type and ZmNLR-7 mutant maize under untreated and pathogen-challenged (48 hpi with *Bipolaris maydis* conidia) conditions. Libraries were prepared using the Illumina TruSeq™ RNA Library Prep Kit and sequenced by Majorbio (Shanghai, China). Raw reads were processed via the Majorbio Cloud Platform for alignment, differential expression analysis, and functional annotation. Total RNA was extracted from tissue samples, with concentration and purity assessed using a Nanodrop 2000 (Thermo Scientific, Waltham, MA, USA) spectrophotometer, integrity verified by agarose gel electrophoresis, and RNA Quality Number (RQN) determined via Agilent 5300 (Agilent, Santa Clara, CA, USA). Library construction required ≥ 1 µg total RNA per sample at concentrations ≥ 30 ng/µL, RQN > 6.5, and $OD_{260/280}$ ratios of 1.8–2.2. Sequencing depth was 5.88 GB clean data/each sample. Differential expression analysis was performed using DESeq2, with the reference genome *Zea mays* (Assembly: Zm-B73-REFERENCE-NAM-5.0; Source: http://plants.ensembl.org/*Zea_mays*/Info/Index, accessed on 6 March 2025), applying significance thresholds of $|\log_2 FC| \geq 1$ and adjusted *p*-value (*p*$_{adj}$) <0.05. The primers for quantitative real-time PCR validation are listed in Table S5.

2.9. Statistical Analysis

Statistical significance of results was calculated using one-way ANOVA, followed by Duncan's test for the least significant difference at a significance level of $p < 0.05$.

3. Results

3.1. Bioinformatics Analysis of the ZmNLR-7 Gene

In the early stage, the research group used the homologous sequence analysis method to integrate the disease resistance-gene information reported in gramineous crops such as wheat (*Triticum aestivum*) and rice (*Oryza sativa*), and screened the homologous genes of maize (*Zea mays*) using the candidate gene association analysis method. The obtained maize homologous genes were verified, and two candidate genes whose expression was significantly upregulated after induction by *Bipolaris maydis* were screened out and named *ZmNLR-7* (*Zm00001d020117*). *ZmNLR-7* is located on chromosome 7 of maize. The CDS sequence of the *ZmNLR-7* gene is 1578 bp in length, encoding 526 amino acids. The molecular mass of the protein encoded by it is predicted to be 67.3 kDa, and the theoretical isoelectric point is 5.42. Further domain analysis showed that the protein encoded by ZmNLR-7 has a typical NB-ARC domain (Figure S1). The presence of these conserved domains suggests that this gene may be involved in the immune response mediated by nucleotide-binding leucine-rich repeat receptors. ZmNLR-7 belongs to the NLR protein family. The phylogenetic evolutionary tree of this protein family was constructed using MEGA7.0 software. Phylogenetic comparison revealed that the *ZmNLR-7* (*Zm00001d020117*) studied in this experiment had a high homology with maize *ZmNLR-3* (*XM 035960670.1*) and *ZmNLR-4* (*XM 008653508.3*), and it was closely related to rice *OsNLR-8* (*CP101152.1*) (Figure 1).

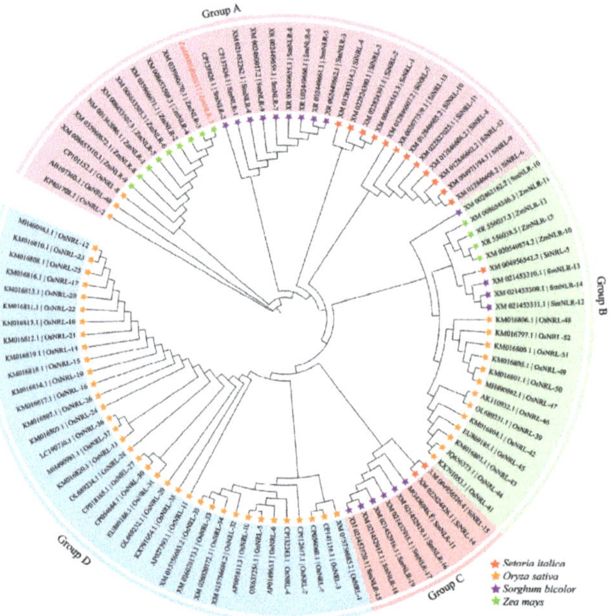

Figure 1. Phylogenetic analysis of ZmNLR-7 from *Triticum aestivum*, *Oryza sativa*, and *Zea mays*. The genes encoding the *ZmNLR-7* were divided into 4 categories. *ZmNLR-7* was marked in red. The species we selected—*Setaria italica*, *Oryza sativa*, *Sorghum bicolor*, and *Zea mays*—are all monocotyledonous plants. Bootstrap values of 1000 replicates are shown as percentages at the branch nodes. Bar = 0.1. The protein-naming method in the figure consisted of autonomously naming with an NCBI number.

3.2. *Analysis of Subcellular Localization of ZmNLR-7*

Based on the analysis of the WoLF PSORT online prediction tool, the ZmNLR-7 protein was predicted to be localized in the plasma membrane system. To verify the biological reliability of this prediction, this study conducted a systematic verification through a dual experimental system. First, the p2300-ZmNLR-7 GFP recombinant plasmid and the empty control p2300 GFP (Figure 2A) were injected into Nicotiana benthamiana leaves using the Agrobacterium-mediated transient expression system, and laser confocal microscopy was performed after culturing in the dark at 22 °C for 48–72 h. The results showed that the positive control GFP signal was widely distributed throughout the cell, while the ZmNLR-7 fusion protein showed clear subcellular localization characteristics, with its green fluorescence signal in the nucleus and plasma membrane (Figure 2B). To further confirm the conservation of this localization pattern in the maize cell system, we used the PEG-mediated protoplast transformation system to introduce p2300-ZmNLR-7: GFP and the control vector p1305-GFP into maize protoplasts, and subcellular localization analysis was performed after culturing in the dark at room temperature for 14–16 h. Consistent with the results of transient expression in tobacco, the ZmNLR-7: GFP fusion protein was also localized in the nucleus and plasma membrane in maize protoplasts (Figure 2C).

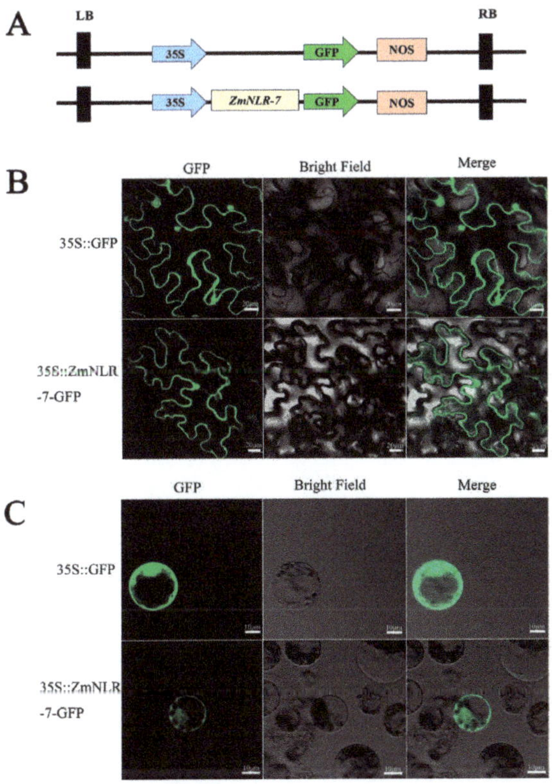

Figure 2. Subcellular localization analysis of ZmNLR-7. (**A**) Schematic representation of the 35S::ZmNLR-7-GFP vector construction for subcellular localization analysis. (**B**) Subcellular localization of ZmNLR-7-GFP in Nicotiana benthamiana leaves; 35S::GFP (empty vector) served as the control. (**C**) Subcellular localization of ZmNLR-7-GFP in maize mesophyll protoplasts. Scale bar: 20 μm, GFP is a green fluorescent protein.

3.3. Analysis of the Expression Pattern of *ZmNLR-7*

In order to explore the expression of the *ZmNLR-7* gene in different tissues of maize, RT-PCR detection revealed that *ZmNLR-7* was expressed to varying degrees in different tissues of maize, showing significant differences (Figure 3A). *ZmNLR-7* had the highest expression level in leaves, while the expression level was significantly reduced in roots, stems, stamens, female ears, filaments, bracts, and flag leaves. In order to analyze the dynamic regulatory characteristics of *ZmNLR-7* in disease-resistance response, its induced expression pattern was further analyzed by the infection experiment of *Bipolaris maydis*. The temporal expression analysis of 0–72 h after pathogen infection showed that the transcription level of *ZmNLR-7* showed a typical time-dependent response: the expression level reached a peak 48 h after infection, which was significantly upregulated by 11.92 times compared with the untreated state (0 h), and then gradually decreased (Figure 3B).

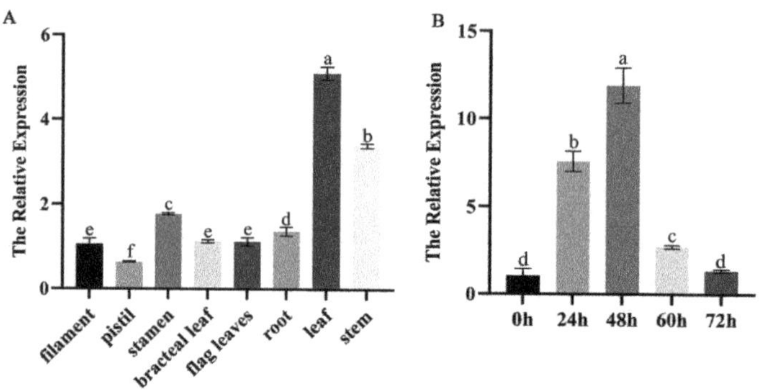

Figure 3. The expression profile of *ZmNLR-7*. (**A**) Tissue-specific expression levels of *ZmNLR-7* in different maize organs. (**B**) Temporal expression dynamics of *ZmNLR-7* following *Bipolaris maydis* inoculation. Significant differences (p-value < 0.05) are indicated with different lowercase letters.

3.4. Obtaining Homozygous Mutants of ZmNLR-7 Maize and Phenotypic Analysis of Response to Infection with Bipolaris Maydis

To verify the genetic stability of *ZmNLR-7*-related EMS mutant lines, this study systematically identified the mutation sites. Using maize B73 wild type and *ZmNLR-7-1* and *ZmNLR-7-2* mutants as materials, leaf tissues were collected at the four-leaf stage to extract genomic DNA. According to the EMS chemical mutagenesis characteristics of the *ZmNLR-7* gene, specific primers were designed in the 350 bp region upstream and downstream of the mutation site (amplification product 700 bp). After PCR amplification and Sanger sequencing analysis (Figure 4A), *ZmNLR-7-1* and *ZmNLR-7-2* had a C→T single base substitution at ChrX:93,688,474 and ChrX:93,694,948, respectively (see Table S1). *Both ZmNLR-7-1* (EMS4-097d06) and *ZmNLR-7-2* (EMS4-097d0c) lines harbor premature termination codons. Subsequent experiments utilized homozygous EMS mutant lines, and the homozygous mutant lines obtained were used for subsequent studies. Field experiments showed that the *ZmNLR-7* mutant exhibited significant growth and development defects: compared with the wild-type B73, the plant height decreased by 11.21–11.33%, and the thousand-grain weight decreased by about 3.7% (Figure S2). To evaluate the disease resistance in the adult stage, a standardized field design (plant spacing of 40 cm) was carried out on the T3 generation homozygous mutant and the wild type at the experimental base of Anhui Agricultural University. The leaves were inoculated with 1×10^5 CFU/mL conidia suspension of *Bipolaris maydis* (*Bipolaris maydis*) at the tasseling stage (n = 15). Seven days after inoculation, the mutant showed a typical disease phenotype: large areas of confluent necrotic spots appeared on the leaves (Figure 4B), and the disease index reached 74.63–84.07, which was significantly higher than that of the wild type (40.74) by 33.89–43.33 (Figure 4C). Further validation through *Bipolaris maydis* inoculation experiments on seedling-stage maize revealed that *ZmNLR-7* mutant lines (*ZmNLR-7-1* and *ZmNLR-7-2*) also exhibited significantly enhanced disease-susceptibility phenotypes (Figure S3), consistent with field observations. The results showed that the loss of *ZmNLR-7* function led to a significant reduction in the resistance of corn to *Bipolaris maydis*.

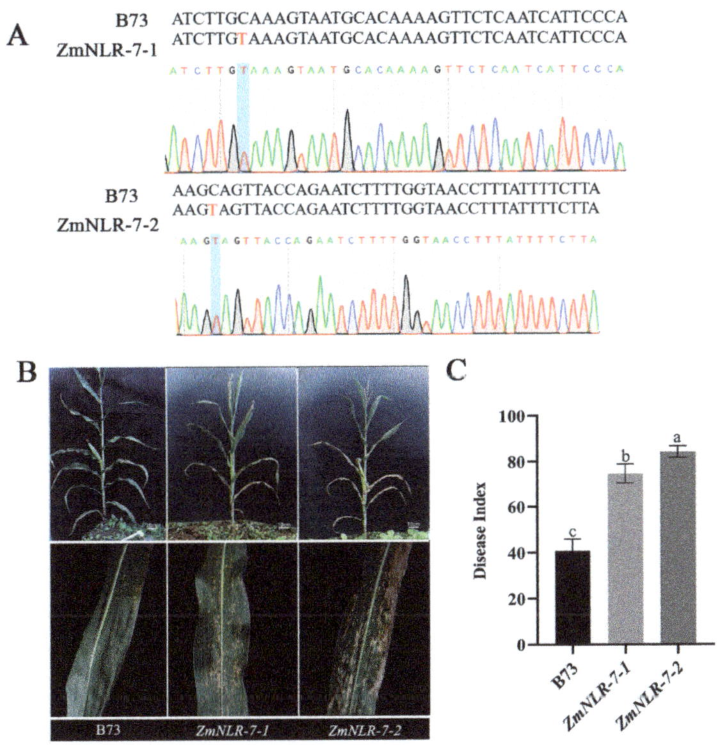

Figure 4. Phenotypic analysis of *ZmNLR-7* mutants in response to *Bipolaris maydis* infection. (**A**) Genotypic identification and sequencing chromatograms of *ZmNLR-7* mutants. (**B**) Disease-resistance phenotypes of wild-type and *ZmNLR-7* mutant lines at 7 days post-inoculation (dpi) with *Bipolaris maydis*. (**C**) Disease index of wild-type and *ZmNLR-7* mutant plants following pathogen challenge. Significant differences (p-value < 0.05) are indicated by different lowercase letters.

3.5. Determination of Physiological Indicators of ZmNLR-7 Maize Mutants in Response to Infection with Bipolaris Maydis

The relevant physiological indicators of mutants and wild-type plants were determined 48 h after inoculation with *Bipolaris maydis* spore suspension. It was found that the SOD activity of *ZmNLR-7-1* and *ZmNLR-7-2* (891.829 U/g and 577.263 U/g) was 30–55% lower than that of wild-type B73 (1269.42 U/g) (Figure 5A), and the POD activity of mutants (20,690 U/g and 19,640 U/g) was 18–22% lower than that of wild-type B73 (25,190 U/g) (Figure 5B). At the same time, the MDA content in the mutant (97.674 nmol/g and 125.597 nmol/g) was 14–46% higher than that in the wild type (85.936 nmol/g) (Figure 5C), indicating that the membrane lipid peroxidation damage was aggravated. The H_2O_2 accumulation of the wild type exceeded that of the mutant, which was 12.51 μmol/g, 8.89 μmol/g, and 10.97 μmol/g, respectively (Figure 5D), revealing that its ROS scavenging ability was impaired. The results showed that the corn *ZmNLR-7* mutant showed significant antioxidant system disorders after infection with *Bipolaris maydis*.

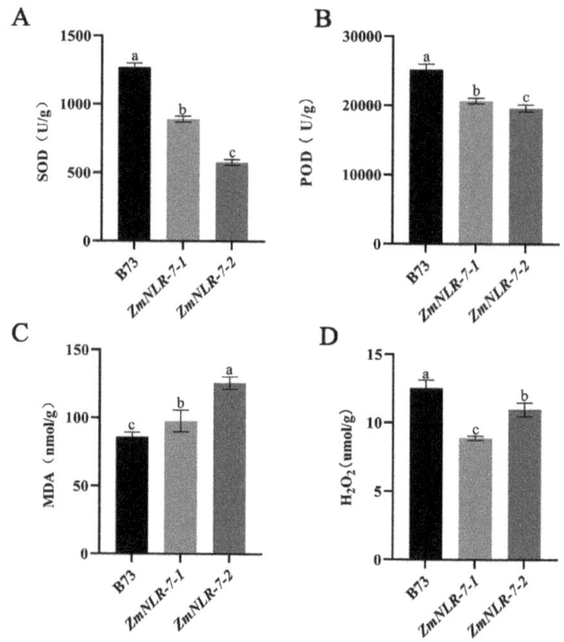

Figure 5. Measurement of disease resistance-related physiological indicators: (**A**) SOD activity; (**B**) POD activity; (**C**) MDA content; and (**D**) H_2O_2 content. Significant differences (p-value < 0.05) are indicated by different lowercase letters.

3.6. Transcriptomic Analysis of ZmNLR-7 Maize Mutant in Response to Bipolaris Maydis Infection

According to principal component analysis (PCA), different time nodes distinguished samples on the first principal axis (PC1) and explained 62.8% of the differences (Figure 6A), and different treatment groups distinguished samples on the second principal axis (PC2) and explained 9.9% of the differences (Figure 6A), which showed that the collected samples were well representative. In addition, the Venn diagram results showed that the mutants induced by the leaf spot fungus and the wild type had 898 common genes and 3983 mutant-specific genes (Figure 6B). In order to further explore the specific response pattern of the mutants compared with the wild type after infection with the *Bipolaris maydis*, the enrichment analysis of the KEGG pathway was performed. First, for the KEGG results, a total of four KEGG pathways were detected to be significantly enriched 48 h after inoculation (Figure 6C). Specifically, the four signaling pathways of plant hormone signal transduction, the MAPK signaling pathway—plant, phenylpropanoid biosynthesis, and starch and sucrose metabolism—were significantly enriched in the mutant *ZmNLR-7* plants (Figure 6C). It is worth noting that the plant hormone signaling pathway dominates the number of enriched genes, so the focus will be on analyzing the regulatory mechanism of this pathway.

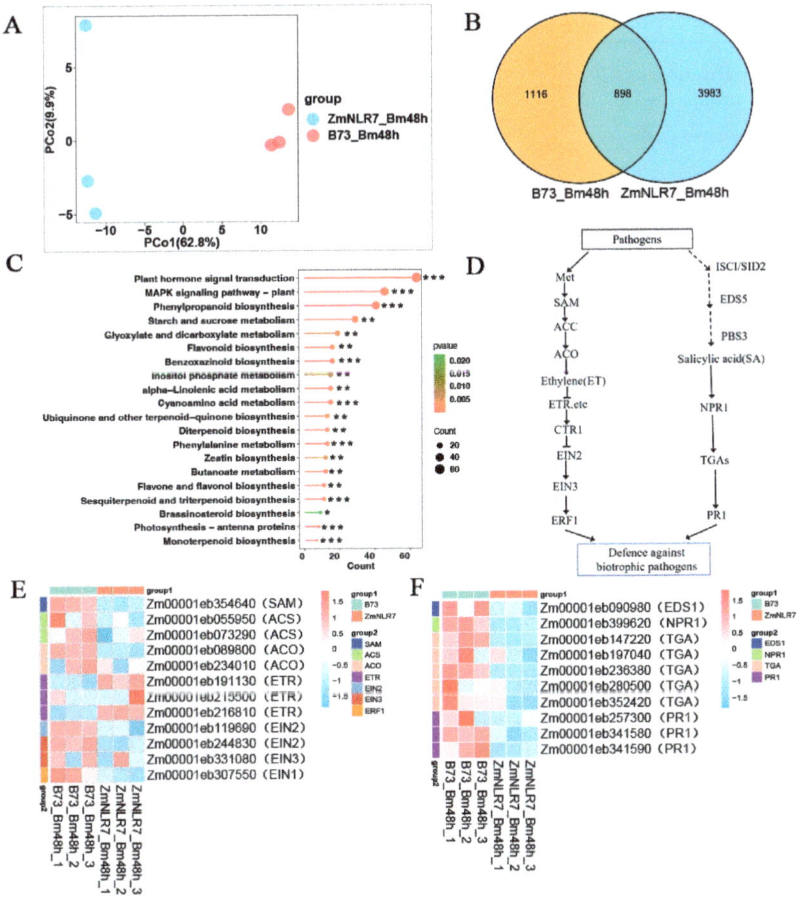

Figure 6. Differential gene expression in maize following *Bipolaris maydis* infection. (**A**) Principal component analysis (PCA) of global transcriptomic profiles across treatments. (**B**) Venn diagram illustrating shared and unique differentially expressed genes (DEGs) among experimental groups (**C**) KEGG pathway enrichment analysis of DEGs in wild type (B73-Bm48 h) versus *ZmNLR-7* mutant (ZmNLR-7-Bm48h) at 48 h post-inoculation with *Bipolaris maydis*. Asterisks indicate significant differences (*: $p < 0.05$; **: $p < 0.01$; ***: $p < 0.001$) (**D–F**) Schematic representation of ethylene (ET) and salicylic acid (SA) signaling pathways, with heatmaps showing expression levels of key genes regulated by *ZmNLR-7* in response to pathogen infection.

In the mutant strain *ZmNLR-7* infected with *Bipolaris maydis* and the wild-type B73 control group (Bm_ZmNLR-7_VS_B73), ethylene pathway analysis of ethylene synthesis-related genes found that the ethylene synthesis-related gene *MetK* (*Zm00001eb354640*) was downregulated by 0.548 times; the *ACS* gene (*Zm00001eb055950* and *Zm00001eb234010*) was downregulated by 0.258 and 0.684 times, respectively; the *ACO* gene (*Zm00001eb073290* and *Zm00001eb131380*) was downregulated by 0.435 and 0.453 times, respectively; and the ethylene receptor-related genes *EIR* and *ERS* (*Zm00001eb191130*, *Zm00001eb234010*, and *Zm00001eb234010*) were downregulated by 0.548 and 0.684 times, respectively. The *Zm00001eb216810* genes were upregulated by 2.843, 2.255, and 1.841 times, respectively; the downstream transcription factor *EIN2* (*Zm00001eb19690*) was downregulated by 0.412 times; the *EIN3* gene (*Zm00001eb244830* and *Zm00001eb331080*) was downregulated by 0.345 and 0.736 times; the *ERF1* gene (Zm00001eb307550) was downregulated by

0.644 times (Table S2); and the *CTR1* gene is a negative regulatory element of the ethylene pathway. The upregulation of *CTR1* leads to the inhibition of the activation state of *EIN2*. The downregulation of *EIN2* further blocks the signal transmission to the core transcription factor *EIN3* and the downstream response gene *ERF1*, resulting in the expression levels of response genes such as *EIN3* and *ERF1* also showing a downward trend. For the analysis of the salicylic acid pathway in Bm_ZmNLR-7_VS_B73 (Table S3), the gene *EDS1* (*Zm00001eb090980*) related to salicylic acid synthesis was downregulated by 0.918-fold; the core receptor *NPR1* (*Zm00001eb399620*) was downregulated by 0.369-fold; and the *TGA* genes (*Zm00001eb147220*, *Zm00001eb197040*, *Zm00001eb236380*, *Zm00001eb280500*, and *Zm00001eb352420*) were downregulated by 0.605, 0.507, 0.522, 0.274, and 0.263 times. The downregulation of the *TGA* genes may lead to the inability to activate the expression of *PR1* and its related genes (*Zm00001eb299340*, *Zm00001eb257300*, *Zm00001eb341580*, and *Zm00001eb341590*). The above results show that under pathogen stress, the synthesis of the salicylic acid signaling pathway and the expression of genes related to the core transduction link of mutant *ZmNLR-7* are inhibited, thus weakening the plant's ability to respond to disease.

This study further demonstrated the expression changes of key genes in the ethylene signaling pathway and salicylic acid signaling pathway in the *ZmNLR-7* mutant 48 h after inoculation with *Bipolaris maydis* by heat maps. After treatment with *Bipolaris maydis*, the expression levels of ethylene synthesis-related genes, *SAM*, *ACO*, and *ACS*, and other genes in the *ZmNLR-7* mutant line decreased; the ethylene receptor *ETR* negatively regulated ethylene signal transduction, resulting in the upregulation of ethylene receptor-related genes *EIR* and *ERS* genes; and for the *CTR1* gene, as a negative regulatory element of the ethylene pathway, its upregulation led to the inhibition of the activation state of *EIN2*, and the downregulation of *EIN2* further blocked the transmission of signals to the core transcription factor *EIN3* and the downstream response gene *ERF1*, resulting in the expression levels of response genes such as *EIN3* and *ERF1* also showing a downward trend (Figure 6E). At the same time, the expression levels of genes related to salicylic acid synthesis, such as *EDS1*, core receptor *NPR1*, and *TGA*, were all downregulated; downregulation of the *TGA* gene may lead to the inability to activate the expression of *PR1* (Figure 6F). At the same time, the genes related to the jasmonic acid (JA) pathway did not change significantly during the infection process (Table S4; Figure S4). These transcriptome results indicate that the *ZmNLR-7* mutant may inhibit the key genes of the ethylene and salicylic acid pathways when infected by the leaf spot fungus, resulting in increased susceptibility of corn to the leaf spot fungus.

3.7. Fluorescence Quantitative Verification of ZmNLR-7 Transcriptomics

To verify the authenticity of the transcriptome data, this study randomly screened several differentially expressed genes from the transcriptome data for verification, including *Zm00001eb310440* (*2′-deoxymugineic-acid 2′-dioxygenase*), *Zm00001eb137930* (*Protein DOWNY MILDEW RESISTANCE 6*), *Zm00001eb042870* (*UDP-glycosyltransferase 73D1-like*), *Zm00001eb230410* (*terpene synthase 2*), *Zm00001eb350770* (*CBL-interacting protein kinase 14*), *Zm00001eb419870* (*probable protein phosphatase 2C 37*), *Zm00001eb341580* (*pathogenesis-related protein PRMS precursor*), *Zm00001eb244830* (*ETHYLENE INSENSITIVE 3-like 5 protein*), *Zm00001eb122500* (*GA 3-oxidase 1*), and *Zm00001eb401290* (*ethylene-responsive transcription factor RAP2-4*). The experimental results showed that although there were certain differences in expression changes compared with RNA-seq gene expression data, the expression trends of these genes remained highly consistent as a whole (Figure 7), which fully demonstrated the authenticity and reliability of the transcriptome data of this study.

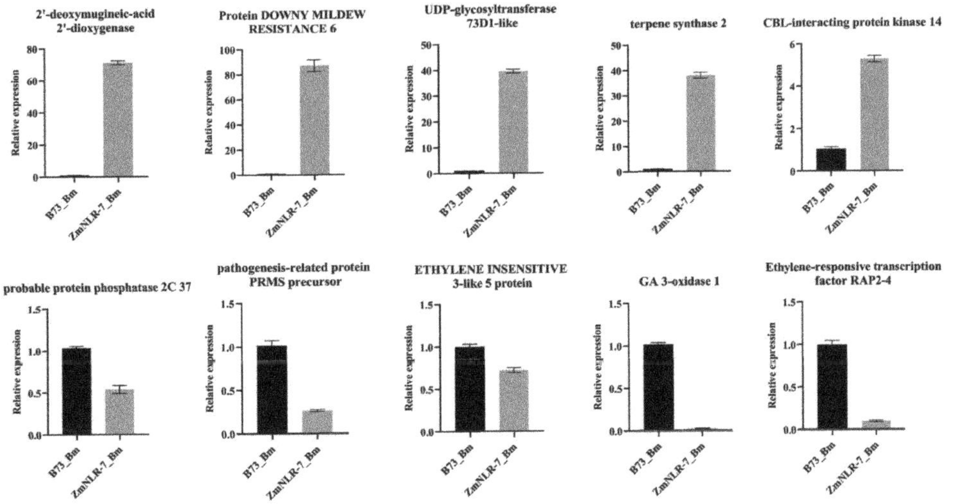

Figure 7. Validation of RNA-Seq data by qRT-PCR.B73-Bm48h and ZmNLR-7-Bm48h denote wild-type and *ZmNLR-7* mutant maize plants, respectively, at 48 h post-inoculation (hpi) with *Bipolaris maydis*. Error bars represent standard deviation (SD) of three independent biological replicates. The left *Y*-axis indicates relative expression levels quantified by quantitative reverse-transcription PCR (qRT-PCR).

4. Discussion

NLR (nucleotide-binding-domain and leucine-rich-repeat-containing) proteins are key receptors in the plant immune system, mainly triggering effector-triggered immune responses (ETIs) by recognizing pathogen effector proteins (effectors) [25]. Plant NLR proteins are the core executors of ETI immunity, and their structural diversity and functional differentiation play a key role in disease resistance. According to the difference in the N-terminal domain, NLR is divided into two categories: CNL and TNL. Monocotyledonous plants (such as corn and wheat) are mainly CNL, while dicotyledonous plants (such as *Arabidopsis*) have both TNL and CNL types [26]. For example, *Arabidopsis* TNL protein RPP1 generates immune signaling molecules (such as 2′cADPR) through the NAD + hydrolysis activity of the TIR domain, activating the EDS1-PAD4-ADR1 complex to drive broad-spectrum resistance [27,28]. The wheat CNL protein TaRPP13L1 relies on the anchor protein TaANK-TPR1 to promote dimerization and enhance resistance to stripe rust [29]. In this study, the maize ZmNLR-7 protein was localized in the nucleus and plasma membrane. The localization of proteins in the cytoplasmic membrane and nucleus is crucial for their functions. The cytoplasmic membrane, as the boundary between the inside and outside of the cell, houses proteins responsible for communication with the external environment, material transport, and intercellular recognition. For example, the wheat protein TaANK-TPR1 targets the NLR protein TaRPP13L1 (which recognizes the pathogen Peronospora Parasitica 13-like 1), promoting its dimerization, thereby triggering intense cell necrosis and ultimately enhancing wheat resistance to stripe rust [29]. In contrast, proteins localized in the nucleus primarily participate in the regulation of gene expression. For instance, the effector protein Pt-1234 from the leaf rust pathogen targets the subdomain C of the wheat transcription factor *TaNAC069*, thereby regulating the role of *TaNAC069* in the defense response against leaf rust in wheat [30]. In this study, through subcellular localization, it was confirmed that ZmNLR-7 can localize in both the cytoplasmic membrane and nucleus. This result suggests that ZmNLR-7 may participate in the plant's response to small-spot-pathogen stress by recognizing the pathogen and through targeted interactions with the pathogen's effector proteins. In subsequent studies, this research group will employ experiments such as yeast two-hybrid and BIFC, using maize ZmNLR-7 as bait to screen for effector proteins from the small-spot pathogen that interact with it.

In addition, NLR-type disease-resistance proteins are one of the most important components of the plant immune system, and their activity is precisely regulated by both intramolecular and intermolecular protein interactions (Fine-Tuning Immunity: Players and Regulators for Plant NLRs). For example, the team led by Wang Guanfeng discovered that the maize ZmVPS23-like protein ZmVPS23L interacts with the CC domain of the maize Rp1-D21 (a mutant generated by intramolecular recombination of two CC-NLR-type disease-resistance genes) and relocates it to endosomes, thereby suppressing the hypersensitive response [31]. The wild emmer wheat NLR genes *TdCNL1/TdCNL5* enhance wheat resistance to powdery mildew by coordinately regulating the MLIW170/PM26 locus [32]. In this study, maize ZmNLR-7 participates in plant defense responses through dual localization in the cytoplasmic membrane and nucleus: the plasma membrane regulates ROS metabolism (SOD/POD activity and H_2O_2 accumulation), and in the nucleus, it suppresses the core transcription factor *ERF1* of the ethylene signaling pathway, specifically downregulating genes in the SA/ET pathway (such as *NPR1* and *PR1*). This reveals that ZmNLR-7 is involved in plant disease-resistance regulation through hormonal crosstalk. However, it remains unclear which related proteins are directly affected by the absence of ZmNLR-7. Therefore, in subsequent studies, the research group will continue to investigate proteins within maize that can interact with ZmNLR-7, aiming to clarify the modules through which ZmNLR-7 participates in plant disease-resistance responses, thus providing new targets for maize disease-resistance breeding.

Low temperature in *Arabidopsis* inhibits ET signaling by enhancing the expression of SA synthesis genes (*ICS1/SID2*), forming an environmental-immune coupling mechanism [33]; while the ethylene response factor *PpERF98* in peach trees inhibits the SA pathway by activating JA/ET signals, increasing sensitivity to pathogens [34]. In this study, upregulation of ET receptor genes (*EIR, ERS*) may inhibit the EIN2-EIN3-ERF1 cascade reaction through *CTR1*, weakening SA-dependent resistance. This SA-ET antagonistic pattern is similar to the hormone interaction under low temperature stress [33]. Current research indicates that NLR immune receptors can perform multiple functions. In addition to activating disease-resistance pathways (such as ETI), some members of the NLR family can also participate in plant growth and development processes by regulating hormone signals [35]. For instance, the deletion of the snc1 gene in Arabidopsis leads to the inhibition of plant growth. In this study, the absence of ZmNLR-7 hinders the activation and expression of defense-related genes ERF1 and PR1 in maize, resulting in the suppression of the ethylene (ET) and salicylic acid (SA) disease-resistance signaling pathways, while having a negligible effect on the expression of genes related to jasmonic acid signaling pathways. This is manifested by a significant increase in the sensitivity of the mutant *ZmCIPK-7* to the small-spot pathogen. Plant hormones such as JA and SA can maintain the balance between plant growth and defense through interactions with other plant-growth regulators (PGRs) [35]. Based on this, the research group speculates that the absence of *ZmNLR-7* disrupts the balance between maize growth and defense, enhancing maize susceptibility while inhibiting plant growth (e.g., reduction in plant height and thousand-grain weight). In subsequent studies, this research group will continue to conduct in-depth expression analyses of genes related to plant growth and development in maize *ZmNLR-7* mutants, as well as protein screening for those that interact with ZmNLR-7, with the aim of clarifying the underlying molecular mechanisms of *ZmNLR-7*'s effects on plant growth. We acknowledge the possibility that the reduced plant height and weight observed in the mutants could be linked to NLR pleiotropy. A potential mechanism for these developmental defects may involve NLR-mediated disruption of GA homeostasis or its interaction with other phytohormones, warranting further analysis [36,37].

This study found that the maize *ZmNLR-7* gene specifically regulates the salicylic acid (SA) and ethylene (ET) disease-resistance signaling pathways (independent of the jasmonic acid pathway), revealing a new mechanism by which NLR proteins achieve directional transmission of immune signals through structural plasticity. Based on this feature, gene-editing technologies (such as CRISPR) can be used to target and modify pathogen response elements, and design disease-resistance modules that are only activated when infected by pathogens. Although the existing EMS mutant (*ZmNLR-7*) has verified the function of the gene, chemical mutagenesis may introduce non-targeted mutations. In the future, the role of the gene as a molecular switch in the disease-resistance–yield trade-off will be systematically analyzed through gene knockout/overexpression, providing theoretical and technical support for the breeding of intelligent disease-resistant varieties.

5. Conclusions

This study systematically analyzed the molecular mechanism of ZmNLR-7, a member of the maize NLR family, in mediating resistance to *Bipolaris maydis*. This gene constructs a multi-level disease-resistance defense system by maintaining ROS metabolic homeostasis (SOD/POD activity regulation) and integrating the SA/ET signal transduction network. The mutant phenotype confirmed that the loss of *ZmNLR-7* function led to increased membrane lipid peroxidation, inhibition of pathogenesis-related genes, and significant decline in systemic resistance. The results provide a theoretical framework and key targets for analyzing broad-spectrum and durable disease-resistant varieties.

Supplementary Materials: The following supporting information can be downloaded at https://www.mdpi.com/article/10.3390/cimb47070573/s1.

Author Contributions: T.D., G.W., and B.S. conceived and designed the experiments; Y.L., R.Z. and S.D. conducted the plant inoculation and RNA extraction; X.Y. conducted the qRT-PCR valid action and statistics analysis on experimental data; H.J. performed all the bioinformatics analysis; B.S. and T.D. provided technical support and edited the manuscript. All authors have read and agreed to the published version of the manuscript.

Funding: This study was funded by the National Natural Science Foundation of China (No. 32172477), the Open Fund of National Engineering Laboratory of Crop Stress Resistance Breeding (NELCOF20230106 and NELCOF20240104), and the Natural Science Research Project of Colleges and Universities in Anhui Province (Nos. 2022AH050907, 2023AH052217, and 2022AH051384).

Institutional Review Board Statement: Not applicable.

Informed Consent Statement: Not applicable.

Data Availability Statement: The authors declare that data supporting the findings of this study are available in the article. If raw data files are required, they are available from the corresponding author upon reasonable request.

Conflicts of Interest: The authors declare no conflicts of interest.

References

1. Jones, J.D.G.; Dangl, J.L. The plant immune system. *Nature* **2006**, *444*, 323–329. [CrossRef] [PubMed]
2. Yu, X.; Niu, H.; Liu, C.; Wang, H.; Yin, W.; Xia, X. PTI-ETI synergistic signal mechanisms in plant immunity. *Plant Biotechnol. J.* **2024**, *22*, 2113–2128. [CrossRef] [PubMed]

3. Zhou, F.; Menke, F.L.; Yoshioka, K.; Moder, W.; Shirano, Y.; Klessig, D.F. High humidity suppresses ssi4-mediated cell death and disease resistance upstream of MAP kinase activation, H2O2 production and defense gene expression. *Plant J.* **2004**, *39*, 920–932. [CrossRef] [PubMed]
4. Yuan, M.; Jiang, Z.; Bi, G.; Nomura, K.; Liu, M.; Wang, Y.; Cai, B.; Zhou, J.-M.; He, S.Y.; Xin, X.-F. Pattern-recognition receptors are required for NLR-mediated plant immunity. *Nature* **2021**, *592*, 105–109. [CrossRef] [PubMed]
5. Ngou, B.P.M.; Ahn, H.-K.; Ding, P.; Jones, J.D.G. Mutual potentiation of plant immunity by cell-surface and intracellular receptors. *Nature* **2021**, *592*, 110–115. [CrossRef] [PubMed]
6. Zhu, B.-S.; Zhu, Y.-X.; Zhang, Y.-F.; Zhong, X.; Pan, K.-Y.; Jiang, Y.; Wen, C.-K.; Yang, Z.-N.; Yao, X. Ethylene activates the EIN2-EIN3/EIL1 signaling pathway in tapetum and disturbs anther development in Arabidopsis. *Cells* **2022**, *11*, 3177. [CrossRef] [PubMed]
7. Duxbury, Z.; Wu, C.-H.; Ding, P. A comparative overview of the intracellular guardians of plants and animals: NLRs in innate immunity and beyond. *Annu. Rev. Plant Biol.* **2021**, *72*, 155–184. [CrossRef] [PubMed]
8. Pan, Q.; Liu, Y.-S.; Budai-Hadrian, O.; Sela, M.; Carmel-Goren, L.; Zamir, D.; Fluhr, R. Comparative genetics of nucleotide binding site-leucine rich repeat resistance gene homologues in the ge-nomes of two dicotyledons: Tomato and Arabidopsis. *Genetics* **2000**, *155*, 309–322. [CrossRef] [PubMed]
9. Pan, Q.; Wendel, J.; Fluhr, R. Divergent evolution of plant NBS-LRR resistance gene homologues in dicot and cereal genomes. *J. Mol. Evol.* **2000**, *50*, 203–213. [CrossRef] [PubMed]
10. Tamborski, J.; Krasileva, K.V. Evolution of plant NLRs: From natural history to precise modifications. *Annu. Rev. Plant Biol.* **2020**, *71*, 355–378. [CrossRef] [PubMed]
11. Jones, D.A.; Takemoto, D. Plant innate immunity–direct and indirect recognition of general and specific pathogen-associated molecules. *Curr. Opin. Immunol.* **2004**, *16*, 48–62. [CrossRef] [PubMed]
12. Martin, G.B.; Bogdanove, A.J.; Sessa, G. Understanding the functions of plant disease resistance proteins. *Annu. Rev. Plant Biol.* **2003**, *54*, 23–61. [CrossRef] [PubMed]
13. Danot, O.; Marquenet, E.; Vidal-Ingigliardi, D.; Richet, E. Wheel of life, wheel of death: A mechanistic insight into signaling by STAND proteins. *Structure* **2009**, *17*, 172–182. [CrossRef] [PubMed]
14. van der Biezen, E.A.; Jones, J.D. The NB-ARC domain: A novel signalling motif shared by plant resistance gene products and regulators of cell death in animals. *Curr. Biol.* **1998**, *8*, R226–R228. [CrossRef] [PubMed]
15. Luo, M.; Xie, L.; Chakraborty, S.; Wang, A.; Matny, O.; Jugovich, M.; Kolmer, J.A.; Richardson, T.; Bhatt, D.; Hoque, M.; et al. A five-transgene cassette confers broad-spectrum resistance to a fungal rust pathogen in wheat. *Nat. Biotechnol.* **2021**, *39*, 561–566. [CrossRef] [PubMed]
16. Hatta, M.A.M.; Arora, S.; Ghosh, S.; Matny, O.; Smedley, M.A.; Yu, G.; Chakraborty, S.; Bhatt, D.; Xia, X.; Steuernagel, B.J. The wheat Sr22, Sr33, Sr35 and Sr45 genes confer resistance against stem rust in barley. *Plant Biotechnol. J.* **2021**, *19*, 273–284. [CrossRef] [PubMed]
17. Dracatos, P.M.; Lu, J.; Sánchez-Martín, J.; Wulff, B.B. Resistance that stacks up: Engineering rust and mildew disease control in the cereal crops wheat and barley. *Plant Biotechnol. J.* **2023**, *21*, 1938–1951. [CrossRef] [PubMed]
18. Zhao, X.; Wang, C.; Zhao, X.; Liu, J.; Cui, T.; Ren, Y.; Niu, Y.; Tang, Z. Research Progress on Pyramiding Breeding of Disease-Resistant Related Genes in Wheat. *Shanxi Agric. Sci.* **2017**, *45*, 308–313.
19. Feng, Y.; Ma, Y.; Feng, F.; Chen, X.; Qi, W.; Ma, Z.; Song, R. Accumulation of 22 kDa α-zein-mediated nonzein protein in protein body of maize endosperm. *New Phytol.* **2022**, *233*, 265–281. [CrossRef] [PubMed]
20. Gu, L.; Cao, Q.; Dong, J.; Qiao, M.; Wang, Z.; Zhang, Z.; Sun, H.; Xie, H.; Ge, M.; Zhang, Y.; et al. Transcription factor ZmNLP8 modulates nitrate utilization by transactivating *ZmNiR1.2* in maize. *Plant J.* **2025**, *122*, e70263. [CrossRef] [PubMed]
21. Hang, T.; Ling, X.; He, C.; Xie, S.; Jiang, H.; Ding, T. Isolation of the zmers4 gene from maize and its functional analysis in transgenic plants. *Front. Microbiol.* **2021**, *12*, 632908. [CrossRef] [PubMed]
22. Mao, L.; Ge, L.; Ye, X.; Xu, L.; Si, W.; Ding, T. ZmGLP1, a germin-like protein from maize, plays an important role in the regulation of pathogen resistance. *Int. J. Mol. Sci.* **2022**, *23*, 14316. [CrossRef] [PubMed]
23. Livak, K.J.; Schmittgen, T.D. Analysis of relative gene expression data using real-time quantitative PCR and the 2− ΔΔCT method. *Methods* **2001**, *25*, 402–408. [CrossRef] [PubMed]
24. Su, B.; Jiang, H.; Song, Z.; Liu, W.; Rao, S.; Jiang, H.; Wu, G.; Ding, T. Overexpression of ZmEREB211 confers enhanced susceptibility to Pseudomonas syringae pv. tomato DC3000 in Arabidopsis. *Plant Sci.* **2025**, *356*, 112482. [CrossRef] [PubMed]
25. Guo, B.-C.; Zhang, Y.-R.; Liu, Z.-G.; Li, X.-C.; Yu, Z.; Ping, B.-Y.; Sun, Y.-Q.; Burg, H.v.D.; Ma, F.-W.; Zhao, T.; et al. Deciphering Plant NLR Genomic Evolution: Synteny-Informed Classification Unveils Insights into TNL Gene Loss. *Mol. Biol. Evol.* **2025**, *42*, msaf015. [CrossRef] [PubMed]
26. Huang, S.; Jia, A.; Ma, S.; Sun, Y.; Chang, X.; Han, Z.; Chai, J. NLR signaling in plants: From resistosomes to second messengers. *Trends Biochem. Sci.* **2023**, *48*, 776–787. [CrossRef] [PubMed]

27. Nishimura, M.T.; Anderson, R.G.; Cherkis, K.A.; Law, T.F.; Liu, Q.L.; Machius, M.; Nimchuk, Z.L.; Yang, L.; Chung, E.-H.; El Kasmi, F.; et al. TIR-only protein RBA1 recognizes a pathogen effector to regulate cell death in *Arabidopsis*. *Proc. Natl. Acad. Sci. USA* **2017**, *114*, E2053–E2062. [CrossRef] [PubMed]
28. Ma, S.; Lapin, D.; Liu, L.; Sun, Y.; Song, W.; Zhang, X.; Logemann, E.; Yu, D.; Wang, J.; Jirschitzka, J.; et al. Direct pathogen-induced assembly of an NLR immune receptor complex to form a holoenzyme. *Science* **2020**, *370*, eabe3069. [CrossRef] [PubMed]
29. Guo, S.; Zhang, F.; Du, X.; Zhang, X.; Huang, X.; Li, Z.; Zhang, Y.; Gan, P.; Li, H.; Li, M.; et al. TaANK-TPR1 enhances wheat resistance against stripe rust via controlling gene expression and protein activity of NLR protein TaRPP13L1. *Dev. Cell* **2025**, *60*, p1702–p1718.e6. [CrossRef] [PubMed]
30. Geng, H.; Zhang, Y.; Qin, Z.; Wang, S.; Liu, C.; Cui, Z.; Liu, D.; Wang, H. Puccinia triticina effector Pt-1234 modulates wheat immunity by targeting transcription factor TaNAC069 via its C subdomain. *Crop J.* **2025**, *13*, 69–78. [CrossRef]
31. Sun, Y.; Ma, S.; Liu, X.; Wang, G.-F. The maize ZmVPS23-like protein relocates the nucleotide-binding leucine-rich repeat protein Rp1-D21 to endosomes and suppresses the defense response. *Plant Cell* **2023**, *35*, 2369–2390. [CrossRef] [PubMed]
32. Zhu, K.; Li, M.; Dong, L.; Zhang, H.; Zhang, D.; Lu, P.; Wu, Q.; Xie, J.; Chen, Y.; Guo, G.; et al. An atypical NLR pair TdCNL1/TdCNL5 from wild emmer confers powdery mildew resistance in wheat. *Nat. Genet.* **2025**, *57*, 1553–1562. [CrossRef] [PubMed]
33. Li, Z.; Liu, H.; Ding, Z.; Yan, J.; Yu, H.; Pan, R.; Hu, J.; Guan, Y.; Hua, J. Low temperature enhances plant immunity via salicylic acid pathway genes that are repressed by ethylene. *Plant Physiol.* **2019**, *182*, 626–639. [CrossRef] [PubMed]
34. Zhang, D.; Zhu, K.; Shen, X.; Meng, J.; Huang, X.; Tan, Y.; Cardinale, F.; Liu, J.; Li, G.; Liu, J. Two interacting ethylene response factors negatively regulate peach resistance to *Lasiodiplodia theobromae*. *Plant Physiol.* **2023**, *192*, 3134–3151. [CrossRef] [PubMed]
35. Li, X.; Liu, C.; Du, J.; Sun, Y.; Hu, R.; Liu, S.; Xu, Q.; He, X.; Tang, C.-X.; Xu, R.; et al. DEAD-box protein SMA1 activates immunity likely through the formation of nuclear condensates with EDS1 in Arabidopsis. *Cell Rep.* **2025**, *44*, 115895. [CrossRef] [PubMed]
36. Wang, Y.; Zhao, J.; Lu, W.; Deng, D. Gibberellin in plant height control: Old player, new story. *Plant Cell Rep.* **2017**, *36*, 391–398. [CrossRef] [PubMed]
37. Zhou, M.; Li, Y.; Cheng, Z.; Zheng, X.; Cai, C.; Wang, H.; Lu, K.; Zhu, C.; Ding, Y. Important Factors Controlling Gibberellin Homeostasis in Plant Height Regulation. *J. Agric. Food Chem.* **2023**, *71*, 15895–15907. [CrossRef] [PubMed]

current issues in molecular biology

MDPI

Article

Physiological and Biochemical Responses and Transcriptome Analysis of *Bangia fuscopurpurea* (Rhodophyta) Under High-Temperature Stress

Minghao Zhao [1], Hongyan Zheng [1,†], Zepan Chen [1] and Weizhou Chen [1,2,*]

1 Marine Science Institute, Shantou University, Shantou 515063, China; 22mhzhao@stu.edu.cn (M.Z.); zhenghy@stu.edu.cn (H.Z.); zepan@stu.edu.cn (Z.C.)
2 Guangdong Provincial Key Laboratory of Marine Biotechnology, Shantou University, Shantou 515063, China
* Correspondence: wzchen@stu.edu.cn
† Current address: College of Biology and Agricultural Resources, Huanggang Normal University, Huangzhou 438000, China.

Abstract

With the advancement of human industrial activities, increased carbon dioxide emissions have made global warming an inescapable trend. Elevated temperatures exert profound effects on the viability of large macroalgae. *Bangia fuscopurpurea* (Rhodophyta) is a commercially important large red alga widely cultivated along the coastal waters of Putian, Fujian Province, China; however, its physiological, biochemical, and molecular responses to heat stress remain unclear. To address this question, we cultured *B. fuscopurpurea* at 15 °C (control) and 28 °C (heat stress) for 7 days, assessed changes in growth and photosynthetic parameters, and performed transcriptome sequencing. Growth analysis revealed that the relative growth rate of *B. fuscopurpurea* at 28 °C was significantly lower than that at 15 °C. After 1 day at 28 °C, the chlorophyll a and carotenoid contents increased significantly; the phycobiliprotein levels rose markedly on days 4 and 7, whereas the *Fv/Fm* ratio decreased significantly on days 1, 4, and 7. Transcriptomic analysis indicated that heat stress up-regulated the majority of differentially expressed genes (DEGs) in *B. fuscopurpurea*. KEGG pathway enrichment analysis revealed that the DEGs were predominantly associated with photosynthesis, carbohydrate and energy metabolism, glycerophospholipid metabolism, and the glutathione cycle. In summary, *B. fuscopurpurea* mitigates the adverse effects of heat stress by up-regulating genes involved in photosynthesis, antioxidant defenses, and glycerophospholipid metabolism. These findings enhance our understanding of the physiological adaptations and molecular mechanisms by which *B. fuscopurpurea* responds to heat stress.

Keywords: gene regulation; *Bangia fuscopurpurea*; macroalgae; transcriptome; photosynthesis; high-temperature stress

1. Introduction

Since the last century, anthropogenic emissions of greenhouse gases such as CO_2 have increased markedly, leading to global climate change [1]. Studies predict that by 2100 the global mean sea surface temperature will increase by 3–7 °C [2]. As a crucial component of the Earth's ecosystem, ocean warming will significantly affect the marine environment, including sea level rise [3], an increased frequency of marine heatwaves (MHWs) [4], and sea ice decline [5]. Macroalgae (including microalgae and large macroalgae), as the main primary producers of marine ecosystems, provide essential matter and energy for other consumers, playing an important role in marine ecosystems. In addition, they are responsible for approximately 50% of Earth's oxygen production and biological carbon dioxide uptake, playing an important role in global oxygen supply, carbon dioxide fixation, and climate change mitigation [6]. The effects of ocean warming on the physiological and biochemical processes of large macroalgae indirectly affect their ecological functions and services in marine ecosystems. Studies have shown that large macroalgae can enhance their survival under high-temperature stress through physiological and molecular mechanisms [7]. High temperatures influence osmotic regulatory substances in macroalgae, such as soluble proteins and soluble sugars, as well as photosynthetic pigments [8]. Short-term high-temperature stress in *Saccharina japonica* triggers a series of physiological and biochemical responses, such as an increased total soluble protein content and enhanced antioxidant enzyme activities [9].

As integral components of marine ecosystems, intertidal macroalgae significantly influence oceanic carbon cycling and energy flux through photosynthesis. Under global climate change, thermal stress substantially affects the photosynthetic performance of macroalgae. In *Kappaphycus alvarezii*, photosynthetic performance and pigment content are significantly impaired by elevated temperatures, as evidenced by a reduction in the maximum photochemical quantum yield (Fv/Fm), indicating temperature sensitivity and damage to the photosynthetic apparatus [10]. Conversely, studies on Gracilaria lemaneiformis demonstrate that short-term thermal exposure stimulates growth and enhances photosynthetic activity [11], with photosynthetic saturation rates and carbon utilization efficiency both increasing with temperature [12]. Similar physiological responses have been documented in Gracilaria bailinae [7].

Bangia species are primarily distributed in the temperate and subtropical regions of the western North Pacific and the North Atlantic and are classified into marine and freshwater taxa [13]. *B. fuscopurpurea* predominantly grows on rocky substrates in the upper intertidal zone; it is a palatable, economically important red alga rich in EPA, free amino acids, polysaccharides, vitamins, and minerals, and among currently cultivated macroalgae, it has the highest EPA content, offering considerable commercial potential [14,15]. Currently, the artificial cultivation of red hair algae in China has been carried out only in Putian, Fujian [16]; however, the physiological responses and molecular mechanisms of marine Bangia thalli under high-temperature stress remain unclear. Therefore, this study investigates the effects of high-temperature stress on marine Bangia thalli in terms of growth and survival, physiological–biochemical parameters, and gene-level changes, aiming to provide further insight into the biological responses of marine Bangia to global warming and to facilitate the breeding of heat-tolerant strains.

2. Materials and Methods

2.1. Seaweed Samples and Temperature Treatment

B. fuscopurpurea samples were collected from aquaculture rafts at Nanri Island, Putian City, Fujian Province (25°13′17.4″ N, 119°28′31.9″ E). The thalli were air-dried and transported back to the laboratory on ice. The samples were first rinsed with seawater filtered through a 0.45 μm membrane to remove debris and epiphytes, then acclimated in an incubator for 2 days. The acclimation conditions were set to 15 °C, a light intensity of 40 μmol photons·m^{-2}·s^{-1}, pH 8.0, a 12 h light/12 h dark photoperiod, and a biomass density of 2 g·L^{-1}; nutrients were supplied as $NaNO_3$-N at 2 mg·L^{-1} and KH_2PO_4-P at 1 mg·L^{-1}, with continuous aeration to maintain suspension and growth. After acclimation, the formal experiments were conducted.

Healthy and morphologically intact specimens of *B. fuscopurpurea* were cultured in 1 L of sterilized seawater medium at 15 °C and 28 °C, supplemented with 2 mg·L^{-1} $NaNO_3$-N and 1 mg·L^{-1} KH_2PO_4-P. The culture conditions were as follows: a light intensity of 40 µmol photons m^{-2}·s^{-1}, seawater salinity of 30 psu, and a 12 h:12 h (light/dark) photoperiod. The experiment lasted 7 days with three replicates per group, and parallel samples were analyzed separately.

The culture conditions for the transcriptome materials were as follows: after the algae were returned to the laboratory and cultivated for two days as described above, they were transferred to sterile seawater containing antibiotics (0.1 g·L^{-1} kanamycin sulfate + 0.1 g·L^{-1} ampicillin + 0.2 g·L^{-1} streptomycin sulfate) for 1 day of cultivation, and then the formal experiments began. At 6 h and 1 d of the experiment, 3 samples of fresh *B. fuscopurpurea* with a weight of 0.3 to 0.5 g each were taken from each treatment group. These samples were rapidly frozen with liquid nitrogen and then stored at −80 °C for use as experimental materials for transcriptome sequencing.

2.2. Determination of Relative Growth Rate and Fluorescence Parameters

After high-temperature stress treatment, the fresh weight of *B. fuscopurpurea* was measured at 0, 1, 3, 5, and 7 days of high-temperature cultivation, and the relative growth rate was calculated using the formula proposed by Yong [17].

The maximum photochemical quantum yield of PSII (*Fv/Fm*) after t days of high-temperature stress was measured using IMAGING-PAM (MAXI, Walz, Nuremberg, Germany), with the formula $F_v = F_m - F_0$. The minimal (F_0) and maximal (F_m) fluorescence was obtained by exposing the thalli, dark-adapted for 20 min, to actinic and saturating light provided by the fluorometer [18].

2.3. Measurement of Chlorophyll a and Carotenoids

Approximately 0.05 g of the algal sample was weighed into a 2 mL grinding tube, and 0.95 mL of anhydrous methanol was added. The sample was ground using a cryogenic grinder (JXFSTPRP-CLN-48, Shanghai Jingxin, Shanghai, China) at 70 Hz and −4 °C for 25 cycles (50 s grinding and 10 s pause per cycle). After grinding, the samples were left to stand at 4 °C for 24 h. The mixture was then centrifuged at 4000 rpm for 10 min at 4 °C. The supernatant was collected for further analysis. Using anhydrous methanol as a blank control, OD values at 480 nm, 510 nm, 652 nm, and 665 nm were measured with a microplate reader. The contents of chlorophyll a and carotenoids were calculated using the formula proposed by Porra [19].

2.4. Algal Phycobiliprotein Determination

Approximately 0.05 g (fresh weight) of algal tissue was placed in a 2 mL grinding tube, to which 0.95 mL of precooled 0.1 mol·L^{-1}, pH 7.0 phosphate buffer was added. Samples were homogenized in a cryogenic grinder under the same conditions as described in Section 2.3. The homogenate was centrifuged at 4 °C and 4000 rpm for 10 min. The supernatant was collected for subsequent analysis. Using 0.1 mol·L^{-1}, pH 7.0 phosphate buffer as the blank, OD values at 651 nm, 645 nm, 618 nm, 614 nm, 595 nm, 592 nm, 564 nm, and 455 nm were measured with a microplate reader (TCP011096, Jet Bio-Filtration, Guangzhou, China). The phycoerythrin and phycocyanin contents were calculated using the formula proposed by Beer [20].

2.5. Library Preparation for Transcriptome Sequencing

A total amount of 1 µg RNA per sample was used as input material for the RNA sample preparations. Sequencing libraries were generated using NEBNext®Ultra™ RNA Library Prep Kit for Illumina® (NEB, USA) following the manufacturer's recommendations, and index codes were added to attribute sequences to each sample. Briefly, mRNA was purified from total RNA using poly-T oligo-attached magnetic beads. Fragmentation was carried out using divalent cations under elevated temperature in NEBNext First Strand Synthesis Reaction Buffer (5×). First-strand cDNA was synthesized using a random hexamer primer and M-MuLV Reverse Transcriptase. Second-strand cDNA synthesis was subsequently performed using DNA polymerase I and RNase H. Remaining overhangs were converted into blunt ends via exonuclease/polymerase activities. After the adenylation of the 3′ ends of DNA fragments, NEBNext Adaptor with a hairpin-loop structure was ligated to prepare for hybridization. In order to select cDNA fragments of preferentially 240 bp in length, the library fragments were purified with the AMPure XP system (Beckman Coulter, Beverly, MA, USA). Then, 3 μL USER Enzyme (NEB, USA) was used with size-selected, adaptor-ligated cDNA at 37 °C for 15 min followed by 5 min at 95 °C before PCR. Then, PCR was performed with Phusion High-Fidelity DNA polymerase, Universal PCR primers, and Index (X) Primer. Finally, the library fragments were purified with AMPure XP system (Beckman Coulter, Beverly, USA), and the library quality was assessed with the Agilent Bioanalyzer 2100 system.

2.6. Quality Control, Transcriptome Assembly, and Gene Function Annotation

Raw data (raw reads) of fastq format were first processed through in-house perl scripts. In this step, clean data (clean reads) were obtained by removing reads containing the adapter, reads containing ploy-N, and low-quality reads from raw data. At the same time, the Q20, Q30, GC content, and sequence duplication level of the clean data were calculated. All the downstream analyses were based on clean data with high quality. The transcriptome was assembled using the Trinity software (2.14.0). Gene function was annotated based on the following databases: NR (NCBI non-redundant protein sequences); Pfam (Protein family); KOG/COG/eggNOG (Clusters of Orthologous Groups of proteins); Swiss-Prot (a manually annotated and reviewed protein sequence database); KEGG (Kyoto Encyclopedia of Genes and Genomes); and GO (Gene Ontology).

2.7. Analysis of Differentially Expressed Genes

The differential expression analysis of the two conditions/groups was performed using the DESeq R package (1.10.1). DESeq provides statistical routines for determining the differential expression in digital gene expression data using a model based on the negative binomial distribution. The resulting p values were adjusted using the Benjamini and Hochberg approach for controlling the false discovery rate. Genes with an adjusted p value < 0.05 found by DESeq were assigned as differentially expressed.

2.8. Statistical Analysis

The RGR, Chl-a, Car, PC, PE, and *Fv/Fm* data were expressed as means ± SD (n = 3). The experimental data were processed and statistically analyzed by Microsoft Office Excel software. The homogeneity of variance was tested by Levene's test. At the significance level of $p < 0.05$, the statistical significance within each group at different time points was tested by One-Way ANOVA, and the statistical significance between groups at the same time point was tested by an independent-sample *T*-test. Graphpad prism 8.0 software was used for data plotting. Cluster trend profile chart, pairwise comparison, and volcano plot analysis was performed using BMKCloud (www.biocloud.net).

3. Results

3.1. Changes in Photosynthetic Pigments and Fluorescence Parameters Under Different Temperatures

B. fuscopurpurea exhibited different relative growth rates when cultured at 15 °C and 28 °C; from day 1 to day 7, the growth rate at 15 °C was consistently higher than at 28 °C, with significant differences ($p < 0.05$). At 15 °C, the maximum relative growth rate (6.95%) was observed on day 2, whereas at 28 °C, the peak rate (0.96%) occurred on day 4. The average relative growth rate was 5.6% at 15 °C and 0.59% at 28 °C (Figure 1).

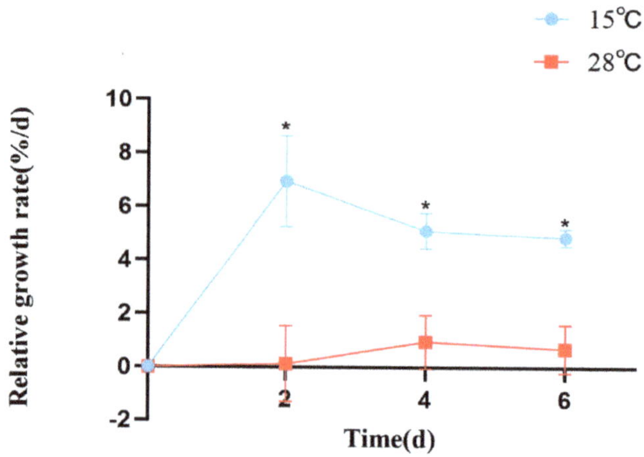

Figure 1. The influence of different temperatures on the relative growth rate of *B. fuscopurpurea*. Note: * means that there is a significant difference between the 15 °C and 28 °C groups at the same time point; * means that there is a statistical difference at the $p < 0.05$ significance level.

The chlorophyll a and carotenoid contents of *B. fuscopurpurea* at 15 °C did not change significantly over time; at 28 °C, the chlorophyll and carotenoid levels initially declined, then increased, and subsequently declined again, reaching their lowest values at 6 h (0.86 and 0.21) and highest values at 1 d (1.21 and 0.29). Moreover, on day 1, the chlorophyll a and carotenoid levels in the 28 °C treatment (1.21 and 0.29) were significantly higher than those in the 15 °C treatment (0.94 and 0.23) ($p < 0.05$) (Figure 2A,B).

At 15 °C, the phycobiliprotein content in *B. fuscopurpurea* did not change significantly over time. At 28 °C, the phycobiliprotein content remained stable initially and then increased, with a significant rise from 3.59 on day 4 to 4.71 on day 7 ($p < 0.05$). Moreover, on days 1, 4, and 7, the phycobiliprotein contents at 28 °C (3.60, 3.59, 4.71) were significantly higher than those at 15 °C (3.06, 2.80, 3.48) ($p < 0.05$). Similarly, at 15 °C, the phycocyanin content in *B. fuscopurpurea* did not exhibit significant changes over time. At 28 °C, the phycocyanin content first decreased and then increased, with a significant rise from 0.37 at 6 h to 0.61 on day 4 ($p < 0.05$) and further to 1.06 on day 7 compared to 0.61 on day 4 ($p < 0.05$). Additionally, on days 4 and 7, the phycocyanin contents in the 28 °C group (0.61, 1.06) were significantly higher than those in the 15 °C group (0.35, 0.50) ($p < 0.05$) (Figure 2C,D). At 15 °C, the Fv/Fm of *B. fuscopurpurea* exhibited an initial increase followed by stabilization: at 6 h, the Fv/Fm (0.48) was significantly higher than at 0 d (0.44) ($p < 0.05$) and thereafter showed no significant change over time. At 28 °C, the *Fv/Fm* showed a rise-and-fall pattern: at 1 d (0.51), it was significantly higher than at 6 h (0.42) ($p < 0.05$), then declined significantly to 0.46 at 4 d compared to 1 d (0.51) and further to 0.38 at 7 d compared to 4 d (0.46) ($p < 0.05$). Moreover, at 28 °C, the *Fv/Fm* values at 6 h,

4 d, and 7 d were 0.42, 0.46, and 0.38, respectively, which were significantly lower than the corresponding values at 15 °C (0.48, 0.51, and 0.50) (Figure 3).

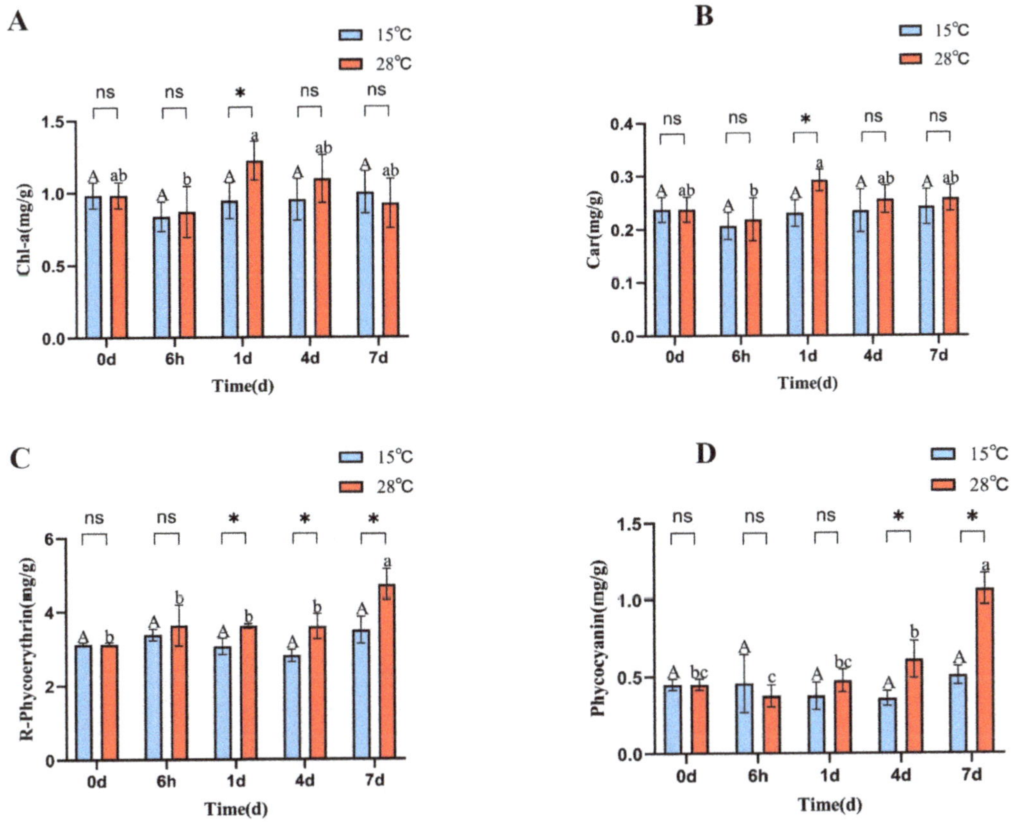

Figure 2. Chl-a, Car, PC, and PE content of *B. fuscopurpurea* at different temperatures. Note: In the figure, A, B, C, D represent the content graphs of chlorophyll, carotenoids, phycocyanin, and phycobilin respectively. The capital letters A, B represent the significant differences in the 15 °C treatment group under different culture times; the lowercase letters a, b, and c represent the significant differences in the 28 °C treatment group under different culture times; ns indicates that there is no significant difference between the 15 °C treatment group and the 28 °C treatment group at the same time point; * represents that there are statistical differences between the 15 °C treatment group and the 28 °C treatment group at the significance level of $p = 0.05$ at the same time point.

3.2. Data Quality

The transcriptome sequencing of twelve samples generated a total of 77.97 Gb of clean data, with each sample yielding at least 5.97 Gb and a Q30 base percentage of ≥85.50% (Supplementary Table S1). Assembly produced 47,609 unigenes in total. Of these, 6945 unigenes exceeded 1 kb in length. The functional annotation of the unigenes yielded annotations for 22,000 entries (Supplementary Table S2).

3.3. Gene Function Annotation

Currently, the whole-genome sequencing of *B. fuscopurpurea* remains incomplete; therefore, unigene sequences were compared against the NR, Swiss-Prot, COG, KOG, eggNOG4.5, and KEGG databases using DIAMOND (v2.0.4). KEGG Orthology assignments were obtained via KOBAS (v2.0), and GO Orthology annotations were derived by analyzing predicted gene sequences with InterProScan (5.34-73.0) against the integrated InterPro database. Following amino acid prediction, HMMER (v3.1b2) searches against the Pfam database were performed to retrieve comprehensive unigene annotation information. By applying BLAST with an E-value cutoff of $\leq 1 \times 10^{-5}$ and HMMER with an E-value threshold of $\leq 1 \times 10^{-10}$, this study ultimately obtained annotation information for 22,000 unigenes.

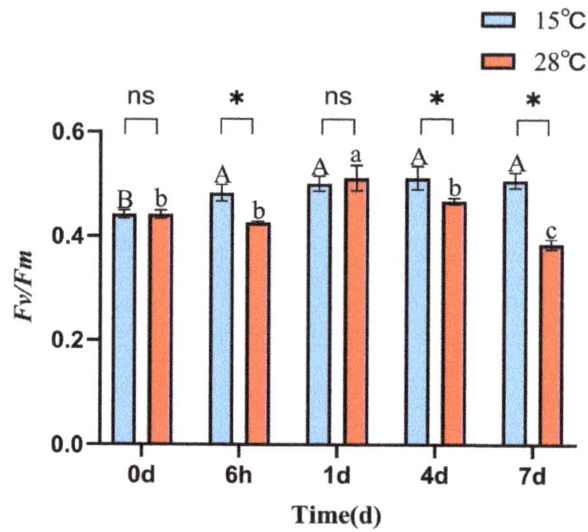

Figure 3. *Fv/Fm* of *B. fuscopurpurea* at different temperatures. Note: The capital letters A, B represent the significant differences in the 15 °C treatment group under different culture times; the lowercase letters a, b, and c represent the significant differences in the 28 °C treatment group under different culture times; ns indicates that there is no significant difference between the 15 °C treatment group and the 28 °C treatment group at the same time point; * represents that there are statistical differences between the 15 °C treatment group and the 28 °C treatment group at the significance level of $p = 0.05$ at the same time point.

In the NR database, the top match was *Porphyra umbilicalis* (54.87%), which, like *B. fuscopurpurea*, belongs to the Bangiaceae family, consistent with their taxonomic relationship (Supplementary Figure S2). A total of 7701 genes were annotated in the COG database and grouped into 25 functional categories: the largest was "Translation, ribosomal structure and biogenesis" (1172), followed by "Posttranslational modification, protein turnover, and chaperones" (880), and the smallest was "RNA processing and modification" (3) (Supplementary Figure S3). The GO database annotated 14,577 genes, which were distributed among three main categories: Cellular Component, Molecular Function, and Biological Process. Within the Cellular Component category, the largest term was "cellular anatomical entity," followed by "intracellular" and then "protein complex." In the Molecular Function category, "binding" was predominant, followed by "catalytic activity" and then "structural molecule activity." In Biological Process, "cellular process" was the most represented, followed by "metabolic process" and then "biological regulation" (Supplementary Figure S4). The number of genes annotated across all databases was 3397 (Supplementary Figure S5).

3.4. Differential Gene Analysis

3.4.1. Overview of Differentially Expressed Genes

Using $\log_2(FC) \geq 2$ and FDR < 0.01 as thresholds for differential gene screening, two comparisons were established, CT-6 h vs. HT-6 h and CT-1 d vs. HT-1 d, to investigate temporal changes in gene expression under heat stress. A total of 3644 DEGs were identified across time points: 1677 DEGs (979 up-regulated and 698 down-regulated) in CT-6 h vs. HT-6 h and 1967 DEGs (1261 up-regulated and 706 down-regulated) in CT-1 d vs. HT-1 d (Figure 4A,B). Among these, 813 DEGs were modulated by heat treatment at both 6 h and 1 d (Figure 4C). In addition, the trend analysis of DEG expression over time under heat stress in *B. fuscopurpurea* yielded eight clustered expression profiles (Figure 5). Profiles A and D (122 and 347 DEGs, respectively) showed expression up-regulation over time at both control and elevated temperatures. Clusters B and E (432 and 647 DEGs, respectively) were characterized by high expression under heat stress and low expression at the control temperature. Clusters C and F (78 and 368 DEGs, respectively) displayed increasing expression over time under heat stress, with minimal change at the control temperature.

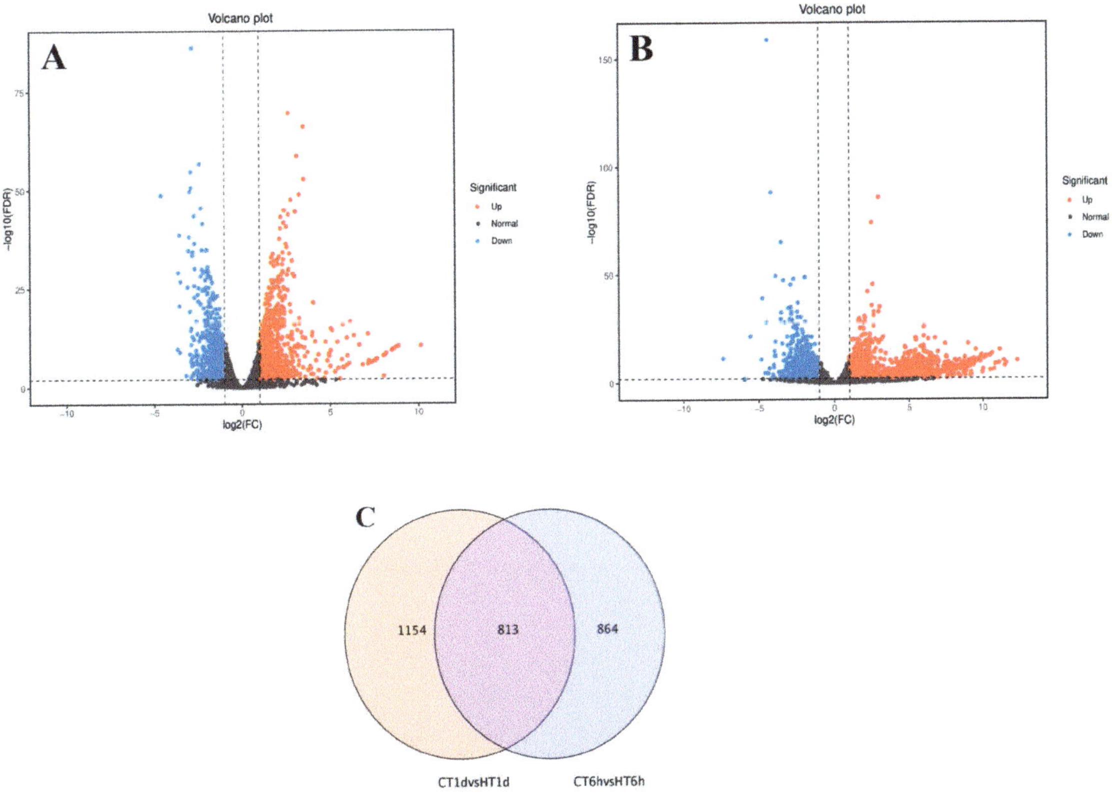

Figure 4. Analysis results of differentially expressed genes between the control group and the high-temperature group (pairwise comparison). Note: (**A**,**B**) are volcano plots of differentially expressed genes in CT group at 6 h compared with HT group at 6 h and CT group at 1 d compared with HT group at 1 d, respectively. The red part represents the number of up-regulated genes, and the blue part represents the number of down-regulated genes. (**C**) is a Venn diagram drawn based on the results of differential expression analysis. The overlapping part refers to the differentially expressed genes that are enriched in both comparison groups. The non-overlapping part refers to the differentially expressed genes that are enriched in each individual group.

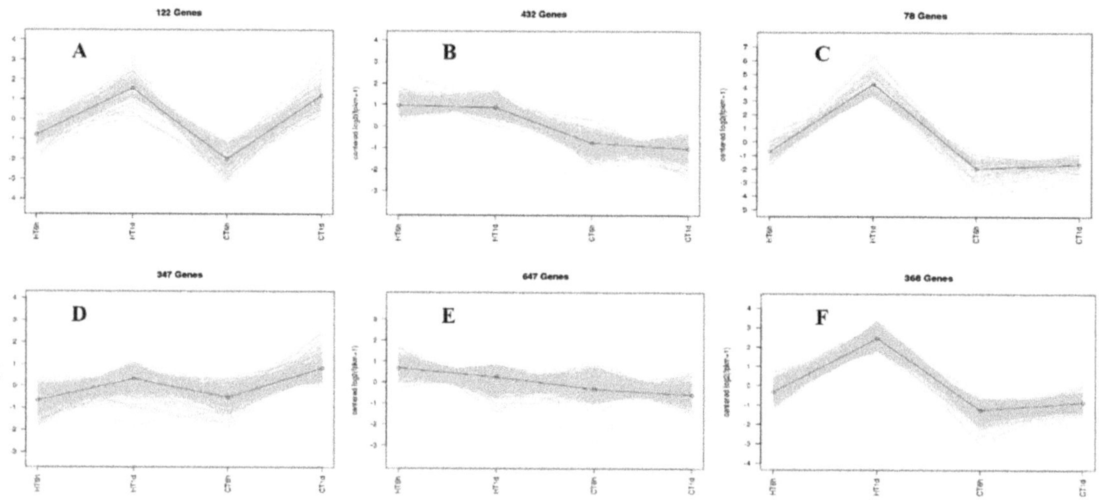

Figure 5. Cluster trend profile chart. Note: The trend clustering curve graph showing similar expression trends for the selected stones (**A–F**) in the figure. In this cluster trend profile, the horizontal axis represents different sample treatment groups, and the vertical axis represents the expression trend of the differentially expressed genes that have been enriched. Positive values indicate an upward trend in expression, while negative values indicate a downward trend in expression.

3.4.2. Analysis of Differentially Expressed Genes Using GO and KEGG Databases

GO and KEGG functional annotations were performed for the DEGs. GO classification revealed that DEGs are involved in three categories, Biological Process, Molecular Function, and Cellular Component, and in both CT-6 h vs. HT-6 h and CT-1 d vs. HT-1 d comparisons, most genes were specifically assigned to metabolic processes, cellular processes, and biological regulation (Supplementary Figure S6). KEGG enrichment analysis showed that at 6 h DEGs were primarily enriched in carbon metabolism, amino acid biosynthesis, and protein processing in the endoplasmic reticulum; at 1 d, they were mainly enriched in ribosome, carbon metabolism, and amino acid biosynthesis (Supplementary Figure S7 and Table 1).

3.4.3. KEGG Differential Expression Pathway Analysis

Based on the KEGG enrichment of DEGs, eight pathways related to energy, carbohydrate, and lipid metabolism were selected to examine the temporal gene expression in *B. fuscopurpurea* under heat stress: photosynthesis (ko00195), photosynthesis antenna proteins (ko00196), carbon fixation in photosynthetic organisms (ko00710), glycolysis/gluconeogenesis (ko00010), the pentose phosphate pathway (ko00030), the Citrate cycle (ko00020), glycerophospholipid metabolism (ko00564), and glutathione metabolism (ko00480). The full gene names used in all heatmaps of DEGs are provided in Supplementary Table S3.

Photosynthesis Pathway

Comparative transcriptome analysis identified 13 DEGs involved in photosynthesis, participating in the assembly of PSII (*PsbA, PsbC, PsbM, PsbO, PsbP, PsbQ, PsbS, PsbU*), PSI (*PsaO*), the cytochrome b6/f complex (*PetC*), and photosynthetic electron transport (*PetF, PetH*) (Figure 6). All of these DEGs were significantly up-regulated at both 6 h and 1 d, with the two down-regulated genes (*PsbA, PsbC*) being enriched at 1 d. In addition, five DEGs related to light-harvesting antenna proteins were identified, involved in the synthesis of the allophycocyanin β subunit (*ApcB*), phycobilisome linker protein (*ApcC*), phycocyanin rod-linker protein (*CpcC*), phycobilin lyase α subunit (*CpcE*), and photosystem I light-harvesting complex (*Lhca1*). All of these antenna-related DEGs were significantly up-regulated at both 6 h and 1 d, with the single down-regulated gene (*ApcB*) enriched at 1 d.

Table 1. Significantly enriched KEGG pathways in CT_6_h vs. HT6h and CT_1_d vs. HT_1_d.

Pathway ID	Pathway	KEGG_B_Class	No. of DEGs	*p* Value
CT6h vs. HT6h				
ko01200	Carbon metabolism	Global and overview maps	29	6.13×10^{-1}
ko01230	Biosynthesis of amino acids	Global and overview maps	20	6.83×10^{-1}
ko00860	Porphyrin and chlorophyll metabolism	Metabolism of cofactors and vitamins	19	2.87×10^{-9}
ko00230	Purine metabolism	Nucleotide metabolism	15	3.25×10^{-2}
ko00710	Carbon fixation in photosynthetic organisms	Energy metabolism	14	3.65×10^{-2}
ko00010	Glycolysis/gluconeogenesis	Carbohydrate metabolism	13	2.72×10^{-1}
ko00630	Glyoxylate and dicarboxylate metabolism	Carbohydrate metabolism	13	3.46×10^{-1}
ko00196	Photosynthesis antenna proteins	Energy metabolism	12	1.68×10^{-8}
ko00195	Photosynthesis	Energy metabolism	11	5.94×10^{-4}
ko00030	Pentose phosphate pathway	Carbohydrate metabolism	11	1.54×10^{-2}
ko00480	Glutathione metabolism	Metabolism of other amino acids	10	8.50×10^{-2}
ko00260	Glycine, serine, and threonine metabolism	Amino acid metabolism	10	1.15×10^{-1}
ko00564	Glycerophospholipid metabolism	Lipid metabolism	9	1.55×10^{-2}
ko00250	Alanine, aspartate, and glutamate metabolism	Amino acid metabolism	9	1.06×10^{-1}
CT1d vs. HT1d				
ko01200	Carbon metabolism	Global and overview maps	34	9.43×10^{-1}
ko01230	Biosynthesis of amino acids	Global and overview maps	29	6.66×10^{-1}
ko00860	Porphyrin and chlorophyll metabolism	Metabolism of cofactors and vitamins	16	6.43×10^{-5}
ko00630	Glyoxylate and dicarboxylate metabolism	Carbohydrate metabolism	16	5.52×10^{-1}
ko00710	Carbon fixation in photosynthetic organisms	Energy metabolism	14	2.87×10^{-1}
ko00010	Glycolysis/gluconeogenesis	Carbohydrate metabolism	14	6.69×10^{-1}
ko00195	Photosynthesis	Energy metabolism	13	1.05×10^{-3}
ko00196	Photosynthesis antenna proteins	Energy metabolism	12	8.25×10^{-7}
ko00480	Glutathione metabolism	Metabolism of other amino acids	12	1.59×10^{-1}
ko00230	Purine metabolism	Nucleotide metabolism	12	6.24×10^{-1}
ko00270	Cysteine and methionine metabolism	Amino acid metabolism	11	6.12×10^{-1}
ko00030	Pentose phosphate pathway	Carbohydrate metabolism	10	2.14×10^{-1}
ko00020	Citrate cycle (TCA cycle)	Carbohydrate metabolism	9	9.25×10^{-1}

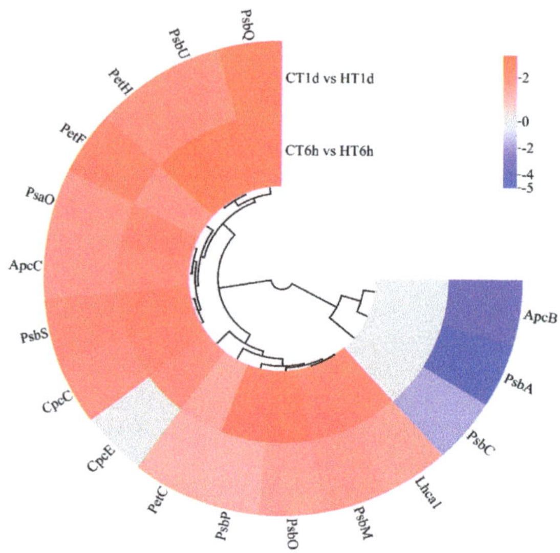

Figure 6. Cluster heatmap of differentially expressed genes in photosynthesis. Note: The logarithmic values of fold change (Log_2FC) of expression levels among the sample groups are shown in the legend, where red indicates significant up-regulation, blue indicates significant down-regulation, and gray indicates no significant difference.

Carbohydrate Synthesis and Energy Metabolism Pathways

Transcriptome analysis revealed that DEGs are significantly enriched in pathways related to carbon and energy metabolism and play roles in the temporal response of *B. fuscopurpurea* to heat stress. In our findings, the Calvin cycle was significantly enriched with 10 DEGs: *SBPase* was uniquely up-regulated at 6 h; *RpiA, TK, TPI,* and *MDH2* were exclusively up-regulated at 1 d; *RbcL* was down-regulated at 1 d; and *ALDO, FBP, PrkB,* and *RPE* were significantly up-regulated at both time points (Figure 7A). The Calvin cycle intermediates glyceraldehyde-3-phosphate and dihydroxyacetone phosphate can funnel into glycolysis/gluconeogenesis for sucrose and starch synthesis, enter the tricarboxylic acid cycle to generate energy, or flow into the pentose phosphate pathway to produce NADPH. In this study, glycolysis/gluconeogenesis was significantly enriched with nine DEGs. *TPI, FrmA,* and ADH were exclusively up-regulated at 1 d; *PK* was significantly down-regulated at both time points; and *FBP, Pgm, ALDO, PdhD,* and *G6PE* were significantly up-regulated at both 6 h and 1 d (Figure 7B). The pentose phosphate pathway was significantly enriched with nine DEGs: *PGD* was down-regulated solely at 6 h; *RpiA, G6PD, PGLS,* and *TK* were up-regulated at both time points; and *ALDO, FBP, Pgm,* and *RPE* were significantly up-regulated at both 6 h and 1 d (Figure 7C). The TCA cycle was significantly enriched with seven DEGs: *SucA* was exclusively up-regulated at 6 h; *CS, ACO, FumC,* and *MDH2* were up-regulated exclusively at 1 d; *SucB* was down-regulated at 1 d; and *PdhD* was significantly up-regulated at both time points (Figure 7D).

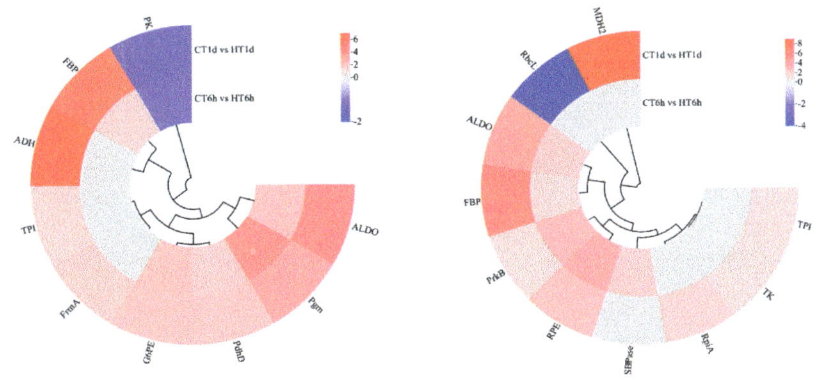

(A,B) Calvin cycle, Citrate cycle

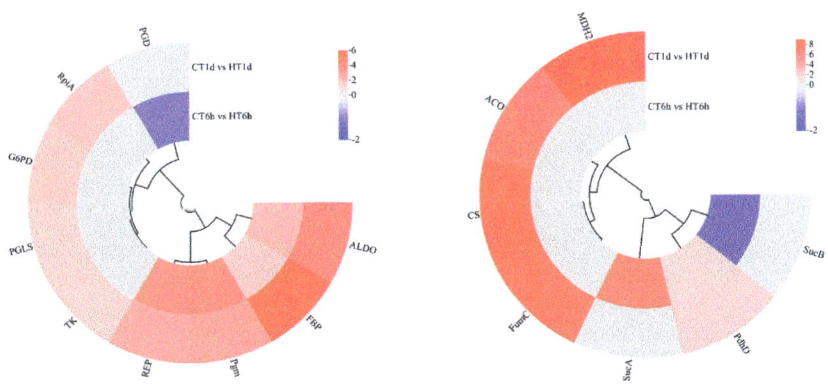

(C,D) Glycolysis/gluconeogenesis, pentose phosphate pathway

Figure 7. Cluster heatmap of differentially expressed genes in Calvin cycle, Citrate cycle, glycolysis/gluconeogenesis, and pentose phosphate pathway. Note: In the figure, A, B, C, and D represent the cluster heatmaps of the Calvin cycle, Citrate cycle, glycolysis/gluconeogenesis, and pentose phosphate pathway.The logarithmic values of fold change (Log_2FC) of expression levels among the sample groups are shown in the legend, where red indicates significant up-regulation, blue indicates significant down-regulation, and gray indicates no significant difference.

Glycerophospholipid Metabolism Pathway

Transcriptome analysis revealed the enrichment of eight DEGs involved in glycerophospholipid metabolism, encoding enzymes for glycerol-3-phosphate dehydrogenase (*GPD1*), another glycerol-3-phosphate dehydrogenase isoform (*GlpA*), diacylglycerol kinase (*DgkA*), acylglycerophosphate acyltransferase 8 (*AGPAT8*), phosphatidylethanolamine N-methyltransferase (*PEMT*), ethanolamine-phosphate transferase (*EPT1*), cytidine diphosphate–ethanolamine synthase (*PCYT2*), and phosphatidylserine decarboxylase (*PSD*). Specifically, *GPD1* and *GlpA* were exclusively down-regulated at 6 h; *PSD* and *PCYT2* showed exclusive up-regulation at 1 d; *AGPAT8*, *DgkA*, and *EPT1* were significantly up-regulated at both time points; and *PEMT* exhibited down-regulation at 6 h followed by up-regulation at 1 d (Figure 8).

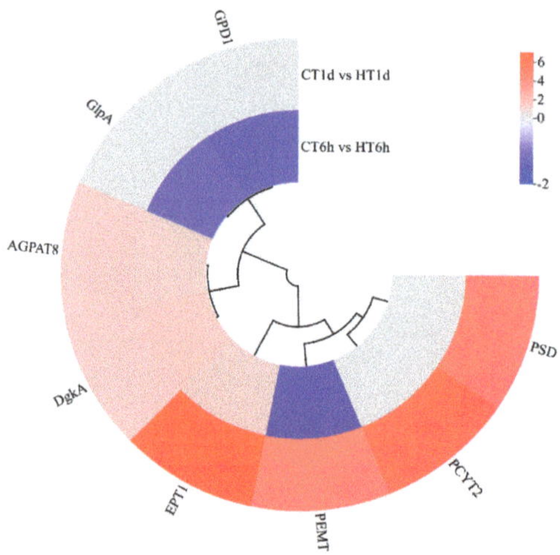

Figure 8. Cluster heatmap of differentially expressed genes in glycerophospholipid metabolism pathway. Note: The logarithmic values of fold change (Log_2FC) of expression levels among the sample groups are shown in the legend, where red indicates significant up-regulation, blue indicates significant down-regulation, and gray indicates no significant difference.

Glutathione Metabolism Pathway

Transcriptome analysis revealed that heat stress induced the differential expression of nine DEGs involved in the glutathione cycle, encoding enzymes such as leucine aminopeptidase (*PepA*), γ-glutamyl transpeptidase (*Ggt*), 6-phosphogluconate dehydrogenase (*PGD*), glutathione peroxidase (*Gpx*), ornithine decarboxylase (*ODC*), glutathione S-transferase (*HPGDS*), glucose-6-phosphate dehydrogenase (*G6PD*), L-ascorbate peroxidase (*APX*), and ribonucleotide reductase subunit M2 (*RRM2*). Among these, *Ggt* was uniquely up-regulated at 6 h; *PepA*, *PGD*, and *ODC* were exclusively down-regulated at 6 h; *G6PD* was specifically up-regulated at 1 d; and *HPGDS*, *Gpx*, *RRM2*, and *APX* were significantly up-regulated at both time points (Figure 9).

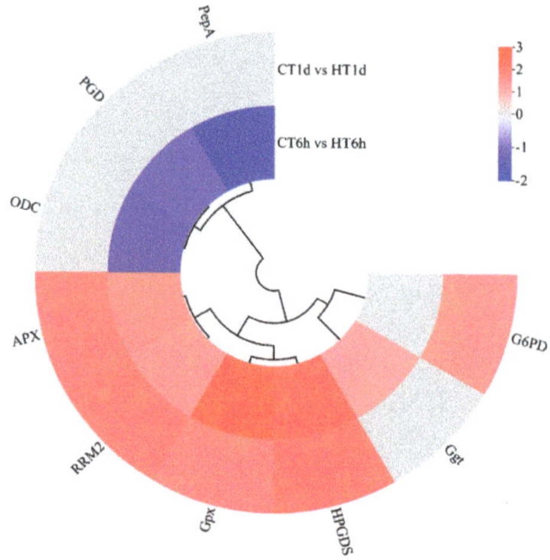

Figure 9. Cluster heatmap of differentially expressed genes in glutathione cycle pathway. Note: The logarithmic values of fold change (Log_2FC) of expression levels among the sample groups are shown in the legend, where red indicates significant up-regulation, blue indicates significant down-regulation, and gray indicates no significant difference.

4. Discussion

As global warming intensifies, large-scale economically cultivated macroalgae are inevitably subjected to the adverse effects of elevated temperatures. Our results demonstrate that *B. fuscopurpurea* grows more favorably at 15 °C compared to at 28 °C. Short-term heat exposure (1 day) significantly increased the chlorophyll a and carotenoid contents while markedly reducing the *Fv/Fm*. Transcriptomic analysis uncovered the molecular mechanisms underlying heat tolerance in this alga. RNA-seq data indicate that these responses involve the regulation of carbohydrate metabolism, photosynthesis, glycerophospholipid metabolism, and the glutathione cycle.

4.1. The Influence of Temperature on Photosynthetic Pigments and Fluorescence Parameters

Photosynthesis and respiration in macroalgae are temperature-dependent, and fluctuations in temperature can alter algal growth performance [21]. Different algal taxa exhibit variable tolerance to temperature shifts, and the growth rate serves as a key metric that directly reflects their physiological condition [22]. Under heat stress, the Atlantic brown alga Fucus vesiculosus exhibited a 65% reduction in growth rate at 26 °C compared to at 12 °C [23]. Similarly, in our study, Bangia fuscopurpurea cultured at 28 °C displayed a significantly suppressed relative growth rate compared to that cultured at 15 °C. Chlorophyll a, the principal light-harvesting pigment, directly determines photosynthetic capacity and indirectly influences the growth rate; carotenoids protect chlorophyll from photodamage, and phycobiliproteins capture and transfer light energy to chlorophyll, all playing crucial roles in photosynthesis [24]. Studies have shown that under heat stress, *Ulva prolifera* attains peak chlorophyll a content after 4 days of cultivation [25]. Likewise, *Gracilaria blodgettii* shows significant increases in chlorophyll a and carotenoid levels under elevated temperatures, which then decline as the temperature rises further [26]. Phycoerythrin (PE), a major light-harvesting pigment found in red algae and cyanobacteria [27], also exhibits potent antioxidant activity by scavenging reactive oxygen species and alleviating oxidative stress [28]. In prolonged-heat-stress experiments with *Neopyropia yezoensis*, both the phycoerythrin and phycocyanin contents were found to increase significantly [29]. Data analysis revealed that in the 28 °C treatment, the chlorophyll a and carotenoid levels in *B. fuscopurpurea* decreased at 6 h and rose at 1 day ($p < 0.05$), while phycoerythrin and phycocyanin exhibited a progressive increase, significantly peaking on day 7 ($p < 0.05$), consistent with previous studies. Photosystem II (PSII) is regarded as one of the most temperature-sensitive components of the photosynthetic apparatus [30,31]. The *Fv/Fm* ratio represents the maximum quantum yield of PSII, reflecting the alga's maximal photosynthetic efficiency and indirectly indicating its growth status [7]. Studies have shown the Fv/Fm to be a valuable metric for assessing macroalgal thermal tolerance [32]. Our results demonstrate that the *Fv/Fm* values in *B. fuscopurpurea* at 28 °C at 6 h, 4 d, and 7 d were significantly lower than those at 15 °C ($p < 0.05$), indicating initial photoinhibition at 6 h, consistent with the findings in *Kappaphycus alvarezii* [33]. Over the extended exposure periods (4 d and 7 d), heat stress progressively impaired the photosynthetic apparatus, further reducing the *Fv/Fm* ($p < 0.05$).

4.2. The Influence of Temperature on the Photosynthetic Pathway

Photosynthesis is one of the most temperature-sensitive biological processes in algae [34], relying on electron transport within the photosystems to supply the energy required for algal growth and development [35]. In red algae, phycobilisomes serve as the principal light-harvesting antennae in the photosynthetic mechanism. Red algal phycobilisomes consist of phycoerythrin (PE), phycocyanin (PC), and allophycocyanin (APC); the red coloration of these algae arises from APC masking the green color of chlorophyll and other pigments [36]. These phycobilisome structures are closely associated with Photosystem II (PSII) and Photosystem I (PSI), ensuring the efficient transfer of excitation energy to their reaction centers and promoting the conversion of light into chemical energy [37]. Studies have shown that in Sargassum horneri, the expression of photosynthesis-related genes is down-regulated during the early phase of heat stress [38]. In this study, genes encoding the PSII dimer subunit PsbM and oxygen-evolving enhancer proteins (*PsbO/P/Q/U*) were rapidly up-regulated to stabilize the oxygen-evolving complex and maintain water-splitting activity, ensuring sustained ATP and NADPH production. Concurrently, genes for photo-protective proteins such as PsbS and Lhca1, along with key electron transport components (*PetC/F/H*), were co-up-regulated, balancing the energy flow between PSII and PSI, preserving non-photochemical quenching and thermal dissipation, and preventing excessive ROS accumulation. At 6 h, the up-regulation of *CpcE* likely reinforces phycobilisome integrity, followed by the moderate down-regulation of core reaction-center proteins (*PsbA/C*, *ApcB*) to reduce light capture and mitigate photoinhibition risk—findings consistent with observations in *Pyropia haitanensis* under heat stress [39].

4.3. The Influence of Temperature on Carbohydrate Synthesis and Energy Metabolism Pathways

Studies have shown that algal metabolic processes are subject to gene-level regulation under stress conditions [40], with carbohydrate metabolism representing a critical component. Algae modulate carbohydrate concentrations and structural composition to optimize the use of endogenous carbon-derived energy and accumulate compatible solutes for molecular protection [41]. However, under excessive environmental stress, algae may experience energy deficits, prompting the up-regulation of intrinsic carbohydrate metabolic routes and the activation of alternative pathways such as glycolysis to sustain ATP supply and carbon skeleton provision for essential processes [42]. For example, *G. lemaneiformis* exhibits a significant accumulation of floridoside and isofloridoside after 1 and 2 days of heat stress [43]. In this study, we conducted an integrated analysis of four carbohydrate and energy metabolism pathways: the Calvin cycle, glycolysis/gluconeogenesis, the pentose phosphate pathway, and the tricarboxylic acid (TCA) cycle. We observed that the majority of Calvin cycle genes were up-regulated, consistent with the response reported in *Saccharina latissima* under elevated temperature [44]. However, *RbcL* was significantly down-regulated at 1 day; the differential temperature sensitivity of Rubisco has been reported across algal taxa [45], and our finding aligns with that of Huang, who observed *RbcS* down-regulation in *G. bailinae* under heat stress [7]. Previous studies suggest that the thermal lability of the electron transport chain may constrain energy metabolism under heat stress [46]. For instance, *P. haitanensis* shows a significant down-regulation of energy metabolism pathways, including the pentose phosphate pathway and TCA cycle, under heat stress [39], whereas *G. lemaneiformis* displays enhanced energy metabolism when exposed to high temperature, mirroring our observations [47]. The enhanced pentose phosphate pathway and glycolysis/gluconeogenesis generate abundant NADPH, supplying energy and supporting antioxidant defenses by maintaining enzyme activities to mitigate oxidative damage. While heat stress has been reported to inhibit the TCA cycle in *Sargassum fusiforme* [48], our study found an overall up-regulation of TCA cycle genes. Key intermediate-synthesizing enzymes (*CS*, *MDH2*, *PdhD*) were significantly induced, indicating that enhanced TCA activity under heat stress bolsters signal transduction, energy supply, and antioxidant defense to counteract thermal damage.

4.4. The Influence of Temperature on Glycerophospholipid Metabolism

Glycerophospholipids, as the principal lipid constituents of biological membranes, are essential for maintaining bilayer fluidity, a property fundamental to the proper function of membrane-bound proteins, ion channels, and receptors [49]. Under heat-stress conditions, the lipid composition and architecture of algal plasma membranes may be altered [50]; the regulation of glycerophospholipid metabolism is therefore critical for preserving membrane fluidity and integrity, which in turn protects cells from thermal damage. Studies have shown that chitosan oligosaccharide treatment enhances thermotolerance in Gracilariopsis lemaneiformis by significantly up-regulating genes involved in glycerophospholipid metabolism, increasing photosynthetic membrane lipid content, and promoting the accumulation of lipid signaling molecules such as phosphatidic acid (PA), thereby improving photosynthetic growth and heat resistance under high-temperature stress [51]. Similarly, exogenous arginine application in heat-stressed Sargassum fusiforme induces the up-regulation of arginine metabolism genes, leading to enhanced phosphatidic acid synthesis and improved thermotolerance [52]. Consistent with these findings, our data show that at 6 h of heat stress, *B. fuscopurpurea* limits excessive lipid synthesis or redirects carbon flux by reducing glycerol-3-phosphate production, while the sustained up-regulation of *DgkA* promotes the conversion of diacylglycerol (DAG) to phosphatidic acid (PA). PA not only fuels phospholipid biosynthesis but also acts as a signaling molecule to orchestrate membrane remodeling and stress-response pathways, facilitating rapid membrane repair and mitigating heat-induced cellular damage.

4.5. The Influence of Temperature on Glutathione Metabolism

Previous studies have demonstrated that plants deploy specific genetic responses to cope with environmental stresses, with glutathione (GSH) and hydrogen peroxide serving as central signaling molecules in both abiotic and biotic stress pathways [53]. The exogenous application of GSH can enhance antioxidant enzyme activities, mitigate oxidative damage, and improve thermotolerance in plants [54]. In macroalgae, the principal mechanism for counteracting excessive reactive oxygen species is the ascorbate–glutathione cycle [55]. Maintaining the dynamic equilibrium between reduced glutathione (GSH) and its oxidized form (GSSG), as well as between ascorbate (ASC) and dehydroascorbate (DHA), is essential for alleviating oxidative stress. Moreover, the activation of antioxidant enzymes—such as glutathione reductase (GR), ascorbate peroxidase (APX), and superoxide dismutase (SOD)—also plays a pivotal role in this defense system [56]. Previous research has shown that the ascorbate–glutathione cycle participates in the low-salinity stress response of *B. fuscopurpurea* [57]. Similarly, in our study, *B. fuscopurpurea* exhibited a significant up-regulation of the γ-glutamyl transpeptidase gene (*Ggt*) at 6 h of heat stress, enhancing GSH synthesis to rapidly scavenge intracellular ROS and ameliorate acute oxidative damage. After 1 day of stress, the thalli had reprogrammed their metabolic networks and engaged additional heat-response pathways, resulting in a normalization of *Ggt* expression. Throughout heat stress, glutathione peroxidase (*GPX*) and L-ascorbate peroxidase (*APX*) enzymes directly detoxifying H_2O_2 and other ROS must remain highly active; accordingly, *GPX* and *APX* genes were significantly up-regulated at both 6 h and 1 d to sustain cellular homeostasis.

4.6. The Gene Regulatory Mechanism Under High-Temperature Stress

Integrating the foregoing pathway analyses, it can be concluded that under heat stress, *B. fuscopurpurea* orchestrates a complex adaptive response network via precise gene regulation. Transcriptome profiling indicates that heat stress induces the modulation of genes involved in energy transfer between PSII and PSI, light harvesting, and heat-shock proteins, thereby mitigating photodamage. Concurrently, it fine-tunes genes in the Calvin cycle, the TCA cycle, glycolysis/gluconeogenesis, and the pentose phosphate pathway to balance ATP production with carbon fixation while triggering antioxidant defenses. In glycerophospholipid metabolism, the sustained up-regulation of *DgkA* drives the conversion of diacylglycerol into phosphatidic acid, promoting membrane remodeling and repair. Moreover, within 6 h of heat exposure, *Ggt* is rapidly induced to sustain the glutathione cycle, followed by the persistent up-regulation of *GPX* and *APX* to scavenge ROS, collectively preserving cellular function and viability. Together, these coordinated regulatory events constitute the comprehensive response strategy of *B. fuscopurpurea* under heat stress (Figure 10).

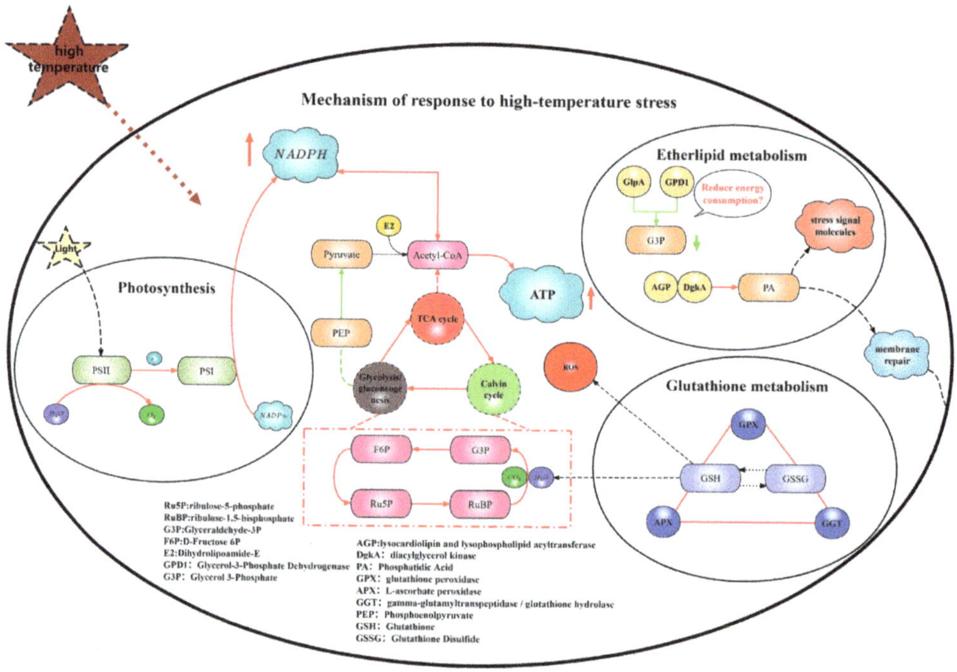

Figure 10. Overall gene regulatory map. Note: In the regulatory diagram of this mechanism, the red arrows indicate that the related regulatory genes are up-regulated, while the green arrows indicate that the related genes are down-regulated.

5. Conclusions

In summary, this study conducted a preliminary investigation into the physiological and molecular mechanisms of *B. fuscopurpurea* under heat stress. Cultivation at 28 °C imposed thermal stress on *B. fuscopurpurea*, resulting in a significant reduction in the relative growth rate and *Fv/Fm* values. The chlorophyll a and carotenoid contents significantly increased on day 1, while the phycobiliprotein levels showed a marked rise on days 4 and 7. Transcriptomic analysis revealed that *B. fuscopurpurea* adapts to high-temperature stress by regulating eight metabolic pathways related to photosynthesis, energy and carbohydrate metabolism, glycerophospholipid metabolism, and the glutathione cycle. The alga mitigates heat-induced damage by enhancing the Calvin cycle, the TCA cycle, and glycolysis/gluconeogenesis pathways, thereby accelerating the synthesis of osmotic regulators and NADPH. Furthermore, this study found that *B. fuscopurpurea* may counteract the adverse effects of heat stress by up-regulating genes involved in glycerophospholipid and glutathione metabolism to alter membrane fluidity and improve the reactive oxygen species scavenging capacity. These findings provide a valuable reference for further research on stress resistance and the development of heat-tolerant cultivars of *B. fuscopurpurea*.

Supplementary Materials: The following supporting information can be downloaded at https://www.mdpi.com/article/10.3390/cimb47070484/s1.

Author Contributions: Conceptualization, W.C.; Methodology, M.Z.; Software, M.Z.; Validation, M.Z.; Formal analysis, M.Z.; Investigation, M.Z., H.Z. and Z.C.; Resources, W.C.; Data curation, M.Z.; Writing—original draft, M.Z.; Writing—review & editing, W.C.; Visualization, M.Z.; Supervision, H.Z. and W.C.; Project administration, W.C.; Funding acquisition, W.C. All authors have read and agreed to the published version of the manuscript.

Funding: This research was supported by the Research on Industrial Innovation Technology for Guangdong Modern Marine Ranching: 2024-MRI-001, China Agriculture Research System (CARS-50), and Science and Technology Project of Shantou City, Guangdong Province: STKJ2024003.

Institutional Review Board Statement: Not applicable

Informed Consent Statement: Not applicable.

Data Availability Statement: Data is contained within the article or Supplementary Material.

Acknowledgments: We sincerely thank all individuals who provided valuable assistance during the course of this study.

Conflicts of Interest: The authors declare no conflict of interest.

References

1. Ji, Y.; Gao, K. Effects of climate change factors on marine macroalgae: A review. *Adv. Mar. Biol.* **2021**, *88*, 91–136. [PubMed]
2. Masson-Delmotte, V.; Zhai, P.; Pirani, A.; Connors, S.L.; Péan, C.; Berger, S.; Caud, N.; Chen, Y.; Goldfarb, L.; Gomis, M. Climate change 2021: The physical science basis. In *Contribution of Working Group I to the Sixth Assessment Report of the Intergovernmental Panel on Climate Change*; Intergovernmental Panel on Climate Change: Geneva, Switzerland, 2021; Volume 2, p. 2391.
3. Shuang, Y.; Kosuke, H.; Qian, A. Acceleration in the Global Mean Sea Level Rise: 2005–2015. *Geophys. Res. Lett.* **2017**, *44*, 11905–11913. [CrossRef]
4. Arno von, K.; Andrew, S.; Gabriele, C.H. Marine heatwaves in global sea surface temperature records since 1850. *Environ. Res. Lett.* **2022**, *17*, 084027. [CrossRef]
5. Polyakov, I.V.; Pnyushkov, A.V.; Alkire, M.B.; Ashik, I.M.; Baumann, T.M.; Carmack, E.C.; Goszczko, I.; Guthrie, J.; Ivanov, V.V.; Kanzow, T.; et al. Greater role for Atlantic inflows on sea-ice loss in the Eurasian Basin of the Arctic Ocean. *Science* **2017**, *356*, 285–291. [CrossRef]
6. Gao, G.; Beardall, J.; Jin, P.; Gao, L.; Xie, S.; Gao, K. A review of existing and potential blue carbon contributions to climate change mitigation in the Anthropocene. *J. Appl. Ecol.* **2022**, *59*, 1686–1699. [CrossRef]
7. Huang, Y.; Cui, J.; Wang, S.; Chen, X.; Liao, J.; Guo, Y.; Xin, R.; Huang, B.; Xie, E. Transcriptome analysis reveals the molecular mechanisms of adaptation to high temperatures in Gracilaria bailinae. *Front. Plant Sci.* **2023**, *14*, 1125324. [CrossRef]
8. Kaori, M.; Yoko, Y.; Kanji, N.; Takato, F.; Naoya, S.; Tatsuo, U.; Yohko, S.-K.; Masayuki, K. Seasonal variation in the chemical composition of a marine brown alga, *Sargassum horneri* (Turner) C. Agardh. *J. Food Compos. Anal.* **2011**, *24*, 231–236. [CrossRef]
9. Pang, S.J.; Jin, Z.H.; Sun, J.Z.; Gao, S.Q. Temperature tolerance of young sporophytes from two populations of *Laminaria japonica* revealed by chlorophyll fluorescence measurements and short-term growth and survival performances in tank culture. *Aquaculture* **2007**, *262*, 493–503. [CrossRef]
10. Kumar, Y.N.; Poong, S.-W.; Gachon, C.; Brodie, J.; Sade, A.; Lim, P.-E. Impact of elevated temperature on the physiological and biochemical responses of *Kappaphycus alvarezii* (Rhodophyta). *PLoS ONE* **2020**, *15*, e0239097. [CrossRef]
11. Liu, L.; Zou, D.; Jiang, H.; Chen, B.; Zeng, X. Effects of increased CO_2 and temperature on the growth and photosynthesis in the marine macroalga *Gracilaria lemaneiformis* from the coastal waters of South China. *J. Appl. Phycol.* **2018**, *30*, 1271–1280. [CrossRef]
12. Zou, D.; Gao, K. Temperature response of photosynthetic light- and carbon-use characteristics in the red *seaweed Gracilariopsis lemaneiformis* (Gracilariales, Rhodophyta). *J. Phycol.* **2014**, *50*, 366–375. [CrossRef]
13. Broom, J.; Farr, T.; Nelson, W. Phylogeny of the Bangia flora of New Zealand suggests a southern origin for Porphyra and Bangia (Bangiales, Rhodophyta). *Mol. Phylogenetics Evol.* **2004**, *31*, 1197–1207. [CrossRef] [PubMed]
14. Ma, J.-H.; Li, S.-J.; Ji, H.-H. An analysis of amino acids and fatty acids in Bangia. *Chin. J. Mar. Drugs* **2002**, *21*, 40–42.
15. Shuijun, L.; Jiahai, M.; Huanhong, J.; Enyi, X. Evaluation of nutrient components of *Bangia* sp. *Acta Oceanol. Sin.* **2003**, *22*, 89–95.
16. Wenjun, W.; Jianyi, Z.; Pao, X.; Jianrong, X.; Xiang-zhi, L.; Chun-Kai, H.; Weibin, S.; Guang, P.; Guangce, W. Characterization of the life history of *Bangia fuscopurpurea* (Bangiaceae, Rhodophyta) in connection with its cultivation in China. *Aquaculture* **2008**, *278*, 101–109. [CrossRef]

17. Yong, Y.S.; Yong, W.T.L.; Anton, A. Analysis of formulae for determination of seaweed growth rate. *J. Appl. Phycol.* **2013**, *25*, 1831–1834. [CrossRef]
18. Li, C.; Nong, Q.; Solanki, M.K.; Liang, Q.; Xie, J.; Liu, X.; Li, Y.; Wang, W.; Yang, L.; Li, Y. Differential expression profiles and pathways of genes in sugarcane leaf at elongation stage in response to drought stress. *Sci. Rep.* **2016**, *6*, 25698. [CrossRef]
19. Porra, R.J. The chequered history of the development and use of simultaneous equations for the accurate determination of chlorophylls a and b. *Photosynth. Res.* **2002**, *73*, 149–156. [CrossRef]
20. Beer, S.; Eshel, A. Determining phycoerythrin and phycocyanin concentrations in aqueous crude extracts of red algae. *Mar. Freshw. Res.* **1985**, *36*, 785–792. [CrossRef]
21. Piñeiro-Corbeira, C.; Barreiro, R.; Cremades, J.; Arenas, F. Seaweed assemblages under a climate change scenario: Functional responses to temperature of eight intertidal seaweeds match recent abundance shifts. *Sci. Rep.* **2018**, *8*, 12978. [CrossRef]
22. Yang, F.; Wei, Z.; Long, L. Transcriptomic and Physiological Responses of the Tropical Reef Calcified Macroalga *Amphiroa fragilissima* to Elevated Temperature1. *J. Phycol.* **2021**, *57*, 1254–1265. [CrossRef] [PubMed]
23. Wilson, K.L.; Kay, L.M.; Schmidt, A.L.; Lotze, H.K. Effects of increasing water temperatures on survival and growth of ecologically and economically important seaweeds in Atlantic Canada: Implications for climate change. *Mar. Biol.* **2015**, *162*, 2431–2444. [CrossRef]
24. Zulfiqar, S.; Sharif, S.; Saeed, M.; Tahir, A. Role of carotenoids in photosynthesis. In *Carotenoids: Structure and Function in the Human Body*; Springer: Berlin/Heidelberg, Germany, 2021; pp. 147–187.
25. Yang, J.J.; Yu, D.C.; Ma, Y.F.; Yin, Y.; Shen, S.D. Antioxidative defense response of *Ulva prolifera* under high or low-temperature stimulus. *Algal Res.* **2019**, *44*, 101703. [CrossRef]
26. Ma, C.; Qin, S.; Cui, H.; Liu, Z.; Zhuang, L.; Wang, Y.; Zhong, Z. Nitrogen enrichment mediates the effects of high temperature on the growth, photosynthesis, and biochemical constituents of *Gracilaria blodgettii* and *Gracilaria lemaneiformis*. *Environ. Sci. Pollut. Res.* **2021**, *28*, 21256–21265. [CrossRef] [PubMed]
27. Ma, J.; Jiangnan, H.; Xinmei, S.; Demei, M.; Yang, R. Phycobiliproteins, the pigment-protein complex form of natural food colorants and bioactive ingredients. *Crit. Rev. Food Sci. Nutr.* **2024**, *64*, 2999–3017. [CrossRef]
28. Kim, E.Y.; Choi, Y.H.; Nam, T.J. Identification and antioxidant activity of synthetic peptides from phycobiliproteins of *Pyropia yezoensis*. *Int. J. Mol. Med.* **2018**, *42*, 789–798. [CrossRef]
29. Gao, T.; Tang, X.; Wang, D.; Yu, Y.; Mao, Y. Morpho-physiological and transcriptomic analyses reveal adaptive responses of *Neopyropia yezoensis* to long-term high temperature. *Plant Stress.* **2025**, *15*, 100778. [CrossRef]
30. Muhammad, I.; Shalmani, A.; Ali, M.; Yang, Q.-H.; Ahmad, H.; Li, F.B. Mechanisms regulating the dynamics of photosynthesis under abiotic stresses. *Front. Plant Sci.* **2021**, *11*, 615942. [CrossRef]
31. Zhang, L.; Chang, Q.; Hou, X.; Wang, J.; Chen, S.; Zhang, Q.; Wang, Z.; Yin, Y.; Liu, J. The effect of high-temperature stress on the physiological indexes, chloroplast ultrastructure, and photosystems of two herbaceous peony cultivars. *J. Plant Growth Regul.* **2023**, *42*, 1631–1646. [CrossRef]
32. Wu, J.; Lian, W.; Liu, Z.; Zeng, X.; Jiang, J.; Wei, Y. High temperature response of chlorophy II fluorescence parameters and heat tolerance evaluation of different grape cultivars. *J. Northwest A F Univ.* **2019**, *47*, 80–88.
33. Borlongan, I.A.G.; Gerung, G.S.; Nishihara, G.N.; Terada, R. Light and temperature effects on photosynthetic activity of *Eucheuma denticulatum* and *Kappaphycus alvarezii* (brown and green color morphotypes) from Sulawesi Utara, Indonesia. *Phycol. Res.* **2017**, *65*, 69–79. [CrossRef]
34. Li, Q.; Zhang, L.; Pang, T.; Liu, J. Comparative transcriptome profiling of *Kappaphycus alvarezii* (Rhodophyta, Gigartinales) in response to two extreme temperature treatments: An RNA-seq-based resource for photosynthesis research. *Eur. J. Phycol.* **2019**, *54*, 162–174. [CrossRef]
35. Qin, F.; Zang, X.; Shui, G.; Wang, Z. Transcriptome analysis of *Gracilariopsis lemaneiformis* at low temperature. *J. Appl. Phycol.* **2021**, *33*, 4035–4050. [CrossRef]
36. Saluri, M.; Kaldmäe, M.; Tuvikene, R. Extraction and quantification of phycobiliproteins from the red alga *Furcellaria lumbricalis*. *Algal Res.* **2019**, *37*, 115–123. [CrossRef]
37. You, X.; Zhang, X.; Cheng, J.; Xiao, Y.; Ma, J.; Sun, S.; Zhang, X.; Wang, H.-W.; Sui, S.-F. In situ structure of the red algal phycobilisome–PSII–PSI–LHC megacomplex. *Nature* **2023**, *616*, 199–206. [CrossRef]
38. Dai, W.; Wang, X.; Zhuang, M.; Sun, J.; Shen, Y.; Xia, Z.; Wu, T.; Jiang, R.; Li, A.; Bi, F.; et al. Responses of photosynthesis-related genes in *Sargassum horneri* to high temperature stress. *Mar. Pollut. Bull.* **2024**, *199*, 115944. [CrossRef] [PubMed]
39. Wang, W.; Lin, Y.; Teng, F.; Ji, D.; Xu, Y.; Chen, C.; Xie, C. Comparative transcriptome analysis between heat-tolerant and sensitive *Pyropia haitanensis* strains in response to high temperature stress. *Algal Res.* **2018**, *29*, 104–112. [CrossRef]
40. Wang, T.; Gao, M.; Song, H.; Wang, C.; He, M. Low temperature modulates the carbon allocation in different metabolic pathways to improve the tolerance of Arctic Chlorella to high light stress. *Algal Res.* **2024**, *80*, 103562. [CrossRef]
41. Barati, B.; Gan, S.Y.; Lim, P.E.; Beardall, J.; Phang, S.M. Green algal molecular responses to temperature stress. *Acta Physiol. Plant.* **2019**, *41*, 26. [CrossRef]

42. Li, X.; Manuel, J.; Slavens, S.; Crunkleton, D.W.; Johannes, T.W. Interactive effects of light quality and culturing temperature on algal cell size, biomass doubling time, protein content, and carbohydrate content. *Appl. Microbiol. Biotechnol.* **2021**, *105*, 587–597. [CrossRef]
43. Lv, Y.; Sun, P.; Zhang, Y.; Xuan, W.; Xu, N.; Sun, X. Response of trehalose, its degrading enzyme, sucrose, and floridoside/isofloridoside under abiotic stresses in *Gracilariopsis lemaneiformis* (Rhodophyta). *J. Appl. Phycol.* **2019**, *31*, 3861–3869. [CrossRef]
44. Li, H.; Monteiro, C.; Heinrich, S.; Bartsch, I.; Valentin, K.; Harms, L.; Glöckner, G.; Corre, E.; Bischof, K. Responses of the kelp *Saccharina latissima* (Phaeophyceae) to the warming Arctic: From physiology to transcriptomics. *Physiol. Plant.* **2020**, *168*, 5–26. [CrossRef] [PubMed]
45. Iñiguez, C.; Galmés, J.; Gordillo, F.J.L. Rubisco carboxylation kinetics and inorganic carbon utilization in polar versus cold-temperate seaweeds. *J. Exp. Bot.* **2019**, *70*, 1283–1297. [CrossRef]
46. Lemieux, H.; Blier, P.U. Exploring thermal sensitivities and adaptations of oxidative phosphorylation pathways. *Metabolites* **2022**, *12*, 360. [CrossRef] [PubMed]
47. Zhang, X.; Hu, C.; Sun, X.; Zang, X.; Zhang, X.; Fang, T.; Xu, N. Comparative transcriptome analysis reveals chitooligosaccharides-induced stress tolerance of *Gracilariopsis lemaneiformis* under high temperature stress. *Aquaculture* **2020**, *519*, 734876. [CrossRef]
48. Liu, L.; Lin, L. Effect of Heat Stress on *Sargassum fusiforme* Leaf Metabolome. *J. Plant Biol.* **2020**, *63*, 229–241. [CrossRef]
49. Gonzalez-Baro, M.R.; Coleman, R.A. Mitochondrial acyltransferases and glycerophospholipid metabolism. *Biochim. Et Biophys. Acta (BBA)-Mol. Cell Biol. Lipids* **2017**, *1862*, 49–55. [CrossRef]
50. Liang, Y.; Huang, Y.; Liu, C.; Chen, K.; Li, M. Functions and interaction of plant lipid signalling under abiotic stresses. *Plant Biol.* **2023**, *25*, 361–378. [CrossRef]
51. Zhang, X.; Wu, C.; Hu, C.; Li, Y.; Sun, X.; Xu, N. Lipid remodeling associated with chitooligosaccharides-induced heat tolerance of marine macroalgae *Gracilariopsis lemaneiformis*. *Algal Res.* **2020**, *52*, 102113. [CrossRef]
52. Zhang, J.; Liu, S.; Hu, C.; Chen, X.; Sun, X.; Xu, N. Physiological and Transcriptome Analysis of Exogenous L-Arginine in the Alleviation of High-Temperature Stress in *Gracilariopsis lemaneiformis*. *Front. Mar. Sci.* **2021**, *8*, 784586. [CrossRef]
53. Pandey, A.K.; Gautam, A. Stress responsive gene regulation in relation to hydrogen sulfide in plants under abiotic stress. *Physiol. Plant.* **2020**, *168*, 511–525. [CrossRef] [PubMed]
54. Suliman, M.S.E.; Elradi, S.B.M.; Zhou, G.; Meng, T.; Zhu, G.; Xu, Y.; Nimir, N.E.A.; Elsiddig, A.M.I.; Awdelseid, A.H.M.; Ali, A.Y.A. Exogenous glutathione protected wheat seedling from high temperature and water deficit damages. *Sci. Rep.* **2024**, *14*, 5304. [CrossRef] [PubMed]
55. Moenne, A.; González, A.; Sáez, C.A. Mechanisms of metal tolerance in marine macroalgae, with emphasis on copper tolerance in Chlorophyta and Rhodophyta. *Aquat. Toxicol.* **2016**, *176*, 30–37. [CrossRef] [PubMed]
56. Celis-Plá, P.S.M.; Moenne, F.; Rodríguez-Rojas, F.; Pardo, D.; Lavergne, C.; Moenne, A.; Brown, M.T.; Huovinen, P.; Gómez, I.; Navarro, N.; et al. Antarctic intertidal macroalgae under predicted increased temperatures mediated by global climate change: Would they cope? *Sci. Total Environ.* **2020**, *740*, 140379. [CrossRef]
57. Niu, C.; Wang, W.; Yao, H.; Liang, Z.; Zhang, P.; Lu, X. Ascorbate−glutathione cycle involving in response of *Bangia fuscopurpurea* (Bangiales, Rhodophyta) to hyposalinity. *Front. Mar. Sci.* **2023**, *10*, 1174472. [CrossRef]

current issues in molecular biology

MDPI

Article

Transcriptomic Analysis Reveals Candidate Hub Genes and Putative Pathways in *Arabidopsis thaliana* Roots Responding to *Verticillium longisporum* Infection

Qiwei Zheng, Yangpujia Zhou and Sui Ni *

School of Marine Sciences, Ningbo University, Ningbo 315211, China; 226003116@nbu.edu.cn (Q.Z.); 2311130071@nbu.edu.cn (Y.Z.)

* Correspondence: nisui@nbu.edu.cn

Abstract

Verticillium longisporum, a soil-borne fungus responsible for Verticillium wilt, primarily colonizes members of the Brassicaceae family. Using *Arabidopsis thaliana* roots as an experimental host, we systematically identify *V. longisporum*-responsive genes and pathways through comprehensive transcriptomic analysis, alongside screening of potential hub genes and evaluation of infection-associated regulatory mechanisms. The GSE62537 dataset was retrieved from the Gene Expression Omnibus database. After performing GEO2R analysis and filtering out low-quality data, 222 differentially expressed genes (DEGs) were identified, of which 184 were upregulated. Gene Ontology and Kyoto Encyclopedia of Genes and Genomes enrichment analyses were performed on these DEGs. A protein–protein interaction network was constructed using the STRING database. CytoHubba and CytoNCA plugins in Cytoscape v3.10.3 were used to analyze and evaluate this network; six hub genes and four functional gene modules were identified. The GeneMANIA database was used to construct a co-expression network for hub genes. Systematic screening of transcription factors within the 14 DEGs revealed the inclusion of the hub gene *NAC042*. Integrative bioinformatics analysis centered on *NAC042* enabled prediction of a pathogen-responsive regulatory network architecture. We report *V. longisporum*-responsive components in *Arabidopsis*, providing insights for disease resistance studies in Brassicaceae crops.

Keywords: *Verticillium longisporum*; Verticillium wilt; *Arabidopsis thaliana*; root; differentially expressed genes; hub genes

1. Introduction

Verticillium wilt, a soil-borne vascular disease caused by *Verticillium* spp., threatens diverse agricultural crops and ornamental plants worldwide [1]. *Verticillium* spp. primarily infect dicotyledonous plants, with common hosts including cotton, potato, tomato, eggplant, lettuce, spinach, alfalfa, strawberry, oilseed rape, sunflower, olive, and woody perennials [2,3]. Symptom presentation of Verticillium wilt varies among host plants, with no strictly uniform manifestation. Common symptoms encompass wilting, stunted growth, chlorosis, vascular discoloration, and premature senescence [2]. The genus comprises diverse ascomycete species exhibiting both sexual and asexual reproduction cycles, of which *V. dahlia* and *V. albo-atrum* are the most extensively studied [4]. As exemplified by *V. dahliae*, this fungus encodes over 700 putative secreted proteins [5]. A substantial subset of these proteins has been experimentally validated to mediate infection-related processes, including plant cell wall degradation, host immune interference, toxic metabolite secretion, oxidative stress neutralization, host nutrient competition, and host microbiota modulation [6].

While originally identified as a variant of *V. dahliae* [7], *V. longisporum* is now recognized to be a unique near-diploid hybrid species. This pathogen evolved through recurrent parasexual hybridization events involving multiple ancestral lineages, leading to its current taxonomic elevation to species rank [8,9]. Morphologically, conidia of *V. longisporum* exhibit dimensions comparable to or exceeding those of *V. dahliae* [7–9]. The unstable morphology of microsclerotia and variable number of sclerotial units per whorl preclude reliable differentiation of hybrid lineages through phenotypic characterization alone, necessitating molecular diagnostic approaches for precise identification [10,11]. Comparative analyses demonstrate differential host range specificity and virulence among the three fungal lineages [7,12]. The infection process parallels *V. dahliae*, commencing with melanized microsclerotia formation. Upon root exudate recognition, these microsclerotia germinate to colonize root hair surfaces before initiating tissue penetration [13,14]. In *Brassica napus*, the fungus further colonizes xylem vessels, producing conidia that induce vascular occlusion and associated pathologies [15]. Ultimately, the pathogen emerges from xylem conduits to form microsclerotia beneath the stem epidermis and within pith tissues. These dormant structures are then released into the soil through host-tissue degradation during plant senescence [16].

Investigating the molecular mechanisms of the interaction between *Verticillium longisporum* and its host plants is critical for developing effective control strategies against this pathogen. The known Verticillium wilt resistance gene, *Ve1*, was initially identified in tomato [17]; however, no authentic resistance gene has been identified specifically for *V. longisporum*. Recent genomic studies on common Brassicaceae crops provide valuable references for resistance breeding [18]. However, considering the differences in genomic and regulatory programs among various crops, and given that *V. longisporum* can still latently colonize Brassicaceae weeds posing potential threats to crops [7], it is feasible to explore more representative auxiliary targets, such as using the model organism *Arabidopsis thaliana* for research. Altered expression or deletion of certain genes has been demonstrated to reduce plant susceptibility to *V. longisporum*, including the following examples: Overexpression of *EWR1* effectively decreases Arabidopsis susceptibility to this pathogen [19], *NPF5.12* and *MLP6* confer resistance to fungal colonization through nitrate competition and participation in suberin barrier synthesis [20], loss of *CRT1a* activates the ethylene signaling pathway and enhances disease resistance [21], and *TPS23/27*, two duplicated monoterpene synthase genes, produce monoterpenes that stimulate *V. longisporum* germination and subsequent invasion [22]. Notably, certain underlying mechanisms of *V. longisporum* preclude direct extrapolation of established resistance-targeting approaches. Ulrich et al. discovered through comparison of the root transcriptome of the *Arabidopsis* coi1 mutant with those of the susceptible JA-Ile-deficient allene oxide synthase mutant and the susceptible wild-type that constitutive expression of COI1—the receptor for the canonical defense hormone JA-Ile—in roots impedes synthesis and transport of defense compounds, thereby paradoxically enhancing *V. longisporum* infection efficiency [23]. Furthermore, screening for *A. thaliana* transcription factors (TFs) via overexpression enables rapid identification of *V. longisporum*-responsive regulators, thereby facilitating discovery of novel resistance targets [24].

V. longisporum is an emerging global threat to Brassicaceae crops, particularly *Brassica napus*. Current Verticillium wilt research has mostly focused on *V. dahlia* and *V. albo-atrum*, with limited transcriptomic analyses available for *V. longisporum*. Using bioinformatics approaches with the susceptible model organism *A. thaliana* [7], we perform comprehensive transcriptomic profiling to identify potential hub genes and associated pathways, and elucidate molecular regulatory networks responsive to *V. longisporum* infection. Our objective is to advance disease-resistant breeding strategies and molecular-targeted interventions in Brassicaceae crops.

2. Materials and Methods

2.1. Biological Materials and Pathogen Inoculation

Wild-type *A. thaliana* Col-0 was used as the experimental host. Surface-sterilized *Arabidopsis* seeds were aseptically inoculated onto Murashige and Skoog (MS) solid medium, stratified at 4 °C for 48 h, then cultivated in growth chambers for 14 d under controlled environmental conditions: 22 °C, a 16-h light/8-h dark photoperiod, and photosynthetic photon flux density of 100–150 $\mu mol \cdot m^{-2} \cdot s^{-1}$. Control group seedlings were maintained on 1/2 MS medium, while treatment group seedlings were cultured on 1/2 MS medium supplemented with *V. longisporum* spores (200 conidia mL^{-1}) for 48 h, followed by root tissue collection. (The *V. longisporum* strain HZM135335 was obtained from HZBio Microbial Conservation Co., Ltd., Wuhan, China) RNA extraction was performed using the E.Z.N.A. Plant RNA Kit (Omega Bio-tek, Inc., Norcross, GA, USA), with cDNA synthesis via HiScript III RT SuperMix (ABclonal Biotechnology Co., Ltd., Wuhan, China).

2.2. Raw Transcriptome Data Collection

Our GSE62537 (https://www.ncbi.nlm.nih.gov/geo/query/acc.cgi?acc=GSE62537; accessed on 6 November 2024) microarray dataset was obtained from the Gene Expression Omnibus (GEO) database [25]. As a public repository, GEO hosts vast amounts of global high-throughput sequencing and microarray data. This dataset was constructed based on the Affymetrix Arabidopsis ATH1 Genome Array (GPL198), containing control and experimental group data for A. thaliana Col-0 and ndr1-1 genotypes, with three biological replicates per group. The gene expression matrix for GSE62537 is available in Table S1.

2.3. Data Processing and Integration

DEGs were analyzed using the GEO2R online tool within the GEO database. GEO2R is an interactive online analytical platform designed to compare two or more sample groups in GEO datasets and identify genes with differential expression under experimental conditions. Visualize the expression profiles of these genes in the form of volcano plots. Probes mapping to multiple genes were rigorously filtered to enhance result accuracy. Selection criteria were set to $p < 0.1$ and absolute log2 fold change ($|\mathrm{logFC}|$) ≥ 1. Based on fold-change analysis of these genes, we selected the top 10 significantly upregulated and downregulated DEGs, respectively. The absolute expression values were normalized via Z-score transformation, followed by heatmap visualization.

2.4. Enrichment Analysis of Differentially Expressed Genes

To elucidate the biological functions of DEGs, GO and KEGG pathway analyses were performed on upregulated and downregulated gene subsets. Enriched terms were filtered using a significance threshold of $p < 0.5$. To reduce redundancy, GO annotations were summarized via the REVIGO database (http://revigo.irb.hr/ accessed on 7 April 2025), with results constrained to medium similarity (0.7) and filtered by false discovery rate (FDR) values and Arabidopsis-specific annotations.

2.5. PPI Network Construction and Module Analysis

PPI networks were constructed by querying the STRING database (https://cn.string-db.org/, version 12.0 accessed on 7 April 2025) to investigate protein relationships [26]. Interactions with composite scores > 0.4 were considered statistically significant. The network was visualized using Cytoscape software v 3.10.3 (http://www.cytoscape.org accessed on 5 December 2024) [27]. Key functional modules were identified via the MCODE plugin in Cytoscape with parameters: K-core = 2, degree cutoff = 2, max depth = 100, and node score cutoff = 0.2. Module genes were subsequently subjected to GO and KEGG analyses.

2.6. Hub Gene Identification and Co-Expression Network Construction

In Cytoscape, network topology was optimized by extracting the largest connected component as the primary network via the NetworkAnalyzer tool v 4.5.0, removing isolated nodes and fragmented subnets. Six topological algorithms (MCC, MNC, Degree, Closeness, Radiality, EPC) from the cytoHubba plugin and the CytoNCA plugin were used to identify consensus hub genes. A co-expression network was then constructed using the GeneMANIA online tool (http://www.genemania.org/ accessed on 9 April 2025) to validate functional associations among these genes [28].

2.7. RT-qPCR Analysis

cDNA samples were prepared following the protocol described in the GSE62537 dataset. RNA samples were assessed for quality using a Nano-300 micro-volume spectrophotometer prior to reverse transcription. Primer pairs targeting hub genes were designed for quantitative real-time PCR (qPCR) analysis (Table S2). SYBR Green I fluorescent dye was used in reactions using Arabidopsis Actin2 as the internal reference. Thermal cycling conditions involved an initial denaturation at 95 °C for 3 min; 40 cycles of 95 °C for 5 s (denaturation), and 60 °C for 30 s (annealing/extension). Relative gene expression was calculated via the $2^{-\Delta\Delta Ct}$ method. Reactions were performed on a Roche LightCycler 480 Instrument II (F. Hoffmann-La Roche Ltd., Basel, Switzerland). SYBR Green I reagent (RK21203) was purchased from ABclonal Biotechnology Co., Ltd. (Wuhan, China).

2.8. Prediction of the NAC042 Regulatory Network

To investigate potential regulatory targets and molecular mechanisms underlying the response to V. longisporum, all A. thaliana TFs were retrieved from the PlantTFDB database (https://planttfdb.gao-lab.org/, version 5.0 accessed on 16 April 2025) [29]. PlantTFDB catalogs plant TFs and provides integrated analytical tools. Venn diagram analysis was performed between TFs and DEGs to identify TF-DEGs, followed by mechanistic exploration. Notably, NAC042 was identified as both a hub gene and a functional TF. As an exploratory approach, upstream miRNAs of NAC042 were predicted using the psRNATarget database (https://www.zhaolab.org/psRNATarget/seq accessed on 17 April 2025) [30], while interacting partners were mined from BioGRID (https://thebiogrid.org/ accessed on 17 April 2025) [31]. psRNATarget specializes in plant miRNA target prediction, whereas BioGRID archives experimentally validated protein–protein genetic and chemical interactions across species. A hypothetical regulatory network was constructed, representing a potential framework for understanding NAC042-mediated defense mechanisms.

3. Results

3.1. Phenotypic Evaluation in Response to V. longisporum Infection

After two weeks of growth followed by 48 h vernalization, *A. thaliana* were subjected to 48 h post-inoculation treatments with either a sterile control or *V. longisporum* for systematic phenotypic evaluation. Control plants exhibited normal root and leaf development, characterized by intact primary root architecture and active lateral root formation. Infected plants had pronounced growth inhibition: rosette leaves were reduced in size, fewer in number, and wilting symptoms were visible, accompanied by suppressed root system development marked by underdeveloped architecture and significantly diminished biomass compared with controls (Figure 1).

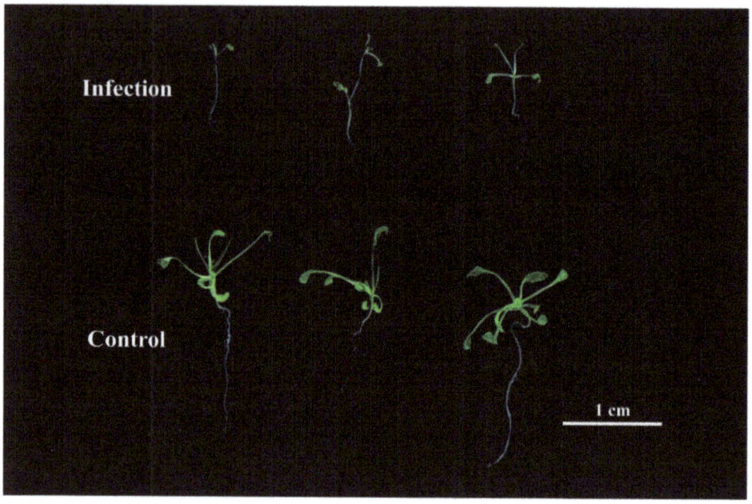

Figure 1. Phenotypic manifestations of *A. thaliana* under control and *V. longisporum* infection treatments at 48 h post-cultivation.

3.2. Identification of DEGs Responsive to V. longisporum Infection

To investigate *A. thaliana* roots' responses to *V. longisporum*, 240 differentially expressed genes (DEGs) were identified between selected sample groups in GSE62837, including 194 that were upregulated (Figure 2A). To enhance quantification accuracy and experimental reproducibility, reads mapping ambiguously to multiple genomic loci were filtered out; 222 high-confidence DEGs were retained, including 184 that were upregulated. The top 10 upregulated and downregulated genes are listed in Table S3. Heatmap analysis reveals pronounced expression divergence between control and infected groups through stark red–blue color demarcation (Figure 2B). Notably, excluded multi-mapped loci may retain biological significance, as exemplified by the tandemly duplicated *TPS23/27* genes, whose monoterpene products have been implicated in promoting *V. longisporum* hyphal growth and proliferation [22].

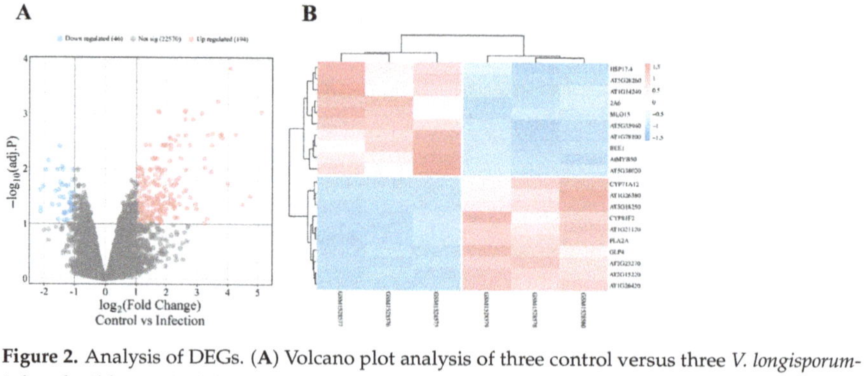

Figure 2. Analysis of DEGs. (**A**) Volcano plot analysis of three control versus three *V. longisporum*-infected wild-type *A. thaliana* samples from the GSE62537 dataset. (**B**) Heatmap analysis of the top 10 most significantly up- and downregulated DEGs (following stringent filtering), showing stark differential expression patterns between treatments.

3.3. Functional Characterization of DEGs

To investigate the roles of DEGs in *A. thaliana* responses to *V. longisporum* infection, Gene Ontology (GO) and Kyoto Encyclopedia of Genes and Genomes (KEGG) enrichment analyses were performed. Table S4 lists the most significantly enriched terms with higher enrichment scores and smaller *p*-values across categories. In the biological process category, the 184 upregulated genes were associated with cell wall macromolecule metabolic processes, indole-containing compound metabolic processes, toxin metabolic processes, cellular responses to hypoxia, secondary metabolic processes, and defense responses to fungi. For cellular components, upregulated genes were primarily linked to secretory vesicles and the dynamic interactions of the plasma membrane–cell wall continuum, including adjacent extracellular regions. Molecular function analysis revealed associations with chitinase activity, transmembrane receptor serine/threonine kinase activity, glutathione transferase activity, FAD-binding, and O-methyltransferase activity (Figure 3A). At a statistical threshold of $p < 0.05$, no significant enrichment was observed for the 38 downregulated genes. KEGG pathway analysis indicated DEG involvement in phenylpropanoid biosynthesis, amino and nucleotide sugar metabolism, biosynthesis of secondary metabolites, and glutathione metabolism (Figure 3B).

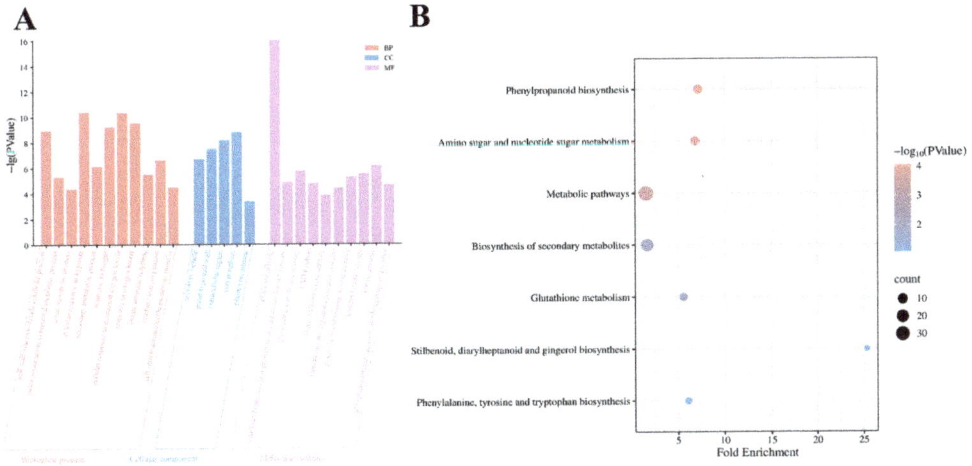

Figure 3. Functional and pathway analysis results of DEGs. (**A**) GO enrichment analysis. (**B**) KEGG pathway enrichment.

3.4. Construction and Module Analysis of the PPI Network

A protein–protein interaction (PPI) network of DEGs was constructed using the STRING database with a confidence score threshold > 0.4. The network consisted of 221 nodes and 269 interactions, with a statistically significant $p < 1.0 \times 10^{-16}$ (Figure 4A). The network was imported into Cytoscape for further analysis. The Molecular Complex Detection (MCODE) plugin in Cytoscape identified four tightly connected gene modules, containing 23 DEGs and 49 interactions (Figure 4B–E). GO analysis demonstrated these module-associated genes to be involved in chitinase activity, FAD-binding, immune system processes, defense response, interspecies interactions, and response to external stimuli. KEGG pathway analysis further indicated their association with metabolic pathways in *A. thaliana* (Figure 4F).

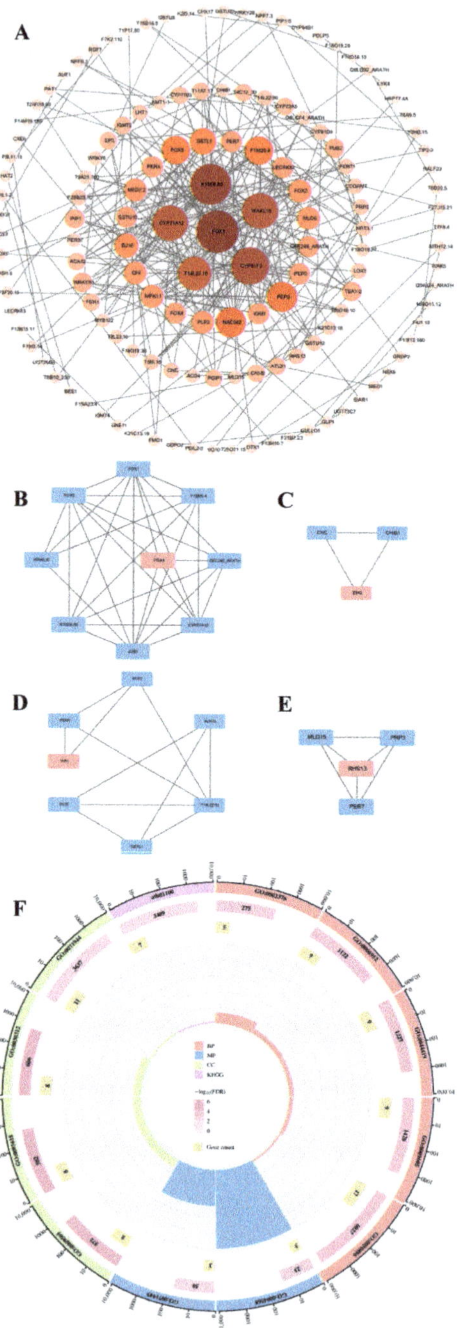

Figure 4. Integrated Protein–Protein Interaction Network Analysis Reveals Functional Modules and Enriched Pathways. (**A**) PPI network constructed using the STRING database. (**B–E**) Four gene modules identified by the MCODE plugin. Node color: red, seed genes; blue, clustered genes. (**F**) Bubble plot of GO and KEGG pathway enrichment for module-associated genes.

3.5. Comprehensive Screening and Analysis of Hub Genes

Hub genes were prioritized through a multi-algorithm consensus approach using six topological methods in the cytoHubba plugin (Degree, MCC, MNC, EPC, Closeness, Radiality) and CytoNCA (Network Centrality Analysis) in Cytoscape. The top 10 candidate genes from each algorithm were collated for cross-validation. To enhance screening specificity, network topology was optimized by extracting the largest connected component via NetworkAnalyzer, which removed isolated nodes and fragmented sub-networks (Table 1). Venn diagram analysis of overlapping candidates across algorithms identified six (*FOX1*, *K15E6.80*, *CYP81F2*, *CYP71A12*, *WAKL10*, and *NAC042*) high-confidence hub genes (Figure 5A). Basic information for these genes is presented in Table 2. Using the GeneMANIA database, co-expression networks and associated functions of hub genes were analyzed. These genes formed complex co-expression networks, wherein functional associations among target genes were mainly mediated by co-expression (98.79%), strongly supporting transcriptional-level functional synergy. This indicates that these genes may be coordinately regulated during *A. thaliana* responses to *V. longisporum* infection. Additionally, co-localization accounted for 1.13%, and predicted interactions for 0.08%; physical interactions fell below the detection threshold (<0.01%). These observations suggest that functional associations among hub genes likely depend on transcriptional coordination or metabolic pathway alignment rather than direct protein–protein interactions. These genes are associated with toxin metabolic processes, secondary metabolic processes, response to hypoxia, innate immune responses, indole-containing compound biosynthesis, sulfur compound metabolic processes, and secondary metabolite biosynthesis (Figure 5B).

Functional annotation of hub genes via the DAVID database identified three pathogen-response genes: *CYP81F2*, *CYP71A12*, and *WAKL10*. Of these, *CYP81F2* and *CYP71A12* were annotated to systemic acquired resistance, monooxygenase activity, oxidoreductase activity acting on paired donors, iron ion binding, and heme binding. *CYP81F2* and *FOX1* were linked to cellular response to hypoxia, and *K15E6.80* and *FOX1* were associated with oxidoreductase activity. *NAC042* received no functional annotation. The repeated mention of *CYP81F2* strongly suggests its critical role in hub gene interactions (Figure 5C).

Table 1. Top 10 hub genes in cytoHubba and CytoNCA.

Degree	MCC	MNC	Closeness	Radiality	EPC	CytoNCA
FOX1	*FOX1*	*FOX1*	*FOX1*	*FOX1*	*T14L22.10*	*FOX1*
FOX5	*K15E6.80*	*CYP81F2*	*K15E6.80*	*F1M20.4*	*FOX1*	*K15E6.80*
F1M20.4	*CYP81F2*	*K15E6.80*	*CYP81F2*	*WAKL10*	*K15E6.80*	*CYP71A12*
NAC042	*CYP71A12*	*T14L22.10*	*WAKL10*	*MLO6*	*CYP81F2*	*NAC042*
CYP71A12	*WAKL10*	*WAKL10*	*CYP71A12*	*CYP71A12*	*CYP71A12*	*FOX5*
WAKL10	*T14L22.10*	*CYP71A12*	*F1M20.4*	*CYP81F2*	*WAKL10*	*T14L22.10*
K15E6.80	*PEP3*	*FOX5*	*T14L22.10*	*K15E6.80*	*NAC042*	*F1M20.4*
CYP81F2	*NAC042*	*NAC042*	*NAC042*	*NAC042*	*FOX5*	*WAKL10*
T14L22.10	*FOX5*	*F1M20.4*	*MLO6*	*MPK11*	*F1M20.4*	*CYP81F2*

3.6. Analysis of RT-qPCR

Quality assessment confirmed that A260/A280 and A260/A230 ratios for RNA samples from both control and experimental groups complied with established quality thresholds. Quantitative real-time polymerase chain reaction (qRT-PCR) was used to measure and compare the expression of identified hub genes between control and *V. longisporum*-infected A. thaliana (Figure 6). Expression levels of all hub genes in infected plants were significantly upregulated compared with controls ($p < 0.05$).

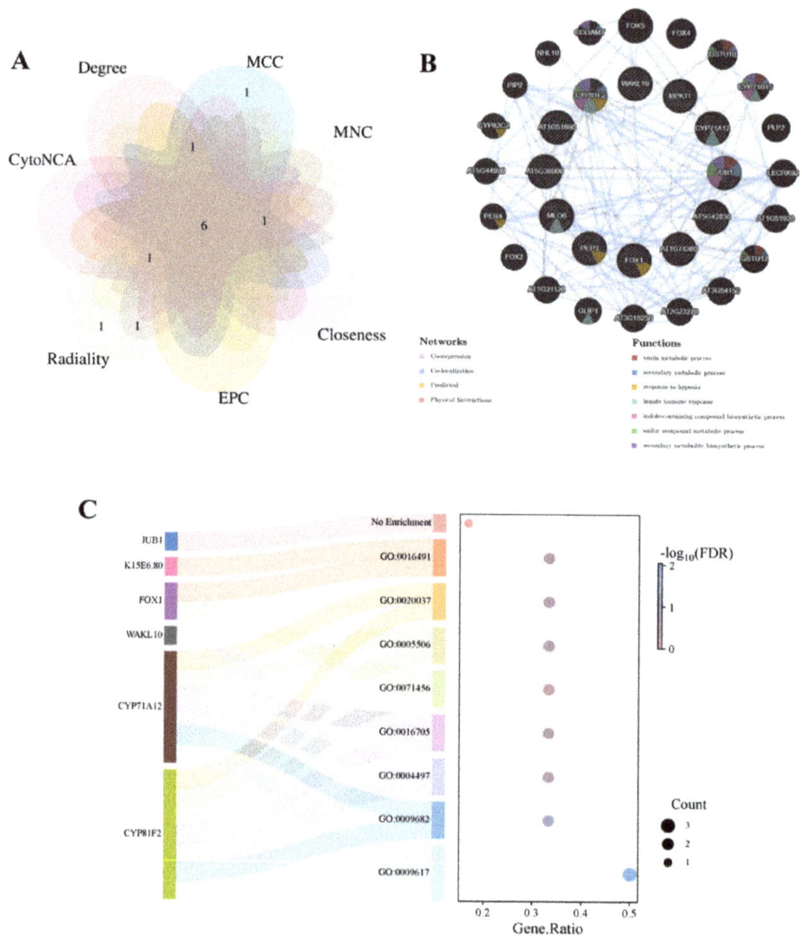

Figure 5. Identification of hub genes, co-expression network analysis, and functional annotation. (**A**) Venn diagram analysis of hub genes identified by cytoHubba and CytoNCA. (**B**) Co-expression network of hub genes predicted by GeneMANIA. (**C**) Mulberry bubble plot of functional annotations for hub genes; bubble size indicates gene counts and color intensity represents false discovery rate.

Table 2. Basic information on hub genes.

TAIR ID	Gene Symbol	Full Name	Function
AT1G26380	*FOX1*	FAD-binding Berberine family protein	Within the biosynthetic pathway of the Arabidopsis cyanogenic phytoalexin 4-hydroxy indole-3-carbonyl nitrile (4-OH-ICN), *FOX1* catalyzes the dehydrogenation of indole cyanohydrin, generating indole carbonyl nitrile.
AT5G38900	*K15E6.80*	Thioredoxin superfamily protein	This enzyme exhibits protein disulfide oxidoreductase activity and participates in defense responses against fungal pathogens during incompatible interactions.

Table 2. *Cont.*

TAIR ID	Gene Symbol	Full Name	Function
AT5G57220	*CYP81F2*	cytochrome P450, family 81, subfamily F, polypeptide 2	As a CYP81F subfamily member participating in glucosinolate metabolism, loss-of-function mutants exhibit compromised fungal resistance, while its mRNA demonstrates intercellular mobility.
AT2G30750	*CYP71A12*	cytochrome P450 family 71 polypeptide	This putative cytochrome P450 cooperates with *CYP71A13* to generate dihydrocamalexic acid (DHCA), the biosynthetic precursor of camalexin. This defense phytoalexin localizes to intercellular spaces and mediates P. syringae resistance in mature Arabidopsis plants via non-antimicrobial pathways.
AT1G79680	*WAKL10*	WALL ASSOCIATED KINASE (WAK)-LIKE 10	The encoded bifunctional kinase/guanylate cyclase protein likely participates in biotic stress response mechanisms through essential cGMP second messenger signaling.
AT2G43000	*NAC042*	NAC domain containing protein 42	This gene encodes an H_2O_2-inducible NAC transcription factor regulating senescence. Its overexpression significantly postpones senescence and confers enhanced tolerance to diverse abiotic stresses.

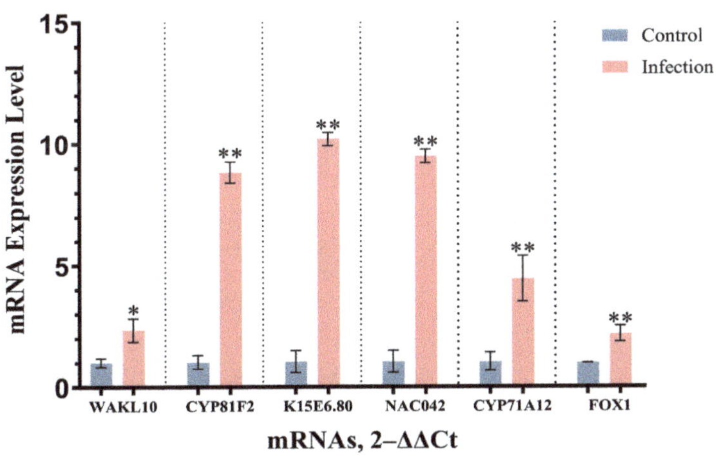

Figure 6. Expression of hub genes in qRT-PCR analysis. * $p < 0.05$, ** $p < 0.01$.

3.7. Identification of TFs in DEGs and Prediction of Upstream miRNAs

The transcription factor (TF) list of *A. thaliana* obtained from PlantTFDB comprised 2296 TFs. Intersection analysis with 184 DEGs revealed 10 to be upregulated and four to be downregulated compared with the control group, including the hub gene *NAC042*. Detailed information on all TF-DEGs is listed in Table 3. Of these, four genes originated from the WRKY family (the highest representation), and three from the MYB family; the remainder were distributed among NAC, bZIP, HD-ZIP, ERF, bHLH, and C2H2 families. The *NAC042* gene was concurrently identified as a hub gene. Integrated analysis of its combined significance across three dimensions (DEGs, TFs, and hub genes) suggests that it may be an important gene in the response of *A. thaliana* roots to *V. longisporum* infection

(Figure 7). The specificity of qPCR amplification was confirmed by melt curve analysis, showing single-peak dissociation curves for all primer pairs (Figure S1).

Table 3. Detailed Information of TF-DEGs.

TAIR ID	Gene Symbol	TF Family	Describe
Upregulated			
AT2G43000	*NAC042*	NAC	See Table 2.
AT1G62300	*WRKY6*	WRKY	Encodes a transcription factor WRKY6. Regulates Phosphate1 (Pho1) expression in response to low phosphate stress. Together with WRKY28 and WRKY41 plays a role redundant to WRKY51 in the suppression of RPW8.1.
AT4G18170	*WRKY28*	WRKY	This gene encodes WRKY6, a transcription factor that modulates Phosphate1 expression under low-Pi stress. Functionally redundant with WRKY28 and WRKY41, it cooperatively suppresses RPW8.1 alongside WRKY51.
AT4G22070	*WRKY31*	WRKY	Classified under Group II-b of the WRKY transcription factor family.
AT4G23810	*WRKY53*	WRKY	As a Group III WRKY transcription factor, it participates in antagonistic regulatory networks with WRKY53 and CRK5, collectively controlling chlorophyll metabolism, senescence progression, and stomatal conductance.
AT1G74650	*MYB31*	MYB	Member of the R2R3 factor gene family; wax regulator associated with reproductive development.
AT1G74080	*MYB122*	MYB	As an R2R3-MYB transcription factor, it modulates cuticular wax biosynthesis during reproductive morphogenesis.
AT1G06850	*bZIP52*	bZIP	This bZIP transcription factor mediates heat stress adaptation by relocating exclusively to the nucleus upon thermal challenge.
AT1G70920	*HB18*	HD-ZIP	Encodes homeodomain-leucine zipper transcription factor HD-Zip 18.
AT3G23230	*TDR1*	ERF	This gene encodes a B-3 subfamily ERF/AP2 transcription factor characterized by a single AP2 domain. The ERF B-3 clade comprises 18 members, with ATERF-1, ATERF-2, and ATERF-5 representing characterized examples.
Downregulated			
AT1G18400	*BEE1*	bHLH	Encodes the brassinosteroid signaling component BEE1 (BR-ENHANCED EXPRESSION 1). Positively modulates the shade avoidance syndrome in Arabidopsis seedlings.
AT5G47370	*HAT2*	HD-ZIP	Homeobox-leucine zipper genes induced by auxin, but not by other phytohormones. Plays opposite roles in the shoot and root tissues in regulating auxin-mediated morphogenesis.
AT3G49930	*AT3G49930*	C2H2	A zinc finger protein belonging to the C2H2/C2HC superfamily characterized by tandem zinc-binding motifs.
AT1G57560	*MYB50*	MYB	As an R2R3-MYB transcription factor, it promotes root cell elongation via transcriptional activation of PECTIN METHYLESTERASE INHIBITOR 8 (PMEI8), while its expression is suppressed by UPB1.

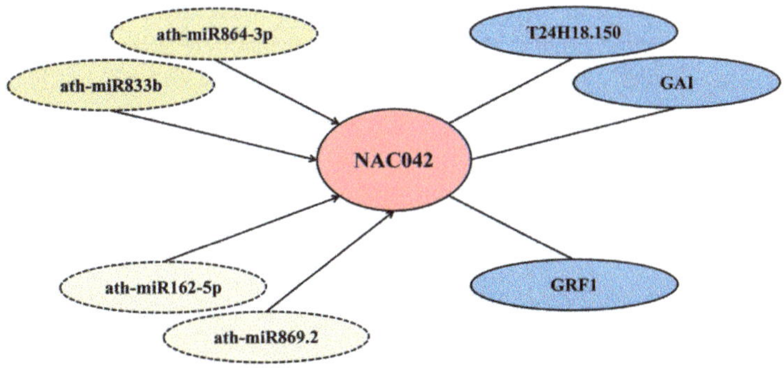

Figure 7. Predicted regulatory network of NAC042. Colors: dark yellow, high-confidence miRNAs; light yellow, moderate-confidence miRNAs; and blue, interacting genes/proteins. Arrows denote putative negative regulation; solid lines indicate experimentally validated interactions.

To investigate potential regulatory mechanisms of the *NAC042* gene in the response of *A. thaliana* to *V. longisporum*, four upstream miRNAs were predicted using psRNATarget. High-confidence candidates included *ath-miR833b* and *ath-miR864-3p* (expectation = 5), with moderate-confidence candidates including *ath-miR162-5p* and *ath-miR869.2* (expectation = 4.5) potentially associated with plant–fungus interaction mechanisms. Protein interaction partners of *NAC042* were retrieved from BioGRID, including *GAI*, *T24H18.150* (validated by yeast two-hybrid assay) and *GRF1* (confirmed via affinity capture-MS). These components were integrated into a putative regulatory network (Figure 7).

4. Discussion

Current management strategies for Verticillium wilt are suboptimal. An integrated control approach is required. For *V. longisporum* specifically, conventional fungicides demonstrate limited efficacy as a standalone management strategy [7]. Resistant breeding is an effective approach for Verticillium wilt management. While multigenic resistance to Verticillium wilt has been identified in several crops [2,32,33], no monogenic resistance genes specifically targeting *V. longisporum* are known. Therefore, studying it through omics and modern molecular biology has more practical significance. By analyzing transcriptome data, we identified DEGs in wild-type *A. thaliana* under *V. longisporum* infection from the GEO database, including 184 that were upregulated and 38 that were downregulated. GO analysis revealed upregulated gene enrichment profiles to align with the established immune response framework in *Arabidopsis*.

Specifically, biological process enrichment related to cell wall macromolecule metabolic processes may correlate with plant responses to pathogen-induced cell wall damage [6]. Furthermore, cell wall remodeling and repair are conserved resistance mechanisms against *V. dahlia* in plants [34]. This suggests a potential conserved mechanism in defense against *V. longisporum*. Glucosinolates, a class of secondary metabolites predominantly found in Brassicaceae plants, are categorized into aliphatic, aromatic, and indolic subtypes [35]. Upregulation of glucosinolates occurs in multiple infected plant species and may contribute to resistance mechanisms against *V. longisporum* [7,36,37]. Furthermore, *V. longisporum* infection induces tryptophan biosynthesis and subsequent tryptophan-derived metabolism. Upregulated genes were enriched in indole-containing compound metabolic processes, suggesting potential crosstalk between enzymatic components regulating indolic glucosinolates, exemplified by the cytochrome P450 family gene *CYP81F2* [38]. While shared precursors such as tryptophan and its derivatives may contribute to this process, Iven et al. [39] demonstrated no significant accumulation of indolic glucosinolate breakdown products or detectable contributions from known tryptophan-derived phytoanticipins in defense against *V. longisporum*. This indicates the involvement of other unidentified indole-derived secondary metabolites in this response. In the cellular component category, the plasma membrane–cell wall continuum serves as the primary interface during early *A. thaliana–V. longisporum* interactions, where effector proteins and enzymes from both organisms are potentially transported via secretory vesicles [40]. Regarding molecular functions, chitinases emerge as critical extracellular hydrolases capable of degrading fungal cell wall components through β-1,4-glycosidic bond cleavage [41]. Serine/threonine kinases constitute core components of the MAPK signaling cascade [42]. The MAPK pathway regulates phenylpropanoid biosynthesis by modulating key enzymatic activities, including FAD-binding and O-methyltransferase functions, thereby driving lignin biosynthesis, energy metabolism, and phytoalexin modifications [43]. Furthermore, MAPK activation induces chitinase expression through WRKY TF-mediated transcriptional reprogramming [44]. KEGG pathway analysis revealed significant enrichment of these genes in phenylpropanoid biosynthesis, amino/nucleotide sugar metabolism, secondary metabolite biosynthesis, and glutathione metabolism pathways. Glutathione, a pivotal reducing agent, scavenges reactive oxygen species (ROS) through redox reactions and modulates defense gene expression via its metabolic flux [45,46]. Intriguingly, the MAPK cascade not only orchestrates glutathione biosynthesis and turnover but it may also be reciprocally regulated by glutathione homeostasis [47–49].

PPI networks enable systematic analysis of functional regulatory mechanisms, cellular signaling pathways, and disease-associated molecular targets through interaction relationship mapping [50]. We constructed a PPI network using a stringent composite score threshold (>0.4), comprising 221 nodes and 269 edges. The observed interaction count significantly exceeded random network expectations ($p < 0.01$, hypergeometric test), indicating strong functional interconnectivity among these proteins. MCODE analysis identified four densely interconnected modules with elevated node clustering coefficients (average 0.72 vs. 0.31 in random networks), containing 23 DEGs and 49 high-confidence interactions. GO and KEGG pathway analyses revealed significant enrichment of these modules in immune regulation (FDR = 2.1×10^{-5}) and metabolic processes (FDR = 4.8×10^{-4}), suggesting coordinated functional modules during pathogen response.

The PPI network was analyzed using cytoHubba (six topological algorithms) and CytoNCA plugins. Intersection analysis identified six high-confidence overlapping hub genes: *FOX1*, *K15E6.80*, *CYP81F2*, *CYP71A12*, *WAKL10*, and *NAC042*. Serving as highly interconnected nodes within the PPI network, hub genes are functionally pivotal to core biological processes and may represent potential therapeutic targets or biomarkers. *CYP71A12*, a putative cytochrome P450 family gene, cooperates with *CYP71A13* to biosynthesize dihydrocamalexic acid, a precursor of phytoalexins [51,52]. This gene, along with *CYP81F2*, is functionally annotated to induce systemic resistance, monooxygenase activity, oxidoreductase activity acting on paired donors, iron ion binding, and heme binding. *WAKL10* encodes a functional guanylate cyclase, although its expression is broadly induced by multiple phytopathogens, indicating limited specificity toward *Verticillium* species [53]. The *NAC042* gene, located on chromosome 2 of *A. thaliana*, functions as a gibberellin-associated TF [54]. It participates in plant defense through DELLA protein accumulation, phytoalexin biosynthesis modulation, and ROS scavenging [55–57]. Co-expression analysis via GeneMANIA revealed that 98.79% of functional associations among hub genes stem from co-expression, with physical interactions below detection thresholds (<0.01%). Network annotations demonstrated enrichment in toxin catabolic processes, secondary metabolic processes, hypoxia response, innate immune response, indole-containing compound biosynthesis, sulfur compound metabolism, and secondary metabolite biosynthesis. qRT-PCR validation confirmed expression patterns consistent with transcriptomic predictions.

We identified 10 significantly differentially expressed TFs through intersection of *A. thaliana* TFs from PlantTFDB with DEGs. The WRKY family contributed four genes, MYB family three genes, and bZIP, HD-ZIP, ERF, and NAC families, one gene each. As one of the largest TF families in plants, WRKY TFs regulate phytohormone signaling, MAPK cascades, ROS/calcium signaling, and phenylpropanoid metabolism to orchestrate plant immunity [58,59]. Multiple phytohormones in *A. thaliana* have been demonstrated to regulate quantitative resistance against *V. longisporum*, including salicylic acid, abscisic acid, and jasmonic acid. Additionally, this study identified that the ERECTA regulatory protein and its pathway also confer resistance to *V. longisporum*. From existing research, it is evident that the molecular regulatory mechanisms of *A. thaliana* in response to *V. longisporum* are complex and may involve further unknown potential associations [60]. All identified TF families have established roles in plant defense responses against biotic and abiotic stresses [61–64]. Notably, *NAC042* was concurrently identified as a DEG, hub gene, and functional TF, suggesting its multifunctional role in the defense of *A. thaliana* against *V. longisporum*. A *NAC042*-centered regulatory network was computationally predicted, providing insights into plant defense mechanisms against *V. longisporum* infection. Finally, all bioinformatics resources and analytical tools used in this study are comprehensively cataloged in Table S5 to facilitate experimental reproducibility and future validation studies.

5. Conclusions

Overall, we performed a comprehensive transcriptomic analysis of the GSE62537 dataset, including the identification of DEGs, performing GO and KEGG analyses, constructing a PPI network and identifying key functional modules and hub genes, analyzing and deriving relevant TF genes, and predicting a potential regulatory network. The TF *NAC042* was identified as a potential target for defense against *V. longisporum*. We acknowledge several research limitations. First, the analysis was based on sequencing data from public databases. Second, certain filtering criteria were applied during some analytical steps, but biological systems cannot be fully captured by simplistic selection rules (e.g., we excluded probe data corresponding to multiple genes and did not retain genes such as *TPS23/27*). Finally, because we did not experimentally validate the regulatory network, the network may contain errors or misleading elements and require quantitative analysis and functional verification.

Supplementary Materials: The following supporting information can be downloaded at: https://www.mdpi.com/article/10.3390/cimb47070536/s1.

Author Contributions: Conceptualization, S.N.; methodology, Q.Z. and Y.Z.; software, Q.Z.; validation, Q.Z. and Y.Z.; formal analysis, Q.Z.; investigation, Q.Z.; resources, S.N.; data curation, Q.Z.; writing—original draft preparation, Q.Z.; writing—review and editing, S.N., Q.Z., and Y.Z.; visualization, Q.Z.; supervision, S.N.; project administration, Q.Z.; funding acquisition, S.N. All authors have read and agreed to the published version of the manuscript.

Funding: This research was funded by Ningbo Public Welfare Science and Technology Plan Project (2024S184).

Institutional Review Board Statement: Not applicable.

Informed Consent Statement: Not applicable.

Data Availability Statement: The datasets analyzed in this study were obtained from the Gene Expression Omnibus (GEO) database (http://www.ncbi.nlm.nih.gov/geo), with accession numbers detailed in the Materials and Methods section.

Acknowledgments: This study was supported by School of Marine Sciences, Ningbo University. The authors would like to express their gratitude to EditSprings (https://www.editsprings.cn) for the expert linguistic services provided.

Conflicts of Interest: The authors declare no conflicts of interest.

References

1. Daayf, F. *Verticillium* wilts in crop plants: Pathogen invasion and host defence responses. *Can. J. Plant Pathol.* **2014**, *37*, 8–20. [CrossRef]
2. Fradin, E.F.; Thomma, B.P.H.J. Physiology and molecular aspects of *Verticillium* wilt diseases caused by *V. dahliae* and *V. albo-atrum*. *Mol. Plant Pathol.* **2006**, *7*, 71–86. [CrossRef]
3. Zhu, Y.T.; Zhao, M.; Li, T.T.; Wang, L.; Liao, C.; Liu, D.; Zhang, H.; Zhao, Y.; Liu, L.; Ge, X.; et al. Interactions between *Verticillium dahliae* and cotton: Pathogenic mechanism and cotton resistance mechanism to *Verticillium* wilt. *Front. Plant Sci.* **2023**, *14*, 1174281. [CrossRef]
4. Sayari, M.; Dolatabadian, A.; El-Shetehy, M.; Daayf, F. Genomic insights into *Verticillium*: A review of progress in the genomics era. *Front. Microbiol.* **2024**, *15*, 1463779. [CrossRef] [PubMed]
5. Klosterman, S.J.; Subbarao, K.V.; Kang, S.; Veronese, P.; Gold, S.E.; Thomma, B.P.H.J.; Chen, Z.; Henrissat, B.; Lee, Y.H.; Park, J.; et al. Comparative genomics yields insights into niche adaptation of plant vascular wilt pathogens. *PLoS Pathog.* **2011**, *7*, e1002137. [CrossRef]
6. Zhang, D.D.; Dai, X.F.; Klosterman, S.J.; Subbarao, K.V.; Chen, J.Y. The secretome of *Verticillium dahliae* in collusion with plant defence responses modulates *Verticillium* wilt symptoms. *Biol. Rev. Camb. Philos. Soc.* **2022**, *97*, 1810–1822. [CrossRef]
7. Depotter, J.R.L.; Deketelaere, S.; Inderbitzin, P.; Von Tiedemann, A.; Höfte, M.; Subbarao, K.V.; Wood, T.A.; Thomma, B.P.H.J. *Verticillium longisporum*, the invisible threat to oilseed rape and other brassicaceous plant hosts. *Mol. Plant Pathol.* **2016**, *17*, 1004–1016. [CrossRef]
8. Ingram, R. *Verticillium dahliae* var. *longisporum*, a stable diploid. *Trans. Br. Mycol. Soc.* **1968**, *51*, 339–341. [CrossRef]
9. Karapapa, V.K.; Bainbridge, B.W.; Heale, J.B. Morphological and molecular characterization of *Verticillium longisporum* comb. nov., pathogenic to oilseed rape. *Mycol. Res.* **1997**, *101*, 1281–1294. [CrossRef]
10. Inderbitzin, P.; Bostock, R.M.; Davis, R.M.; Usami, T.; Platt, H.W.; Subbarao, K.V. Phylogenetics and taxonomy of the fungal vascular wilt pathogen *Verticillium*, with the descriptions of five new species. *PLoS ONE* **2011**, *6*, e28341. [CrossRef]
11. Inderbitzin, P.; Davis, R.M.; Bostock, R.M.; Subbarao, K.V. Identification and Differentiation of *Verticillium* Species and *V. longisporum* Lineages by Simplex and Multiplex PCR Assays. *PLoS ONE* **2013**, *8*, e65990. [CrossRef] [PubMed]
12. Inderbitzin, P.; Davis, R.M.; Bostock, R.M.; Subbarao, K.V. The Ascomycete *Verticillium longisporum* Is a Hybrid and a Plant Pathogen with an Expanded Host Range. *PLoS ONE* **2011**, *6*, e18260. [CrossRef] [PubMed]
13. Zhou, L.; Hu, Q.; Johansson, A.; Dixelius, C. *Verticillium longisporum* and *V. dahliae*: Infection and disease in *Brassica napus*. *Plant Pathol.* **2006**, *55*, 137–144. [CrossRef]
14. Eynck, C.; Koopmann, B.; Grunewaldt-Stoecker, G.; Karlovsky, P.; von Tiedemann, A. Differential interactions of *Verticillium longisporum* and *V. dahliae* with *Brassica napus* detected with molecular and histological techniques. *Eur. J. Plant Pathol.* **2007**, *118*, 259–274. [CrossRef]
15. Kamble, A.; Koopmann, B.; von Tiedemann, A. Induced resistance to *Verticillium longisporum* in *Brassica napus* by β-aminobutyric acid. *Plant Pathol.* **2013**, *62*, 552–561. [CrossRef]
16. Heale, J.B.; Karapapa, V.K. The *Verticillium* threat to Canada's major oilseed crop: Canola. *Can. J. Plant Pathol.* **1999**, *21*, 1–7. [CrossRef]
17. Fradin, E.F.; Abd-El-Haliem, A.; Masini, L.; van den Berg, G.C.M.; Joosten, M.H.; Thomma, B.P. Interfamily transfer of tomato Ve1 mediates *Verticillium* resistance in *Arabidopsis*. *Plant Physiol.* **2011**, *156*, 2255–2265. [CrossRef]
18. Wang, Y.; Fredua-Agyeman, R.; Yu, Z.; Hwang, S.F.; Strelkov, S.E. Genome-wide association study of *Verticillium longisporum* resistance in *Brassica* genotypes. *Front. Plant Sci.* **2024**, *15*, 1436982. [CrossRef]
19. Yadeta, K.A.; Valkenburg, D.J.; Hanemian, M.; Marco, Y.; Thomma, B.P.H.J. The Brassicaceae-specific *EWR1* gene provides resistance to vascular wilt pathogens. *PLoS ONE* **2014**, *9*, e88230. [CrossRef]
20. Dölfors, F.; Ilbäck, J.; Bejai, S.; Fogelqvist, J.; Dixelius, C. Nitrate transporter protein NPF5.12 and major latex-like protein MLP6 are important defense factors against *Verticillium longisporum*. *J. Exp. Bot.* **2024**, *75*, 4148–4164. [CrossRef]
21. Pröbsting, M.; Schenke, D.; Hossain, R.; Häder, C.; Thurau, T.; Wighardt, L.; Schuster, A.; Zhou, Z.; Ye, W.; Rietz, S.; et al. Loss of function of CRT1a (calreticulin) reduces plant susceptibility to *Verticillium longisporum* in both *Arabidopsis thaliana* and oilseed rape (*Brassica napus*). *Plant Biotechnol. J.* **2020**, *18*, 2328–2344. [CrossRef]
22. Roos, J.; Bejai, S.; Mozūraitis, R.; Dixelius, C. Susceptibility to *Verticillium longisporum* is linked to monoterpene production by TPS23/27 in *Arabidopsis*. *Plant J.* **2015**, *81*, 572–585. [CrossRef]

23. Ulrich, L.; Schmitz, J.; Thurow, C.; Gatz, C. Coronatine Insensitive 1-mediated repression of immunity-related genes in *Arabidopsis* roots is lifted upon infection with *Verticillium longisporum*. *J. Exp. Bot.* **2025**, *76*, 2356–2372. [CrossRef] [PubMed]
24. Fröschel, C.; Iven, T.; Walper, E.; Bachmann, V.; Weiste, C.; Dröge-Laser, W. A Gain-of-Function Screen Reveals Redundant ERF Transcription Factors Providing Opportunities for Resistance Breeding Toward the Vascular Fungal Pathogen *Verticillium longisporum*. *Mol. Plant Microbe Interact.* **2019**, *32*, 1095–1109. [CrossRef] [PubMed]
25. Edgar, R.; Domrachev, M.; Lash, A.E. Gene Expression Omnibus: NCBI gene expression and hybridization array data repository. *Nucleic Acids Res.* **2002**, *30*, 207–210. [CrossRef] [PubMed]
26. Szklarczyk, D.; Gable, A.L.; Lyon, D.; Junge, A.; Wyder, S.; Huerta-Cepas, J.; Simonovic, M.; Doncheva, N.T.; Morris, J.H.; Bork, P.; et al. STRING v11: Protein-protein association networks with increased coverage, supporting functional discovery in genome-wide experimental datasets. *Nucleic Acids Res.* **2019**, *47*, D607–D613. [CrossRef]
27. Smoot, M.E.; Ono, K.; Ruscheinski, J.; Wang, P.L.; Ideker, T. Cytoscape 2.8: New features for data integration and network visualization. *Bioinformatics* **2011**, *27*, 431–432. [CrossRef]
28. Warde-Farley, D.; Donaldson, S.L.; Comes, O.; Zuberi, K.; Badrawi, R.; Chao, P.; Franz, M.; Grouios, C.; Kazi, F.; Lopes, C.T.; et al. The GeneMANIA prediction server: Biological network integration for gene prioritization and predicting gene function. *Nucleic Acids Res.* **2010**, *38*, W214–W220. [CrossRef]
29. Jin, J.; Tian, F.; Yang, D.C.; Meng, Y.Q.; Kong, L.; Luo, J.; Gao, G. PlantTFDB 4.0: Toward a central hub for transcription factors and regulatory interactions in plants. *Nucleic Acids Res.* **2017**, *45*, D1040–D1045. [CrossRef]
30. Dai, X.; Zhuang, Z.; Zhao, P.X. psRNATarget: A plant small RNA target analysis server (2017 release). *Nucleic Acids Res.* **2018**, *46*, W49–W54. [CrossRef]
31. Oughtred, R.; Rust, J.; Chang, C.; Breitkreutz, B.J.; Stark, C.; Willems, A.; Boucher, L.; Leung, G.; Kolas, N.; Zhang, F.; et al. The BioGRID database: A comprehensive biomedical resource of curated protein, genetic, and chemical interactions. *Protein Sci.* **2021**, *30*, 187–200. [CrossRef] [PubMed]
32. Kemmochi, I.; Kobayashi, I.; Tsuchiya, M.; Sakai, H.; Shimizu, M. Breeding materials for resistance to *Verticillium* wilt in Japanese cabbage (*Brassica oleracea* L. var. *capitata*). *J. Jpn. Soc. Hortic Sci.* **2000**, *69*, 483–491. [CrossRef]
33. Rygulla, W.; Snowdon, R.J.; Friedt, W.; Happstadius, I.; Cheung, W.Y.; Chen, D. Identification of quantitative trait loci for resistance against *Verticillium longisporum* in oilseed rape (*Brassica napus*). *Phytopathology* **2008**, *98*, 215–221. [CrossRef]
34. Song, R.; Li, J.; Xie, C.; Jian, W.; Yang, X. An Overview of the Molecular Genetics of Plant Resistance to the *Verticillium* Wilt Pathogen *Verticillium dahliae*. *Int. J. Mol. Sci.* **2020**, *21*, 1120. [CrossRef]
35. Wittstock, U.; Halkier, B.A. Glucosinolate research in the *Arabidopsis* era. *Trends Plant Sci.* **2002**, *7*, 263–270. [CrossRef]
36. Njoroge, S.M.; Vallad, G.E.; Park, S.Y.; Kang, S.; Koike, S.T.; Bolda, M.; Burman, P.; Polonik, W.; Subbarao, K.V. Phenological and phytochemical changes correlate with differential interactions of *Verticillium dahliae* with broccoli and cauliflower. *Phytopathology* **2011**, *101*, 523–534. [CrossRef]
37. Witzel, K.; Hanschen, F.S.; Klopsch, R.; Ruppel, S.; Schreiner, M.; Grosch, R. *Verticillium longisporum* infection induces organ-specific glucosinolate degradation in *Arabidopsis thaliana*. *Front. Plant Sci.* **2015**, *6*, 508. [CrossRef] [PubMed]
38. Guengerich, F.P. Cytochrome P450 research and The Journal of Biological Chemistry. *J. Biol. Chem.* **2019**, *294*, 1671–1680. [CrossRef]
39. Iven, T.; König, S.; Singh, S.; Braus-Stromeyer, S.A.; Bischoff, M.; Tietze, L.F.; Braus, G.H.; Lipka, V.; Feussner, I.; Dröge-Laser, W. Transcriptional Activation and Production of Tryptophan-Derived Secondary Metabolites in *Arabidopsis* Roots Contributes to the Defense against the Fungal Vascular Pathogen *Verticillium longisporum*. *Mol. Plant* **2012**, *5*, 1389–1402. [CrossRef]
40. Yun, H.S.; Sul, W.J.; Chung, H.S.; Lee, J.H.; Kwon, C. Secretory membrane traffic in plant-microbe interactions. *New Phytol.* **2023**, *237*, 53–59. [CrossRef]
41. Xu, J.; Xu, X.; Tian, L.; Wang, G.; Zhang, X.; Wang, X.; Guo, W. Discovery and identification of candidate genes from the chitinase gene family for *Verticillium dahliae* resistance in cotton. *Sci. Rep.* **2016**, *6*, 29022. [CrossRef]
42. Zhang, M.; Su, J.; Zhang, Y.; Xu, J.; Zhang, S.Q. Conveying endogenous and exogenous signals: MAPK cascades in plant growth and defense. *Curr. Opin. Plant Biol.* **2018**, *45*, 1–10. [CrossRef] [PubMed]
43. Dixon, R.A.; Achnine, L.; Kota, P.; Liu, C.J.; Reddy, M.S.S.; Wang, L. The phenylpropanoid pathway and plant defence-a genomics perspective. *Mol. Plant Pathol.* **2002**, *3*, 371–390. [CrossRef] [PubMed]
44. Yang, L.; Ye, C.; Zhao, Y.; Cheng, X.; Wang, Y.; Jiang, Y.Q.; Yang, B. An oilseed rape WRKY-type transcription factor regulates ROS accumulation and leaf senescence in *Nicotiana benthamiana* and *Arabidopsis* through modulating transcription of *RbohD* and *RbohF*. *Planta* **2018**, *247*, 1323–1338. [CrossRef]
45. Forman, H.J.; Zhang, H.; Rinna, A. Glutathione: Overview of its protective roles, measurement, and biosynthesis. *Mol. Asp. Med.* **2009**, *30*, 1–12. [CrossRef] [PubMed]
46. Wingate, V.P.M.; Lawton, M.A.; Lamb, C.J. Glutathione causes a massive and selective induction of plant defense genes. *Plant Physiol.* **1988**, *87*, 206–210. [CrossRef]
47. Boro, P.; Chattopadhyay, S. Crosstalk between MAPKs and GSH under stress: A critical review. *J. Biosci.* **2022**, *47*, 71. [CrossRef]

48. Liu, Y.; He, C. A review of redox signaling and the control of MAP kinase pathway in plants. *Redox Biol.* **2017**, *11*, 192–204. [CrossRef]
49. Zhao, S.P.; Kamran, M.; Rizwan, M.; Ali, S.; Yan, L.; Alwahibi, M.S.; Elshikh, M.S.; Riaz, M. Regulation of proline metabolism, AsA-GSH cycle, cadmium uptake and subcellular distribution in *Brassica napus* L. under the effect of nano-silicon. *Environ. Pollut.* **2023**, *335*, 122321. [CrossRef]
50. Ngounou Wetie, A.G.; Sokolowska, I.; Woods, A.G.; Roy, U.; Deinhardt, K.; Darie, C.C. Protein-protein interactions: Switch from classical methods to proteomics and bioinformatics-based approaches. *Cell. Mol. Life Sci.* **2014**, *71*, 205–228. [CrossRef]
51. Müller, T.M.; Böttcher, C.; Morbitzer, R.; Götz, C.C.; Lehmann, J.; Lahaye, T.; Glawischnig, E. TRANSCRIPTION ACTIVATOR-LIKE EFFECTOR NUCLEASE-Mediated Generation and Metabolic Analysis of Camalexin-Deficient *cyp71a12 cyp71a13* Double Knockout Lines. *Plant Physiol.* **2015**, *168*, 849–858. [CrossRef]
52. Pastorczyk, M.; Kosaka, A.; Piślewska-Bednarek, M.; López, G.; Frerigmann, H.; Kułak, K.; Glawischnig, E.; Molina, A.; Takano, Y.; Bednarek, P. The role of CYP71A12 monooxygenase in pathogen-triggered tryptophan metabolism and *Arabidopsis* immunity. *New Phytol.* **2020**, *225*, 400–412. [CrossRef]
53. Meier, S.; Ruzvidzo, O.; Morse, M.; Donaldson, L.; Kwezi, L.; Gehring, C.; Newbigin, E. The *Arabidopsis* wall associated kinase-like 10 gene encodes a functional guanylyl cyclase and is co-expressed with pathogen defense related genes. *PLoS ONE* **2010**, *5*, e8904. [CrossRef] [PubMed]
54. Marín-de la Rosa, N.; Sotillo, B.; Miskolczi, P.; Gibbs, D.J.; Vicente, J.; Carbonero, P.; Oñate-Sánchez, L.; Holdsworth, M.J.; Bhalerao, R.; Alabadí, D.; et al. Large-scale identification of gibberellin-related transcription factors defines group VII ETHYLENE RESPONSE FACTORS as functional DELLA partners. *Plant Physiol.* **2014**, *166*, 1022–1032. [CrossRef]
55. Shahnejat-Bushehri, S.; Nobmann, B.; Devi Allu, A.; Balazadeh, S. JUB1 suppresses *Pseudomonas syringae*-induced defense responses through accumulation of DELLA proteins. *Plant Signal. Behav.* **2016**, *11*, e1181245. [CrossRef] [PubMed]
56. Saga, H.; Ogawa, T.; Kai, K.; Suzuki, H.; Ogata, Y.; Sakurai, N.; Shibata, D.; Ohta, D. Identification and characterization of *ANAC042*, a transcription factor family gene involved in the regulation of camalexin biosynthesis in *Arabidopsis*. *Mol. Plant Microbe Interact.* **2012**, *25*, 684–696. [CrossRef] [PubMed]
57. Zhang, Z.; Liu, C.; Li, K.; Li, X.; Xu, M.; Guo, Y. CLE14 functions as a "brake signal" to suppress age-dependent and stress-induced leaf senescence by promoting JUB1-mediated ROS scavenging in *Arabidopsis*. *Mol. Plant* **2022**, *15*, 179–188. [CrossRef]
58. Rushton, P.J.; Somssich, I.E.; Ringler, P.; Shen, Q.J. WRKY transcription factors. *Trends Plant Sci.* **2010**, *15*, 247–258. [CrossRef]
59. Javed, T.; Gao, S.J. WRKY transcription factors in plant defense. *Trends Genet.* **2023**, *39*, 787–801. [CrossRef]
60. Häffner, E.; Karlovsky, P.; Splivallo, R.; Traczewska, A.; Diederichsen, E. ERECTA, salicylic acid, abscisic acid, and jasmonic acid modulate quantitative disease resistance of *Arabidopsis thaliana* to *Verticillium longisporum*. *BMC Plant Biol.* **2014**, *14*, 85. [CrossRef]
61. Dubos, C.; Stracke, R.; Grotewold, E.; Weisshaar, B.; Martin, C.; Lepinie, L. MYB transcription factors in *Arabidopsis*. *Trends Plant Sci.* **2010**, *15*, 573–581. [CrossRef] [PubMed]
62. Dröge-Laser, W.; Snoek, B.L.; Snel, B.; Weiste, C. The *Arabidopsis* bZIP transcription factor family-an update. *Curr. Opin. Plant Biol.* **2018**, *45(Pt A)*, 36–49. [CrossRef]
63. Sharif, R.; Raza, A.; Chen, P.; Li, Y.; El-Ballat, E.M.; Rauf, A.; Hano, C.; El-Esawi, M.A. HD-ZIP Gene Family: Potential Roles in Improving Plant Growth and Regulating Stress-Responsive Mechanisms in Plants. *Genes* **2021**, *12*, 1256. [CrossRef] [PubMed]
64. Feng, K.; Hou, X.L.; Xing, G.M.; Liu, J.X.; Duan, A.Q.; Xu, Z.S.; Li, M.Y.; Zhuang, J.; Xiong, A.S. Advances in AP2/ERF super-family transcription factors in plant. *Crit. Rev. Biotechnol.* **2020**, *40*, 750–776. [CrossRef] [PubMed]

current issues in molecular biology

MDPI

Article

Molecular Regulation of Antioxidant Defense and Metabolic Reprogramming in Xiaozhan Rice Genotypes: Differential Roles of Salicylic Acid and Melatonin Under Salt Stress

Yang Wu [1,†], Yongbo Duan [2,†], Xifan Luo [1], Mingjun Li [1], Hengjie Gao [1], Wei Zhu [3], Fei Zhao [1], Jian Liu [1,*] and Wenzhong Zhang [4,*]

1 Tianjin Key Laboratory Intelligent Breeding Major Crops, College of Agronomy & Resources and Environment, Tianjin Agricultural University, Tianjin 300392, China; wuyang991103@163.com (Y.W.); luoxifan9902@163.com (X.L.); mingjunli0425@163.com (M.L.); hengjie.gao@foxmail.com (H.G.); 13920357546@126.com (F.Z.)

2 Anhui Provincial Engineering Laboratory for Efficient Utilization of Featured Resource Plants, College of Life Sciences, Huaibei Normal University, Huaibei 235026, China; yboduan@163.com

3 Key Laboratory of Hybrid Japonica Rice Genetic Breeding, Ministry of Agriculture, National Hybrid Rice Engineering and Technology Research Centre, Tianjin Sub-Centre, Tianjin 300350, China; zwzswzw@126.com

4 Rice Research Institute, Shenyang Agricultural University, Shenyang 110866, China

* Correspondence: 13821757926@163.com (J.L.); zwzhong1@syau.edu.cn (W.Z.); Tel.: +86-022-2378-1298 (J.L.); +86-024-8848-7135 (W.Z.)

† These authors contributed equally to this work.

Abstract: Against the background of increasing global soil salinity, exogenous salicylic acid (SA) and melatonin (MT) have attracted much attention for their potential in regulating plant stress tolerance and have become an important research direction for the development of green and sustainable agriculture. In this study, the alleviating effects of different concentrations of SA (100–900 μM) and MT (100–900 μM) on salt stress (50 mM NaCl) and their physiological mechanisms were systematically investigated using the Tianjin specialty rice, Xiaozhan rice, as the research object. The results showed that salt stress significantly inhibited the germination and seedling growth of the two varieties, in which the salt-sensitive variety Jinchuan No. 1 showed significantly higher decreases in root length, plant height, and biomass (54.7–69.1%) than the salt-tolerant variety Jindao 919 (4.0–28.9%). Exogenous SA and MT were effective in mitigating salt stress injury, but there were genotypic differences in their pathways of action. For the first time in *japonica* rice, the genotype specificity of the SA/MT response was clearly revealed: SA dominated the response of salt-tolerant varieties by enhancing antioxidant defences, whereas MT optimized the overall performance of the salt-sensitive varieties through scavenging of reactive oxygen species, and in addition, it was further determined that SA and MT exhibited optimal mitigating effects on both varieties in the 300–700 μM concentration range, showing the best mitigation effect for both varieties. This finding provides an important theoretical basis and technological paradigm for precision stress tolerance cultivation of saline rice, and the application of appropriate concentrations of SA/MT according to genotype specificity to reduce the dependence on agrochemicals is of practical value in promoting green and sustainable production in saline agriculture.

Keywords: Xiaozhan rice; salt stress; salicylic acid; melatonin; genotype interactions

1. Introduction

Climate change has continuously exacerbated soil salinization, leading to a growing global footprint in terms of both extent and severity [1,2]. According to the latest data from the Food and Agriculture Organization (FAO), approximately 415 million hectares of global land area are affected by salinization, constituting 6.2% of the world's total terrestrial surface. China ranks among the most severely impacted nations, boasting an estimated 99.13 million hectares of saline-alkali soils. These are primarily distributed across three key regions: the arid northwest, the Songnen Plain in Northeast China, and the coastal areas of the Huang-Huai-Hai Plain [3,4]. Additionally, roughly 20% of the world's arable land and 33% of irrigated agricultural areas face varying intensities of salt stress, a trend that poses a serious threat to global food security [5]. In China, the severity of this issue is exemplified by the Tianjin Binhai New Area and the Jinghai area, where long-term accumulation of salinity (low sea level) directly affects the yield of local crops. Concurrently, it triggers a series of secondary environmental consequences, such as soil structural deterioration and decreased water infiltration capacity [6,7].

As a specialty agricultural product of Tianjin, 'Xiaozhan rice' functions not only as a carrier of the living heritage of regional farming culture but also as a model of practising eco-friendly agriculture [8]. As of 2024, the planting area of Xiaozhan rice accounted for over 80% of the arable land in Tianjin. This rice variety has also been gradually extended to the Beijing-Tianjin-Hebei urban agglomeration, emerging as a benchmark for the green transformation of regional agriculture. However, the widespread soil salinity problem in the coastal area of Tianjin has severely restricted the large-scale cultivation of Xiaozhan rice and the increase in its yield. Therefore, there is an urgent demand to develop eco-friendly approaches for managing saline lands [9].

Fortunately, multiple chemicals, reagents, hormones, and growth regulators have been developed to mitigate the detrimental effects of soil salinization on rice [10]. Among them, SA and MT, as endogenous plant metabolites, offer the advantages of natural origin and environmental compatibility. Moreover, their large-scale production cost is only about \$150/tonne (purity $\geq$ 99%), which is significantly lower than that of the traditional chemical modifiers [11–14]. Notably, these compounds degrade easily in soil and do not pose a bioaccumulation risk, aligning with the global trend towards drug reduction and low-carbon agriculture. Studies have indicated that SA can significantly enhance salt stress tolerance in rice by activating antioxidant enzyme systems (e.g., superoxide dismutase (SOD) and catalase (CAT)) and regulating ion transport proteins [15,16], such as SA is involved in the cold tolerance mechanism of citrus by regulating metabolite accumulation through the *TGA2-P5CS1/ICS1* module [17]. In contrast, MT plays a crucial role in maintaining cellular homeostasis by scavenging reactive oxygen species (ROS) and promoting the synthesis of osmoregulatory substances (e.g., soluble sugars and proline) [18–20], as in the case of sweetpotato, MT upregulated expression of sodium hydrogen exchanger 2 (*NHX2*), K^+ transporter 1 (*AKT1*), cation/H^+ exchanger (*CHX*), CBL-interacting protein kinase 1 (*CIPK1*) and antioxidant enzyme genes are expressed to enhance the adaptability to salt stress [21]. However, existing conclusions are mainly based on studies on other crops or specific rice genotypes (*Nipponbare*), among other things, SA regulates the expression of antioxidant genes through npr1-dependent/non-dependent pathways (e.g., SA-activated *MAPK* cascade pathway revealed by Molecular Plant), and MT enhances salt tolerance by regulating ion channels (e.g., *HKT1*) and the ABA signalling pathway [22–24]. There are still large gaps in the mechanisms of SA/MT interactions in different rice genotypes (especially local varieties), such as genotypic differences in signalling pathways, specificity of transporter protein expression, and environmental synergistic effects. In summary, in-depth exploration of the SA/MT interaction mechanisms of specialty varieties, such as 'Xiao

Zhan Rice,' in combination with genotype-specific pathways is needed, thus facilitating the precise application of SA/MT in regional agriculture [25].

Therefore, salt-tolerant and salt-sensitive types were hereby selected from the above four Xiaozhan rice extension varieties for the study. Moreover, the differences in salt stress tolerance were systematically investigated between exogenous SA and MT, and attempts were made to elucidate the intrinsic mechanism of the two compounds to enhance salt tolerance through the effects of the two compounds on seed germination, seedling growth, and physiological metabolism under controlled conditions. Overall, the outcomes of this study are anticipated to provide actionable insights for optimizing the targeted breeding strategy and the precise application of hormones in rice within the salinized agro-ecosystems in Tianjin. Furthermore, these findings will propel the advancement of agro-ecosystems in salinized areas towards a more environmentally friendly and efficient direction.

2. Materials and Methods

2.1. Experimental Site and Materials

The experiment was conducted from September to December 2024 in a climate-controlled growth chamber at Tianjin Agricultural University (38°96′ N, 117°12′ E). The artificial climate chamber was commissioned to be constructed by Nanjing Lithgow Instrument and Equipment Co., Ltd. (Nanjing, China), and the controlled environmental conditions were as follows: temperature 26 °C, relative humidity 75%, light cycle 12 h/darkness 12 h, LED cold light source, photosynthetic photon flux density of 150 μmol m^{-2} s^{-1}, light homogeneity > 90%, three-layer matrix design and edge light decay < 15%. The experimental materials were Jinyuan U99, Tianlongyou 619, Jinchuan No. 1, and Jindao 919, provided by the Xiaozhan Rice Science and Technology Yard (Baodi District, Tianjin, China). Germination boxes (12 cm × 12 cm) and 96-well hydroponic culture boxes (127 mm × 87 mm × 114 mm, aperture diameter 5.0 mm) were purchased from Haimen Yurong Experimental Analysis Instrument Factory (Haimen, China). NaCl was obtained from Yingda Rare Chemical Reagent Factory (Tianjin, China). SA, MT, and Hoagland's nutrient solution [26] were purchased from Solarbio Science & Technology Co., Ltd. (Beijing, China). All other common chemicals were of analytical grade.

2.2. Experimental Design

Intact rice seeds of uniform size were selected for the experiment. The seeds were surface-sterilized with 75% (v/v) ethanol for 1 min, followed by immersion in 1% (v/v) NaClO for 15 min, and subsequently rinsed five times with sterile distilled water to eliminate contaminants. Under aseptic conditions on a clean bench maintained at 25 °C, the sterilized seeds were evenly spread and air-dried for subsequent use.

The germination test was conducted according to the standardized method specified in Germination Test (GB/T 3543.4-1995) [27]. Each rice germination box was lined with two layers of filter paper, and 10 mL (containing 50 mmol/L NaCl and exogenous substances, Table 1) of treatment solution was added. Salt concentrations were predetermined in preliminary experiments, while volumes were added to ensure that waterlogging did not occur in the system. Three replicates were prepared with 50 seeds per box without contacting each other. All boxes were incubated in an artificial climate chamber under controlled conditions. During cultivation, the appropriate solution was supplemented as needed to maintain moisture. Germinated seeds were counted and recorded daily during the period, and destructive sampling was carried out on the 14th day after sowing to determine the indices.

Table 1. Experimental treatment scheme.

Code	Treatment	Code	Treatment
CK	0 mM NaCl + 0 μM SA/MT	NaCl	50 mM NaCl
S1	50 mM NaCl + 100 μM SA	M1	50 mM NaCl + 100 μM MT
S2	50 mM NaCl + 300 μM SA	M2	50 mM NaCl + 300 μM MT
S3	50 mM NaCl + 500 μM SA	M3	50 mM NaCl + 500 μM MT
S4	50 mM NaCl + 700 μM SA	M4	50 mM NaCl + 700 μM MT
S5	50 mM NaCl + 900 μM SA	M5	50 mM NaCl + 900 μM MT

To simulate a saline environment, surface-sterilized seeds were soaked in distilled water for 24 h and then transferred to 96-well hydroponic boxes containing half-strength nutrient solution to ensure normal growth until the one-leaf-one-heart stage (96 wells are filled). The solution was then replaced with full-strength nutrient solution containing 50 mM NaCl and exogenous compounds (SA/MT) (Table 1). The nutrient solution was renewed every 3 days to maintain optimal nutrient balance and support healthy rice growth. After 14 days of cultivation, destructive sampling was performed to evaluate target parameters, with three replicates per treatment.

2.3. Measurement of Rice Emergence Growth Parameters

During the germination test, seeds were considered germinated when the radicle length reached ≥ 1× seed length and the coleoptile length attained ≥ 1/2 seed length. The number of germinated seeds was recorded daily in the box until termination at 14 DAS. The following parameters were calculated: GP (1), GR (2), GI (3), and VI (4).

$$GP\% = \frac{N_5}{N} \times 100\% \quad (1)$$

Germination Potential (GP, %) represents the germination percentage at 5 DAS, N_5 is the number of seeds with normal germination (radicle ≥ 1× seed length) at 5 DAS, and N is the total number of seeds tested.

$$GR\% = \frac{N_{14}}{N} \times 100\% \quad (2)$$

Germination rate (GR, %) represents the cumulative germination percentage at 14 DAS, N_{14} is the number of seeds with normal germination (radicle length ≥ 1× seed length and coleoptile length ≥ 1/2 seed length) at 14 DAS, and N is the total number of seeds tested.

$$GI = \sum \frac{G_t}{D_t} \quad (3)$$

G_t denotes the number of seeds germinated at a specific time point (day t), while D_t represents the corresponding germination day number associated with G_t.

$$VI = GI \times S \quad (4)$$

S represents the mean seedling height (cm) at the termination of the germination test.

In two different experiments, 10 uniform seedlings were randomly selected from each group for index measurement. After surface moisture was blotted dry with filter paper, Seedling Length (SL) and Seedling Root Length (SRL) were measured using a vernier calliper (0~300 mm, ± 0.02 mm accuracy). Root length was defined as the linear distance from the root-shoot junction to the tip of the primary root, while plant height was measured as the vertical distance from the stem base to the apical meristem. The fresh weight of the whole plant (SFW) was measured using a precision balance (accuracy: 0.001 g). Each sample was then wrapped in aluminum foil and placed in an oven at 105 °C for 30 min to deactivate enzymes. Subsequently, the samples were dried at 80 °C for 72 h until a constant weight was achieved, after which the dry weight (SDW) was recorded.

2.4. Measurement of Rice Seedling Growth Parameters in Simulated Real Stresses

Of the replicate treatments in the hydroponics trial, rice leaves were randomly excised using scissors, ensuring that all samples had identical fresh weights (2.00 ± 0.1 g). The collected leaf samples were immediately wrapped in aluminum foil and flash-frozen in a foam box containing liquid nitrogen for subsequent enzyme activity assays. Subsequent measurements, including plant height (PH), root length (RL), fresh weight (FW), dry weight (DW), and stem base diameter (SD), were conducted following the protocol described in Section 2.3.

2.5. Measurement of Physiological and Biochemical Indicators

To analyze physiological and biochemical changes in plant tissues, fresh leaf samples were homogenized in liquid nitrogen. The following parameters were then measured using commercial assay kits (Solarbio Science & Technology Co., Ltd., Beijing, China) according to the manufacturer's instructions: POD activity [28], SOD activity [29], CAT activity [30], MDA content [31], GSH content [32], soluble sugar content [33] and lignin content [34]. All physiological parameters were quantified using three biological replicates with three technical replicates per assay. Mean values were derived from nine data points (three biological × three technical) unless otherwise noted.

2.6. Statistical Analysis

Data processing was performed using Microsoft Excel 2019 (Microsoft Corporation, Redmond, WA, USA). Statistical analyses, including one-way analysis of variance (ANOVA) and significance testing (Tukey's honestly significant difference [HSD] test, $p < 0.05$), were conducted with IBM SPSS Statistics 26 (IBM Corp., Armonk, NY, USA). The salt tolerance of four rice varieties was evaluated using a fuzzy membership function method. The membership function value was calculated as follows Equation (5):

$$\mu(X_j) = (X_j - X_{\min})/(X_{\max} - X_{\min}) \tag{5}$$

where $\mu(X_j)$ represents the membership function value, X_j denotes the measured value of the j indicator, and $X_{\max}$ and $X_{\min}$ indicate the maximum and minimum values of the j indicator. The coefficient of variation V_j for each indicator was determined using Equation (6):

$$V_j = \frac{\sqrt{\frac{\sum_{i=1}^{n}(X_{ij}-\overline{X_j})^2}{n-1}}}{\overline{X}_j} \tag{6}$$

where $\overline{X_j}$ represents the average value of the j indicator across different growth stages, X_{ij} denotes the value of the affiliation function of indicator j for fertility period i, and V_j is the standard deviation of the membership function values for the j indicator. The weight coefficient W_j for each indicator was calculated as follows Equation (7):

$$W_j = \frac{V_j}{\sum_{j=1}^{n} V_j} \tag{7}$$

The comprehensive evaluation value (D) was determined using Equations (8) and (9):

$$D = \sum_{j=1}^{n} [\mu(X_j \times W_j)] \quad (8)$$

$$\Delta D = D_{CK} - D_{NaCl} \quad (9)$$

Higher ΔD values indicated weaker salt tolerance of the evaluated varieties, and the two genotypes with the highest and weakest salt tolerance were selected for correlation analysis and cluster visualization using the 'correlation plot' and 'heat map with dendrogram' applications in Origin 2024 (OriginLab Corporation, Northampton, MA, USA). All data were normalized using Equation (5) and then analyzed by Principal Component Analysis (PCA) using IBM SPSS Statistics 26 in order to reduce the dimensionality of the evaluation indicators. The contribution rates of communalities were calculated and adopted as weight coefficients (W_j) for corresponding indicators. These weighted parameters were subsequently integrated with fuzzy membership function analysis (using only Equation (8), where W_j represents principal component-derived weights) to comprehensively evaluate salt stress mitigation capacities under different exogenous treatments.

3. Results

3.1. Analysis of Growth Differences and Mechanism in Warble Xiaozhan Rice Under Salt Stress

Salt stress significantly inhibited the growth status (D-value, Supplementary Materials S1) of the experimental rice, with the salt-sensitive cultivar Jinchuan No. 1 exhibiting the largest decline (Δ = 0.242), while the salt-tolerant cultivar Jindao 919 showed the smallest reduction (Δ = 0.122) (Table 2). Comparative analysis revealed these differential responses stemmed from fundamentally distinct physiological regulation strategies: Jindao 919 maintained ionic homeostasis through enhanced antioxidant enzyme activities, whereas Jinchuan No. 1 displayed higher salt sensitivity due to insufficient oxidative damage repair capacity. Notably, the substantial difference in salt tolerance (63.8%) between Tianlongyou 619 (Δ = 0.216) and Jinyuan U99 (Δ = 0.132) further corroborated the critical influence of genetic background on salt stress responses. Based on the comprehensive consideration of significant phenotypic differences and representative mechanisms, Jinchuan No. 1 and Jindao 919 were selected for the comparison of physiological regulation strategies under salt stress.

Table 2. Comparison of antioxidant enzyme activities and D values of different rice varieties under salt stress.

Cultivar	Treatment	SOD Activity ($U\ g^{-1}$ FW)	POD Activity ($U\ g^{-1}$ FW)	CAT Activity ($U\ g^{-1}$ min)	D Value	ΔD
Jinyuan U99	CK	214.125 ± 16.25	215.66 ± 41.90	776.49 ± 237.76	0.609	0.132
	NaCl	272.275 ± 30.40	324.33 ± 47.92	978.725 ± 97.43	0.477	
Tianlongyou 619	CK	250.37 ± 29.66	272.00 ± 35.38	380.465 ± 120.81	0.656	0.216
	NaCl	278.575 ± 39.23	386.33 ± 23.65	567.615 ± 174.79	0.440	
Jindao 919	CK	141.31 ± 21.68	319.33 ± 34.48	624.79 ± 310.25	0.462	0.122
	NaCl	217.655 ± 59.46	383.66 ± 24.60	828.075 ± 180.73	0.340	
Jinchuan No.1	CK	157.50 ± 26.49	402.00 ± 35.38	442.18 ± 132.73	0.552	0.242
	NaCl	186.41 ± 30.05	459.00 ± 14.28	927.53 ± 136.88	0.310	

3.2. Effects of Different Exogenous Substances on Seed Germination Characteristics of Rice Under Salt Stress

As shown in Table 3, the germination characteristics of the salt-tolerant variety Jindao 919 and the salt-sensitive variety Jinchuan No. 1 showed significant differentiation under salt stress. The GR and VI of Jindao 919 remained relatively stable under NaCl treatment, while several germination parameters of Jinchuan No. 1 were significantly suppressed, revealing that its salt-sensitive genotypes were more responsive to ionic toxicity at the early germination stage. Interestingly, the vigour index of Jinchuan No. 1 was generally higher than that of Jindao 919 under exogenous treatments, but its stress recovery threshold was significantly lagged, speculating that salt-tolerant varieties may achieve damage repair through rapid mobilization of preexisting metabolites.

Table 3. Exogenous SA and MT affect rice seed germination under salt stress (%).

Cultivar	Treatment	Germination Potential (GP)	Germination Rate (GR)	Germination Index (GI)	Vigour Index (VI)
Jinchuan No. 1	CK	95.33 ± 1.15 a	98.66 ± 1.15 a	14.88 ± 0.24 a	140.49 ± 2.26 a
	NaCl	79.33 ± 3.05 b	93.33 ± 1.15 b	10.32 ± 0.31 d	63.00 ± 0.88 e
	S1	90.00 ± 4.00 ab	96.66 ± 1.15 ab	13.39 ± 0.12 b	121.08 ± 1.13 b
	S2	95.33 ± 1.15 a	98.00 ± 2.00 a	14.09 ± 0.69 ab	128.34 ± 6.28 ab
	S3	96.66 ± 1.15 a	99.33 ± 1.15 a	15.02 ± 0.28 a	143.16 ± 2.73 a
	S4	93.33 ± 3.05 a	96.66 ± 1.15 ab	12.96 ± 0.22 c	139.85 ± 2.44 a
	S5	90.00 ± 4.00 a	94.66 ± 1.15 b	11.72 ± 0.27 cd	82.56 ± 1.93 d
	M1	82.66 ± 3.05 bc	94.66 ± 1.15 b	10.62 ± 0.06 d	68.29 ± 0.24 e
	M2	85.33 ± 5.03 bc	94.66 ± 1.15 b	10.94 ± 0.25 d	82.47 ± 1.95 d
	M3	89.33 ± 3.05 ab	95.33 ± 2.30 ab	11.50 ± 0.22 cd	109.25 ± 2.14 c
	M4	93.33 ± 1.15 a	98.66 ± 1.15 a	12.11 ± 0.31 c	77.14 ± 1.98 de
	M5	86.00 ± 2.00 bc	94.00 ± 0.00 b	10.70 ± 0.31 d	47.45 ± 1.07 f
Jindao 919	CK	97.33 ± 1.15 a	99.33 ± 1.15 a	12.87 ± 0.28 a	109.41 ± 2.43 a
	NaCl	72.00 ± 3.46 d	90.66 ± 1.15 e	9.58 ± 0.19 j	73.94 ± 1.67 de
	S1	90.66 ± 3.05 ab	96.00 ± 2.00 abcd	11.08 ± 0.54 cd	86.56 ± 4.29 c
	S2	91.33 ± 4.16 ab	98.00 ± 2.00 ab	11.45 ± 0.39 bc	97.62 ± 3.40 b
	S3	98.00 ± 2.00 a	99.33 ± 1.15 a	12.08 ± 0.18 ab	110.17 ± 1.73 s
	S4	90.00 ± 4.00 ab	96.66 ± 1.15 abcd	10.95 ± 0.28 cde	95.26 ± 2.43 b
	S5	79.33 ± 4.16 cd	94.00 ± 2.00 cd	10.21 ± 0.20 gh	71.01 ± 1.43 e
	M1	74.00 ± 2.00 d	93.33 ± 1.15 d	9.95 ± 0.06 hi	55.75 ± 0.37 g
	M2	80.00 ± 2.00 cd	96.00 ± 0.00 abcd	10.27 ± 0.03 fgh	79.12 ± 0.23 d
	M3	83.33 ± 3.05 bc	97.33 ± 2.30 abc	10.68 ± 0.25 def	62.67 ± 1.46 f
	M4	84.66 ± 1.15 bc	95.33 ± 1.15 bcd	10.38 ± 0.23 efg	51.25 ± 1.16 h
	M5	76.00 ± 2.00 d	92.66 ± 3.05 d	9.78 ± 0.19 i	32.60 ± 0.66 i

The data in the table are the mean ± standard deviation, $n = 3$. Different lowercase letters in the same column indicate significant differences at $p < 0.05$; the same letter indicates no significant difference ($p > 0.05$).

The regulatory effect of exogenous SA was dose-specific, with the S3 treatment (500 μM SA) showing a significant gain in both varieties: an 18.6% increase in the germination rate of Jindao 919 and a 33.5% increase in the vigour index of Jinchuan No. 1. In contrast, MT treatment showed a genotype-dependent 'dose-effect difference': 500 μM MT increased biomass accumulation of Jinchuan No. 1 by 22.4%. whereas Jindao 919 showed significant gains in response to a gradient of MT concentration. In addition, the response of Jindao 919 to the MT concentration gradient was flat and only showed a weak promotion trend in the M3 treatment, and it was speculated that there might be a compensatory inhibition mechanism of the melatonin signalling pathway in its germination regulatory network.

3.3. Effects of Different Exogenous Substances on Seeding Growth of Rice Under Salt Stress

As shown in Table 4, the salt-sensitive variety Jinchuan No. 1 and the salt-tolerant variety Jindao 919 exhibited significant growth differences under 50 mM NaCl stress, but the physiological vulnerability of Jinchuan No. 1 was particularly prominent: its root length, seedling height, Seedling Fresh Weight, and dry weight decreased sharply by 54.7%, 69.1%, 52.4%, and 13.4%, compared with the control, whereas the corresponding reductions in Jindao 919 were only 28.9%, 16.4%, 14.2%, and 4.0%, which was a visual confirmation that the salt-tolerant genotypes are seen to have strong ionic homeostatic and osmoregulatory functions. The differential restorative capacity was particularly pronounced: SA maximized root regeneration in Jinchuan No. 1 (S4, +192.3%), whereas MT optimized photosynthetic partitioning in Jindao 919. This organ-specific efficacy highlights SA's superiority in root signalling and MT's role in shoot resource allocation.

Table 4. Effects of SA and MT on growth in rice seedlings under salt stress.

Cultivar	Treatment	Seedling Root Length (SRL, cm)	Seedling Length (SL, cm)	Seedling Fresh Weight (SFW, mg)	Seedling Dry Weight (SDW, mg)
Jinchuan No. 1	CK	10.38 ± 1.53 bc	8.94 ± 0.93 bc	96.65 ± 8.70 ab	19.63 ± 0.73 bc
	NaCl	4.70 ± 1.75 e	2.76 ± 0.15 f	46.00 ± 4.58 e	17.00 ± 1.00 de
	S1	8.57 ± 1.15 cd	9.04 ± 0.57 bc	74.80 ± 6.62 cd	18.43 ± 0.57 cd
	S2	8.93 ± 1.40 cd	9.11 ± 0.54 bc	79.21 ± 6.75 bcd	19.06 ± 0.38 bcd
	S3	10.31 ± 1.36 bc	9.53 ± 0.88 ab	84.01 ± 8.51 abc	19.56 ± 1.50 bc
	S4	14.24 ± 1.17 a	10.79 ± 0.73 a	87.53 ± 11.76 ab	21.02 ± 0.18 a
	S5	7.38 ± 0.80 d	7.04 ± 0.91 de	74.68 ± 10.22 cd	17.46 ± 0.87 de
	M1	8.03 ± 1.61 cd	4.13 ± 0.96 ef	66.66 ± 13.79 de	19.00 ± 2.64 bcd
	M2	10.13 ± 1.72 bc	7.53 ± 1.84 cde	69.00 ± 7.21 de	19.66 ± 2.08 abc
	M3	11.76 ± 1.78 ab	9.50 ± 0.75 ab	104.33 ± 12.50 a	20.00 ± 1.00 ab
	M4	9.13 ± 2.28 cd	6.36 ± 0.32 de	98.33 ± 15.04 ab	20.66 ± 1.52 a
	M5	7.70 ± 2.10 d	3.43 ± 0.51 ef	55.66 ± 11.93 e	15.00 ± 4.00 e
Jindao 919	CK	8.82 ± 1.19 bcd	8.50 ± 0.92 abc	101.08 ± 8.53 abc	19.03 ± 0.29 bcd
	NaCl	6.27 ± 1.37 e	7.11 ± 0.56 de	86.71 ± 3.78 cd	18.28 ± 0.15 de
	S1	7.28 ± 1.11 de	7.81 ± 0.45 bcd	87.05 ± 5.26 cd	18.60 ± 0.28 cde
	S2	10.47 ± 1.85 ab	8.52 ± 0.83 abc	100.71 ± 10.84 abc	19.31 ± 1.06 abcd
	S3	8.44 ± 1.51 cd	9.12 ± 0.67 a	105.05 ± 8.21 ab	21.32 ± 0.46 a
	S4	8.30 ± 0.77 cd	8.70 ± 0.67 ab	101.58 ± 13.34 abc	20.44 ± 0.28 ab
	S5	8.03 ± 0.60 d	6.95 ± 0.64 def	90.88 ± 7.6 b bc	18.26 ± 1.41 de
	M1	8.90 ± 0.60 bcd	5.60 ± 0.45 fg	102.66 ± 11.50 abc	18.66 ± 2.08 cde
	M2	10.20 ± 0.20 abc	7.70 ± 0.78 cd	115.66 ± 11.59 a	20.00 ± 3.60 abc
	M3	10.93 ± 0.80 a	5.86 ± 0.77 ef	88.33 ± 8.38 cd	21.33 ± 3.05 a
	M4	10.43 ± 0.66 ab	4.93 ± 0.41 g	84.00 ± 8.18 d	19.33 ± 5.85 abcd
	M5	6.23 ± 1.23 e	3.33 ± 0.92 h	73.00 ± 13.22 e	17.66 ± 2.51 e

The data in the table are the mean ± standard deviation, $n = 3$. Different lowercase letters in the same column indicate significant differences at $p < 0.05$; the same letter indicates no significant difference ($p > 0.05$).

The restorative effect of exogenous SA treatment on salt-sensitive varieties was even more significant, with Jindao 919 reaching its peak growth at S3, whereas Jinchuan No. 1 demonstrated a wonderful restorative effect at higher concentrations of S4, with root length increasing by 192.3% to 14.24 cm and dry weight even exceeding that of the control by 8.3%, revealing that SA can break through the salt damage threshold by activating the root regeneration pathway. It is noteworthy that the root development of both varieties responded better to SA than MT treatment, suggesting that SA may occupy a core regulatory node in root-crown signalling. In contrast, the effect of MT showed a typical genotype-concentration interaction pattern: the salt-tolerant variety Jindao 919 showed a 41.6%

increase in root elongation and biomass under the M2-M3 gradient, suggesting that the medium concentration of MT optimized its resource capture strategy by enhancing ROS scavenging efficiency, whereas the salt-sensitive variety Jinchuan No. 1 showed a 37.2% increase in plant height and an 18.4% increase in dry weight under the high concentrations of M3-M4, suggesting that the above-ground part of the root crown signalling was enhanced by MT-mediated redistribution of photosynthesis products.

3.4. Effects of Different Exogenous Substances on Plant Morphogenesis of Rice Under Salt Stress

Salt stress showed significant genotype-specific physiological inhibition in the two rice varieties (Table 5). In the salt-sensitive variety Jinchuan No. 1, the root length dropped sharply from 14.89 cm to 4.52 cm (a 69.6% decrease) when exposed to 50 mM NaCl, which was a much larger drop compared to the salt-tolerant variety Jindao 919, which only decreased by 54.7% (from 10.29 cm to 4.66 cm). Plant height, fresh weight, and dry weight indicators simultaneously corroborated this trend: the corresponding indicators of Jinchuan No. 1 decreased by 69.1%, 52.4%, and 13.4%, while those of Jindao 919 decreased by only 16.4%, 14.2%, and 4.0%, which highlighted the innate resistance advantage of salt-tolerant genotypes through ionic compartmentalisation and osmoregulation mechanisms.

Table 5. Effects of SA and MT on plant morphogenesis of rice under salt stress.

Cultivar	Treatment	Root Length (RL, cm)	Plant Height (PH, cm)	Fresh Weight (FW, mg)	Dry Weight (DW, mg)	Stem Diameter (SD, mm)
Jinchuan No. 1	CK	14.89 ± 1.37 a	17.78 ± 0.38 b	218.66 ± 4.61 a	22.00 ± 0.00 a	1.72 ± 0.14 cd
	NaCl	4.52 ± 0.84 cd	13.10 ± 1.76 efg	110.25 ± 37.62 ef	14.66 ± 0.57 e	1.66 ± 0.13 d
	S1	8.36 ± 0.70 b	15.44 ± 1.07 cde	128.00 ± 5.56 cd	19.66 ± 0.57 b	1.92 ± 0.12 bcd
	S2	10.63 ± 0.46 ab	15.69 ± 1.09 cde	135.66 ± 1.52 c	20.33 ± 0.57 ab	2.60 ± 0.20 a
	S3	5.36 ± 0.28 cd	16.97 ± 0.13 bc	141.33 ± 3.21 bc	22.33 ± 0.57 a	2.89 ± 0.18 a
	S4	4.79 ± 0.89 cd	14.61 ± 0.82 def	113.33 ± 4.16 ef	18.00 ± 1.00 bc	3.21 ± 0.34 a
	S5	4.05 ± 0.27 d	12.12 ± 0.65 fg	92.33 ± 24.58 fg	16.00 ± 1.00 cd	1.36 ± 0.13 e
	M1	9.67 ± 1.88 ab	14.23 ± 1.98 def	154.00 ± 19.16 b	17.00 ± 1.00 cd	1.97 ± 0.09 bc
	M2	5.82 ± 0.61 cd	23.10 ± 2.15 a	129.25 ± 26.66 cd	16.66 ± 0.57 cd	2.05 ± 0.21 ab
	M3	5.23 ± 0.97 cd	16.00 ± 1.74 bcd	119.00 ± 9.30 de	15.33 ± 0.57 de	2.51 ± 0.24 a
	M4	4.81 ± 0.66 cd	14.55 ± 2.27 def	111.75 ± 18.57 ef	13.00 ± 1.00 ef	2.48 ± 0.13 a
	M5	4.25 ± 1.02 d	10.86 ± 2.71 g	90.25 ± 20.12 g	11.33 ± 1.15 f	1.78 ± 0.14 cd
Jindao 919	CK	10.29 ± 0.56 a	13.82 ± 0.42 cd	149.66 ± 17.03 a	21.33 ± 2.51 ab	1.85 ± 0.09 f
	NaCl	4.66 ± 0.14 d	11.87 ± 1.20 ef	99.00 ± 16.69 e	14.66 ± 1.52 f	1.85 ± 0.08 f
	S1	7.14 ± 0.46 b	13.70 ± 0.46 cd	122.66 ± 8.14 cd	20.00 ± 1.00 bc	2.54 ± 0.18 de
	S2	7.49 ± 0.54 b	15.13 ± 0.36 b	138.33 ± 5.13 ab	21.00 ± 1.00 ab	3.05 ± 0.22 bc
	S3	4.97 ± 0.70 d	16.92 ± 0.60 a	147.00 ± 7.54 a	23.00 ± 1.00 a	3.44 ± 0.21 ab
	S4	4.73 ± 0.26 d	14.48 ± 0.60 bc	130.00 ± 9.53 bc	18.13 ± 0.80 cd	2.72 ± 0.21 cd
	S5	4.42 ± 0.21 d	11.04 ± 0.64 fg	122.66 ± 4.72 cd	17.66 ± 2.08 de	1.84 ± 0.17 f
	M1	4.83 ± 1.02 d	12.49 ± 1.63 de	113.00 ± 7.48 d	16.33 ± 1.52 e	2.64 ± 0.38 cd
	M2	6.47 ± 0.43 c	12.60 ± 1.59 de	121.00 ± 13.24 cd	17.33 ± 0.57 de	2.69 ± 0.30 cd
	M3	7.37 ± 1.15 b	16.63 ± 0.47 a	147.00 ± 16.79 a	18.00 ± 1.00 cd	3.14 ± 0.82 abc
	M4	4.86 ± 0.37 d	10.48 ± 0.64 g	119.75 ± 15.08 cd	18.66 ± 1.15 cd	3.70 ± 0.21 a
	M5	4.58 ± 0.04 d	7.36 ± 2.26 h	105.25 ± 35.96 de	14.66 ± 0.57 f	2.24 ± 0.11 e

The data in the table are the mean ± standard deviation, $n = 3$. Different lowercase letters in the same column indicate significant differences at $p < 0.05$; the same letter indicates no significant difference ($p > 0.05$).

The exogenous SA treatments showed distinct varietal suitability: The salt-tolerant variety Jindao 919 grew taller and heavier, reaching 16.92 cm and 23.00 mg under the S3 treatment, which was the highest among all treatments. Meanwhile, the salt-sensitive Jinchuan No. 1 returned its dry weight to the normal level of 22.33 mg and increased its stem thickness to 2.89 mm under the same treatment. This indicates that SA greatly improves its ability to adapt by activating certain metabolic processes, and that SA helps root growth more effectively than MT in both varieties, possibly due to its role in regulating signalling. In contrast, MT treatment showed a complicated interaction between concentration and genotype. In contrast, MT treatment showed a complex concentration-genotype interaction pattern. The salt-tolerant variety Jindao 919 synergistically increased plant height (16.63 cm) and stem thickness (3.14 mm) at M3 concentration, while Jinchuan No. 1 plant height soared to 23.10 cm at M2 concentration (e.g., 300 μM MT), which was 28.6% higher than that of other treatment groups. However, MT showed a weak improvement in root length and fresh weight with significant concentration fluctuations, which may be limited by genotype-specific melatonin receptor abundance or membrane transport efficiency.

3.5. Regulatory Effects of Different Exogenous Substances on Physiological Metabolism in Salt-Stressed Rice Seedlings

3.5.1. Antioxidant Enzyme System Response

Salt stress markedly suppressed antioxidant enzyme activities in both salt-tolerant Jindao 919 and salt-sensitive Jinchuan No. 1, though SA and MT exhibited mechanistically distinct remediation patterns across cultivars (Figure 1). Under salinity, Jindao 919 showed 18.7%, 22.1%, and 15.3% reductions in SOD, Peroxidase (POD), and CAT activities, while Jinchuan No. 1 suffered greater declines (SOD: −26.5%, POD: −31.2%, CAT: −20.8%), confirming heightened vulnerability of stress-sensitive genotypes' antioxidant systems.

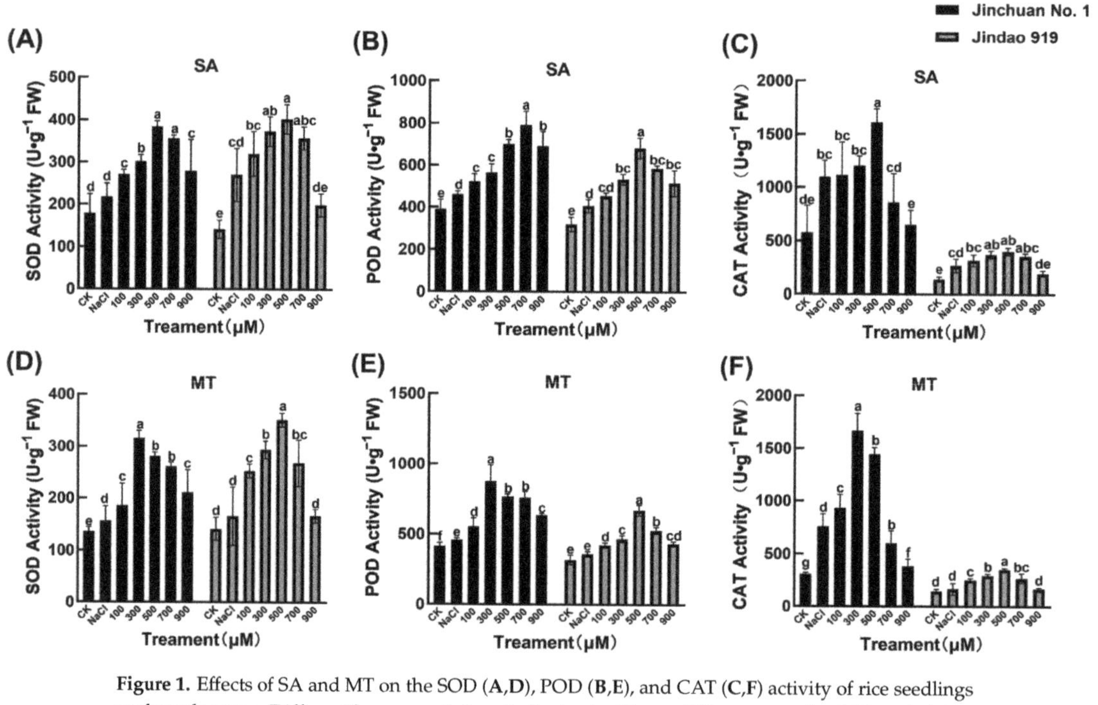

Figure 1. Effects of SA and MT on the SOD (**A**,**D**), POD (**B**,**E**), and CAT (**C**,**F**) activity of rice seedlings under salt stress. Different lowercase letters indicate significant differences at the 0.05 probability level ($p < 0.05$), determined by one-way analysis of variance (ANOVA) and Tukey's HSD post hoc test for significance. The vertical bar chart represents the mean ± standard deviation (SD) calculated from three repetitions.

The physiological indices of Jinchuan No. 1 and Jindao 919 exhibited significant differences under SA treatment (Figure 1A–C). Several indices of Jinchuan No. 1 showed a tendency of increasing and then decreasing, suggesting that low concentrations of SA may activate its defence or metabolism pathway, whereas high concentrations of SA triggered the inhibitory effect. In contrast, some indicators of Jindao 919 consistently increased or stayed the same as SA concentration went up, showing a more cautious way of adapting to SA. The differences between the two varieties may stem from the differences in genotypic sensitivity to the SA signalling pathway. In the MT treatment group (Figure 1D–F), Jinchuan No. 1 showed a stronger dynamic response, such as obvious fluctuations in CAT after MT treatment. In contrast, Jindao 919 was relatively stable with small changes in various indexes, such as the CAT level, indicating that its cellular homeostatic mechanism has a high inhibitory capacity against the disturbance of MT, and this difference may be related to the basal activity levels of antioxidant enzyme systems (e.g., SOD, POD) or melatonin receptor expression of the two species. Critically, antioxidant regulation diverged by genotype-hormone interaction: SA preferentially enhanced SOD in Jindao 919 (+76.9%), while MT maximally activated CAT in Jinchuan No. 1 (+37.2%). This indicates SA's dominance in reinforcing endogenous defences of tolerant varieties, whereas MT compensates for defective ROS scavenging in sensitive lines.

3.5.2. GSH and Soluble Sugar Content Dynamics

Salinity stress-induced metabolic constraints further elucidated the mechanistic divergence of SA and MT interventions across rice cultivars (Figure 2A–D). The salt-tolerant Jindao 919 exhibited a 41.5% reduction in glutathione (GSH) content under stress, whereas the salt-sensitive Jinchuan No. 1 suffered greater depletion (53.2%), confirming compromised antioxidant buffering capacity in stress-vulnerable genotypes.

The exogenous SA significantly enhanced the GSH biosynthesis capacity of the salt-tolerant variety Jindao 919 by targeting the activation of key glutathione synthesis genes and the GSH content of the salt-tolerant variety Jindao 919 was restored to 89.3% of the control level under the treatment of S3, which was much higher than that of the salt-sensitive variety Jinchuan No. 1, which was 45.3%, revealing the molecular advantage of salt-tolerant genotypes to maintain the redox homeostasis through the highly efficient antioxidant system. advantage. Complementarily, MT preferentially alleviated osmotic stress in Jinchuan No. 1 through the soluble sugar synthesis pathway, and its soluble sugar content surged to 104.33 mg g^{-1} (126.8% higher than the control) under M5 treatment, while that of Jindao 919 increased by only 33.4%, suggesting that the sensitive cultivar responded rapidly to osmotic imbalance through reprogramming of sugar metabolism, but the high SA concentration exposed the osmotic stress in Jinchuan No. 1 to a higher level of osmosis. However, the high concentration of SA exposed the metabolic vulnerability of Jinchuan No. 1: its soluble sugar content fell back by 11.1% from the peak, which was more than the 13.5% decrease in Jindao 919, suggesting that the buffering capacity of the sensitive genotypes against exogenous interventions was limited by the insufficient redundancy of metabolic networks.

3.5.3. Lignin Biosynthesis and MDA Accumulation

Salt stress-induced membrane lipid peroxidation (MDA) and lignification processes showed significant genotypic differentiation (Figure 2E–H), with the MDA content of the salt-tolerant variety Jindao 919 elevated by 76.9% under salt stress, whereas the salt-sensitive variety Jinchuan No. 1 showed an increase of 92.4% in MDA.

Exogenous regulatory pathway analysis revealed that MT and SA repaired the damage through differentiated mechanisms. SA preferentially alleviated cell wall lignification in Jindao 919, and the S3 treatment suppressed its lignin content to 1.2-fold of the salt stress level (23.1% reduction), far exceeding the 8.7% improvement in Jinchuan No. 1. The ability of MT to directly scavenge ROS was prominent in Jinchuan No. 1, and the M5 treatment reduced its MDA content by 55.7%, which was significantly better than the 42.3% reduction in Jindao 919. In addition high-dose exogenous treatments exposed the metabolic vulnerability boundary between the two varieties: the S5 concentration triggered a rebound of lignin content in Jinchuan No. 1 to 1.1-fold of the salt-stress level (10.2% increase), whereas in Jindao 919 it only slightly increased by 4.3%; similarly the genotypic differences in the efficacy of ROS scavenging by excess MT (M5) continued to be amplified, with MDA inhibition in Jindao 919 attenuating to 28.5%, whereas Jinchuan No. 1 still maintained a significant improvement of 44.6%, confirming that the redox homeostasis of salt-tolerant varieties is more susceptible to disturbances by dose overload.

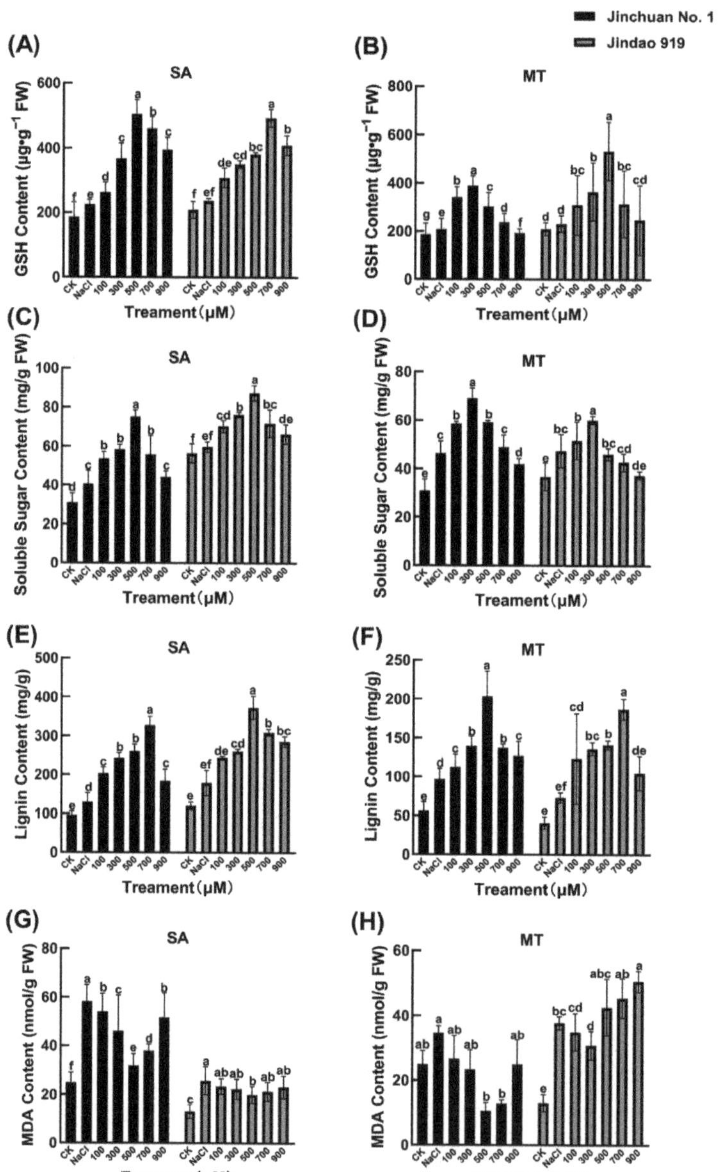

Figure 2. Effects of SA and MT on the GSH (**A**,**B**), soluble sugar (**C**,**D**), lignin (**E**,**F**), and MDA (**G**,**H**) content of rice seedlings under salt stress. Different lowercase letters indicate significant differences at the 0.05 probability level ($p < 0.05$), determined by one-way analysis of variance (ANOVA) and Tukey's HSD post hoc test for significance. The vertical bar chart represents the mean ± standard deviation (SD) calculated from three repetitions.

3.6. Correlation Data and Principal Component Analysis

Correlation analysis revealed significant positive correlations ($p < 0.05$) between physiological parameters and exogenous SA/MT treatments in Jinchuan No. 1, indicating effective alleviation of salt stress inhibition through hormonal regulation (Figure 3A). In contrast, Jindao 919 exhibited weaker correlations across most parameters, with only marginal positive associations between certain antioxidant enzymes and MT treatments, suggesting limited salt tolerance enhancement (Figure 3B). Cluster analysis further distinguished the response patterns between cultivars: Jinchuan No. 1′s parameters aggregated into a high-response cluster with strong SA-MT treatment linkage, demonstrating robust synergistic regulation, whereas Jindao 919′s indicators dispersed in low-response clusters, showing blurred treatment boundaries, reflecting inefficient hormonal signalling (Figure 3C,D).

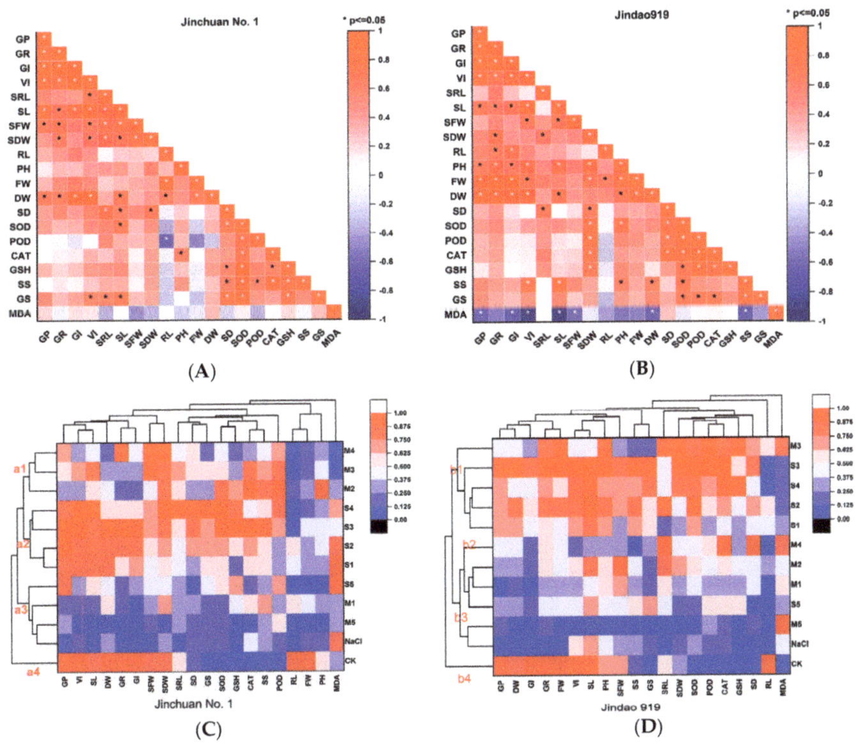

Figure 3. Correlation analysis (**A**,**B**) and cluster analysis (**C**,**D**) of the seedling growth rate, seedling biomass, plant morphogenesis, and physiological indexes of rice seeds. * indicates significant differences at $p < 0.05$ levels. Different red lowercase letters indicate different categories in cluster analysis.

Notably, SA preferentially enhanced osmoregulatory substance accumulation, while MT predominantly strengthened antioxidant defence systems, exhibiting complementary mitigation effects. By comparison, Jindao 919 displayed attenuated responses to both hormones, particularly showing non-significant improvements in osmoregulation. These findings align with the cultivars' differential salt tolerance capacities, revealing that SA/MT-mediated mitigation efficacy fundamentally depends on genetic predisposition to stress adaptation.

3.7. Comprehensive Evaluation Methodology Analysis

PCA systematically elucidated cultivar-specific differences in physiological and morphological regulation pathways underlying salt tolerance in rice varieties (Jindao 919 and Jinchuan No. 1) under exogenous SA and MT treatments (Supplementary Materials S2, Tables S1 and S2). The salt-tolerant Jindao 919 exhibited superior performance, with its first principal component (PC1, 51.37% variance) predominantly governed by high-loading germination parameters (germination potential [GP, 0.850] and germination rate [GR, 0.901]) and antioxidant enzymes (SOD 0.741, POD 0.603, CAT 0.592), demonstrating core salt tolerance mechanisms through synergistic seed vigour and endogenous antioxidant systems. The secondary component (PC2, 21.75%) featured Stem Diameter (SD, 0.669) and glutathione (GSH, 0.650), indicating auxiliary regulation via morphological development and redox homeostasis. In contrast, the salt-sensitive Jinchuan No. 1 showed compromised tolerance, with its PC1 (44.70% variance) dominated by VI (0.896), FW (0.804), and osmolytes (proline, 0.558), suggesting greater reliance on biomass accumulation and osmotic adjustment. PC2 (22.80%) displayed significant loadings for RL (0.875) and MDA (−0.083), highlighting root impairment and membrane peroxidation as key limiting factors under salinity. PCA resolved hormone-genotype specificity: Jindao 919 achieved peak mitigation via SA-activated antioxidant genes, whereas Jinchuan No. 1 relied on MT-driven osmoprotection. This quantifies the core dichotomy: antioxidant reinforcement dominates tolerant varieties, and osmotic adjustment prevails in sensitive lines.

Exogenous treatment showed different regulatory pathways: Jindao 919 exhibited higher SA sensitivity by directly activating antioxidant enzyme genes (such as SOD and CAT), and the comprehensive index reached the maximum value at S3. On the contrary, Jinchuan No.1 has a priority response to MT (optimal M5) and shows enhanced tolerance through the elimination of ROS and proline biosynthesis. Although SA has a certain improvement effect on the RL of Jinchuan No. 1, its effect is still not as good as that of MT. Meanwhile, Jindao 919 shows a limited morphological response to MT and mainly relies on its inherent antioxidant capacity to alleviate salinity.

A comprehensive evaluation resolved that salt stress significantly suppressed the overall performance of both varieties, but there were significant differences in the intensity of their genotypic responses (Table 6). The ranking of the D value of the salt-sensitive variety Jinchuan No. 1 plummeted from seventh to 11th place in the control (CK) under NaCl treatment, while the salt-tolerant variety Jindao 919 was suppressed (ranked from 3rd to 11th place in CK), but the decrease was relatively moderate, confirming the stability of its endogenous salt-tolerance mechanism. The exogenous SA and MT treatments showed a dose-dependent regulation: Jinchuan No. 1 was the top D-value in the M3 treatment, with a synergistic increase in plant height and biomass, while Jindao 919 reached the peak D-value in the S3 treatment (ranked first), with a significant optimization of photosynthetic efficiency and ionic homeostatic indexes. Under the control condition, Jindao 919 showed basic salt tolerance (D value ranked third), and its cellular osmoregulation and antioxidant system preconditioning capacity were the key; whereas, the D value of Jinchuan No. 1 (ranked seventh) declined precipitously under salt stress, revealing the inherent defects of its ion efflux and ROS scavenging mechanism. After salt stress, Jindao 919 maintained ion homeostasis by rapidly up-regulating the expression of salt-sensitive genes, whereas the lag in the activation of the SOS pathway in Jinchuan No. 1 resulted in the accumulation of ion toxicity, and this genotypic difference provides a key phenotypic marker for breeding for salt tolerance.

Table 6. Comprehensive index evaluation of exogenous substances' capacity to alleviate salt stress.

Cultivar	Treatment	Membership Function				D Value	Rank
Jinchuan No. 1	CK	0.636	0.000	0.877	0.366	0.452	7
	NaCl	0.157	0.552	0.013	0.623	0.240	11
	S1	0.565	0.511	0.196	0.131	0.432	8
	S2	0.785	0.620	0.414	0.405	0.611	3
	S3	1.000	0.770	0.102	0.130	0.701	2
	S4	0.733	0.804	0.043	0.521	0.588	4
	S5	0.405	0.661	0.000	0.397	0.376	10
	M1	0.294	0.704	0.422	0.763	0.400	9
	M2	0.535	0.691	0.489	1.000	0.543	5
	M3	0.656	1.000	1.000	0.595	0.708	1
	M4	0.411	0.958	0.781	0.068	0.515	6
	M5	0.000	0.725	0.505	0.000	0.217	12
Jindao 919	CK	0.620	1.000	0.230	0.406	0.571	3
	NaCl	0.003	0.251	0.796	0.651	0.215	11
	S1	0.583	0.423	0.965	0.500	0.518	4
	S2	0.758	0.396	0.908	0.507	0.584	2
	S3	1.000	0.173	0.808	0.618	0.640	1
	S4	0.912	0.076	0.662	0.022	0.504	5
	S5	0.340	0.185	1.000	0.201	0.330	8
	M1	0.292	0.325	0.329	0.761	0.317	9
	M2	0.584	0.000	0.105	1.000	0.371	6
	M3	0.662	0.102	0.010	0.299	0.349	7
	M4	0.500	0.249	0.000	0.000	0.280	10
	M5	0.000	0.196	0.541	0.100	0.117	12

Among the SA treatments, the S3 treatment produced broad-spectrum gains in both varieties: Jinchuan No. 1 rose to the 2nd place in D value, with increased root vigour and proline content, while Jindao 919 strengthened the cell wall barrier through SA-mediated lignin deposition, with a stable first place in D value. In contrast, MT treatment showed a 'genotype-dose' interaction polarity: Jinchuan No. 1 relied on the MT-activated SOD/POD system for ROS scavenging in M3 treatment, while Jindao 919 had a discrete response to MT and even inhibited GSH synthesis at high concentrations (e.g., M5), suggesting that there was a saturating threshold of melatonin signalling in its antioxidant network. SA is superior to MT in terms of cross-species generalization, while MT is more adapted to the emergency repair needs of sensitive genotypes.

4. Discussion

Under the ongoing impact of global climate change, soil salinization is deteriorating steadily, presenting a major threat to agricultural productivity and food security. In contrast to previous work focusing on the universal salt tolerance effect of SA or MT [35,36], this study took Xiaozhan rice in the saline-alkaline land of Tianjin Binhai as the research object. For the first time, exogenous SA and MT were applied to the salt-tolerant variety, Jindao 919, and the salt-sensitive variety, Jinchuan No. 1, to systematically reveal the differential responses of these two varieties to SA and MT. This discovery not only offers novel insights into the mechanisms underlying plant salt tolerance but also establishes an important theoretical foundation for the sustainable cultivation of Xiaozhan rice in saline and alkaline environments. It holds immense economic and ecological significance [37].

When delving into the effects of exogenous SA and MT on root and root tip development of different rice varieties during the seedling growth stage, many findings were found to be related to previous studies yet showed differences and innovations, such as

SA performed better in restoring root length in the salt-sensitive variety Jinchuan No. 1. By contrast, MT performed better in increasing fresh weight of the salt-tolerant variety Jindao 919 [38,39]. Concurrently, while numerous researchers have studied the mitigation of abiotic stresses in plants by exogenous substances, fewer in-depth studies have been carried out on its specific expression among different varieties [40,41]. However, this group's observations reveal for the first time that SA enhances salt stress adaptation in sensitive genotypes by activating root developmental regulators, whereas MT attenuates biomass loss in sensitive genotypes by indirectly stimulating the biosynthesis of osmoprotectants. Additionally, through gradient experiments, the optimal mitigation concentration of SA and MT (500 μM) was clarified. Meanwhile, the potential risks of high-concentration treatments (e.g., the soluble sugar content of salt-sensitive varieties decreased by 11.1% at a high concentration of SA) were also revealed. This finding is distinct from previous studies that mainly focused on the effective concentration. It provides a dosage guideline for the safe and efficient field application of exogenous substances [41]. By doing so, it avoids the waste of resources or secondary stress due to blind application. Ultimately, it contributes to achieving the sustainable development objectives of green agriculture [42,43].

Salt stress exhibited significant inhibitory effects on the growth and development of different rice cultivars during the plant morphogenesis stage. Notably, the stress-relieving effects of exogenous SA and MT showed a distinct genotype dependence, breaking through the limitations of the traditional single-hormone or generic regulatory strategies. For example, the SOD activity of the salt-tolerant cultivar Jindao 919 was significantly increased by 76.9% after SA treatment—a phenomenon that is consistent with the mechanism by which SA and MT can alleviate oxidative stress by enhancing the activities of antioxidant enzymes, such as SOD and POD, in *Zea mays*, *barley* and other crops [24,44]. However, a fluctuating rebound effect of lignin and MDA content was observed in the salt-sensitive variety Jinchuan No. 1 at high doses of SA/MT (900 μM), whereas no similar phenomenon was observed in Jindao 919. In-depth analyses showed that Jinchuan No. 1 relied more on the MT-mediated inhibition of membrane lipid peroxidation to enhance salt tolerance (e.g., 55.7% reduction in MDA content) due to a weaker basal antioxidant system [45]. This is similar to the mechanism by which MT maintains membrane stability by inhibiting MDA accumulation in crops such as *cotton* and *maize*, also fits Foyer's theory of redox homeostasis in plants, in which exogenous MT preferentially maintains the integrity of the membrane system by specifically inhibiting the lipid peroxidation chain reaction when endogenous antioxidant capacity is insufficient [46–48]. In addition, Yin Yanling et al. demonstrated that MT can compensate for the defective antioxidant enzyme network of salt-sensitive genotypes by activating relateds [49]. In contrast, the enhancement of antioxidant defence capacity of the salt-tolerant variety Jindao 919 after SA treatment was more efficient, which is highly consistent with the physiological mechanism of 'SA alleviates stress by enhancing plant osmoregulation' proposed by Nasrin and Lei [50,51]. In conclusion, SA and MT exhibit complementary functions in mitigating salt stress by differentially regulating hormone homeostasis, the antioxidant system, and ion homeostasis. This not only challenged the traditional 'universal' regulatory paradigm but also provided a theoretical cornerstone for precision breeding and hormone-targeted application of genotypic characteristics [35].

This study further revealed the genotype-specific regulation of SA and MT by Principal Component Analysis. For instance, considerable studies confirmed that SA and MT could enhance antioxidant capacity, without quantifying their contribution weights in different genotypes [52,53]. SA dominated the integrated response of salt-tolerant varieties (D-value of 0.640), whereas MT dominated the ROS scavenging capacity of salt-sensitive varieties (D-value of 0.708). This finding forges a theoretical foundation for devising a synergistic multi-hormone regulatory strategy. By comparing the physiological metabolic differences between salt-tolerant and salt-sensitive varieties (e.g., the root regeneration capacity of salt-sensitive varieties was increased by 192.3%), the study echoed the idea of lignin's involvement in plant mechanical support put forward by Daqiu Zhao [54], which revealed the central role of genotype-hormone interactions in salt domestication. Moreover, herein, the analysis of the affiliation function was conducted, and the results revealed that SA and MT showed synergistic effects in salt-sensitive varieties but antagonistic in salt-tolerant varieties. This endeavour fills the gap in the study of salt tolerance mechanisms in *Japonica* rice by using SA/MT and for plant adversity biology (e.g., SA-mediated lignin deposition).

Given the escalating challenges of global soil salinity, improving crop salt tolerance has become a key priority [55,56]. The stress-relieving effects of exogenous SA and MT provide theoretical support for the advancement of saline precision agriculture management strategies [14,57–59]. The strategic application of these exogenous substances can effectively improve the growth performance and yield stability of Xiaozhan rice in saline environments, thereby ensuring food security [60]. In addition, this study provides an innovative method to study the mechanism of plant salt tolerance. Future studies should concentrate on deciphering the molecular regulatory networks of SA and MT during salt acclimation, especially their interactions with other stress signalling pathways [61,62]. While these studies will lay the foundation for breeding superior salt-tolerant crop varieties through targeted genetic manipulation, there are still some limitations warranting further consideration. First, this study focused on the physiological effects of exogenous SA and MT under short-term salt stress, while their mechanisms and effects under long-term salt stress were not explored. Long-term salt exposure may induce more complex physiological and molecular adaptations in plants, and further studies are necessitated to elucidate the long-term responses [63]. Second, the controlled climate chamber conditions used for salt stress simulation in saline soils differ significantly from the actual field environment. Field-specific variables, including soil texture, microbial communities, and climatic factors, may exert substantial impacts on plant growth and salt tolerance [64]. Thus, field validation of the results of SA and MT applications is required.

In conclusion, this study systematically revealed the differential mechanisms of exogenous SA and MT in alleviating salt stress and promoting seedling growth in Xiaozhan rice. These findings not only expand the theoretical framework of plant salt tolerance physiology but also propose novel strategies for addressing the issue of soil salinity in agricultural production. In future research endeavours, the emphasis should be placed on exploring the application potential of these exogenous substances across diverse environmental conditions and crop varieties. By doing so, it will facilitate the advancement of sustainable agricultural practices and contribute to green agriculture and environmental protection.

5. Conclusions

In the context of global climate change, the problem of soil salinity is becoming more and more serious, posing a serious threat to agricultural productivity and food security. In this study, we comprehensively investigated the differential mechanisms of exogenous SA and MT on the alleviation of salt stress in Xiaozhan rice seedlings by using a salt-tolerant variety (Jindao 919) and a salt-sensitive variety (Jinchuan No. 1). The results showed that 500 μM SA treatment significantly increased the germination rate, antioxidant enzyme activities (SOD elevated by 76.9%, CAT elevated by 47.1%), and the accumulation of protective substances (GSH increased by 124.7%, and soluble sugars increased by 86.1%) in the salt-sensitive variety, Jinchuan No. 1. Meanwhile, the same concentration of MT showed outstanding protective effects on cell membranes, reducing MDA content by 69.5%. These findings highlight the different but complementary roles of SA and MT in the enhancement of salt tolerance: SA improves plant salt tolerance mainly by enhancing the antioxidant defence system, whereas MT attenuates the damage caused by salt stress in plants mainly by enhancing membrane protection. This study provides practical green solutions for agricultural practices in saline environments, e.g., 500 μM SA soaked seeds or foliar sprays can not only significantly improve salt tolerance in Xiaozhan rice, but also reduce the use of chemical fertilizers and pesticides, which in turn reduces the negative impacts on the environment. Further research on the synergistic effects of salt stress and MT and their integration with molecular breeding methods to develop more salt stress-tolerant rice varieties will contribute to sustainable agricultural development in salinity-affected areas, ensure food security stability in the face of the challenge of increasing soil salinity, and make an important contribution to green agriculture and environmental protection.

Supplementary Materials: The following supporting information can be downloaded at: https://www.mdpi.com/article/10.3390/cimb47060432/s1.

Author Contributions: Y.W. and Y.D. designed the experiment; Y.W., H.G. and M.L. conducted the experiment; Y.W., Y.D. and X.L. wrote the manuscript; F.Z. and W.Z. (Wei Zhu) provided laboratory and other technical support; F.Z., W.Z. (Wenzhong Zhang) and J.L. critically reviewed and edited the manuscript. All authors have read and agreed to the published version of the manuscript.

Funding: This research was funded by the project supported by the National Key Research and Development Program of China (No. 2024YFD1201005), the Major Research Plan of Tianjin (No. 21YFSNSN00100).

Institutional Review Board Statement: Not applicable.

Informed Consent Statement: Not applicable.

Data Availability Statement: The data presented in this study are available in the figures and tables provided in the manuscript.

Acknowledgments: We appreciate and thank the anonymous reviewers for helpful comments that led to an overall improvement of the manuscript. We also thank the Journal Editor Board for their help and patience throughout the review process.

Conflicts of Interest: The authors declare no conflicts of interest.

Abbreviations

The following abbreviations are used in this manuscript:

SA	Salicylic acid
MT	Melatonin
GP	Germination potential
GR	Germination rate
GI	Germination Index
VI	Vigour index
SRL	Seedling Root Length
SL	Seedling Length
SFW	Seedling Fresh Weight
SDW	Seedling Dry Weight
RL	Root length
PH	Plant height
FW	Fresh weight
DW	Dry weight

SD	Stem Diameter
SOD	Superoxide dismutase
POD	Peroxidase
CAT	Catalase
GSH	Reduced glutathione
GS	Lignin
SS	Soluble sugar
MDA	Malondialdehyde
ROS	Reactive oxygen species
PCA	Principal Component Analysis
P5CS1	Delta-1-pyrroline-5-carboxylate synthetase 1
TGA1	TGACG sequence-specific binding proteins 1
ICS1	ISOCHORISMATE SYNTHASE1
HKT1	High-affinity K transporter 1

References

1. Qadir, M.; Quillérou, E.; Nangia, V.; Murtaza, G.; Singh, M.; Thomas, R.J.; Drechsel, P.; Noble, A.D. Economics of salt-induced land degradation and restoration. *Nat. Resour. Forum* **2014**, *38*, 282–295. [CrossRef]
2. Rengasamy, P. World salinization with emphasis on Australia. *J. Exp. Bot.* **2006**, *57*, 1017–1023. [CrossRef]
3. FAO. *Global Status of Salt-Affected Soils—Main Report*; FAO: Rome, Italy, 2024. [CrossRef]
4. Shao, H.; Chu, L.; Lu, H.; Qi, W.; Chen, X.; Liu, J.; Kuang, S.; Tang, B.; Won, V. Towards sustainable agriculture for the salt-affected soil. *Land Degrad. Dev.* **2019**, *30*, 574–579. [CrossRef]
5. Mohanavelu, A.; Naganna, S.R.; Al-Ansari, N. Irrigation Induced Salinity and Sodicity Hazards on Soil and Groundwater: An Overview of Its Causes, Impacts and Mitigation Strategies. *Agriculture* **2021**, *11*, 983. [CrossRef]
6. Li, J.G.; Pu, L.J.; Han, M.F.; Zhu, M.; Zhang, R.S.; Xiang, Y.Z. Soil salinization research in China: Advances and prospects. *J. Geogr. Sci.* **2014**, *24*, 943–960. [CrossRef]
7. Guo, Z.D.; Zhang, X.J.; Liu, Y.R. Researches on the coal fly ash applied to saline soil improvement in rural roads. In Proceedings of the 2nd International Conference on Civil, Architectural and Hydraulic Engineering (ICCAHE 2013), Zhuhai, China, 27–28 July 2013; p. 2616.
8. Chen, S.; Guo, J.; Zhao, Y.; Li, X.; Liu, F.; Chen, Y. Evaluation and grading of climatic conditions on nutritional quality of rice: A case study of Xiaozhan rice in Tianjin. *Meteorol. Appl.* **2021**, *28*, e2021. [CrossRef]
9. Zhao, H.X.; Gu, B.J.; Chen, D.C.; Tang, J.J.; Xu, X.L.; Qiao, Z.; Wang, J.Q. Physicochemical properties and salinization characteristics of soils in coastal land reclamation areas: A case study of China-Singapore Tianjin Eco-City. *Heliyon* **2022**, *8*, e12629. [CrossRef]
10. Rhaman, M.S.; Imran, S.; Rauf, F.; Khatun, M.; Baskin, C.C.; Murata, Y.; Hasanuzzaman, M. Seed Priming with Phytohormones: An Effective Approach for the Mitigation of Abiotic Stress. *Plants* **2021**, *10*, 37. [CrossRef] [PubMed]
11. Waadt, R.; Seller, C.A.; Hsu, P.-K.; Takahashi, Y.; Munemasa, S.; Schroeder, J.I. Plant hormone regulation of abiotic stress responses. *Nat. Rev. Mol. Cell Biol.* **2022**, *23*, 680–694. [CrossRef]
12. Nasircilar, A.G.; Erkaymaz, T.; Ulukapi, K. Reflection of the synergistic/antagonistic effects of melatonin and salicylic acid on the biochemical profile of *Allium cepa* L. under drought stress. *South Afr. J. Bot.* **2024**, *166*, 1–13. [CrossRef]
13. Zulfiqar, F.; Moosa, A.; Ferrante, A.; Nafees, M.; Darras, A.; Nazir, M.M.; AlShaqhaa, M.A.; Elsaid, F.G. Melatonin and salicylic acid synergistically improve arsenic induced oxidative stress tolerance in ornamental sword lily. *Sci. Hortic.* **2023**, *322*, 112389. [CrossRef]
14. Karimi, M.R.; Sabokdast, M.; Korang Beheshti, H.; Abbasi, A.R.; Bihamta, M.R. Seed priming with salicylic acid enhances salt stress tolerance by boosting antioxidant defense in Phaseolus vulgaris genotypes. *BMC Plant Biol.* **2025**, *25*, 489. [CrossRef]
15. Hasanuzzaman, M.; Bhuyan, M.; Zulfiqar, F.; Raza, A.; Mohsin, S.M.; Al Mahmud, J.; Fujita, M.; Fotopoulos, V. Reactive Oxygen Species and Antioxidant Defense in Plants under Abiotic Stress: Revisiting the Crucial Role of a Universal Defense Regulator. *Antioxidants* **2020**, *9*, 681. [CrossRef]
16. Banik, S.; Dutta, D. Membrane Proteins in Plant Salinity Stress Perception, Sensing, and Response. *J. Membr. Biol.* **2023**, *256*, 109–124. [CrossRef]
17. Xiao, W.; Zhang, Y.; Wang, Y.; Zeng, Y.; Shang, X.; Meng, L.; Zhang, Y.; Fang, T.; Xiao, P.; Qu, J.; et al. The transcription factor TGA2 orchestrates salicylic acid signal to regulate cold-induced proline accumulation in Citrus. *Plant Cell* **2024**, *37*, koae290. [CrossRef]
18. Andronachi, V.-C.; Simeanu, C.; Matei, M.; Radu-Rusu, R.-M.; Simeanu, D. Melatonin: An Overview on the Synthesis Processes and on Its Multiple Bioactive Roles Played in Animals and Humans. *Agriculture* **2025**, *15*, 273. [CrossRef]

19. Spolaor, B.O.; Bertoli, S.C.; Sukert, D.S.; Sala, H.R.; Picoli de Oliveira, B.F.; de Freitas, I.R.; Lima-Moro, A. Exogenous melatonin induces tolerance to drought stress damage in seedlings and soybean plants. *Chil. J. Agric. Res.* **2022**, *82*, 515–526. [CrossRef]
20. Ahmad, J.; Hayat, F.; Khan, U.; Ahmed, N.; Li, J.; Ercisli, S.; Iqbal, S.; Javed, H.U.; Alyas, T.; Tu, P.; et al. Melatonin: A promising approach to enhance abiotic stress tolerance in horticultural plants. *South Afr. J. Bot.* **2024**, *164*, 66–76. [CrossRef]
21. Kumar, S.; Liu, Y.; Wang, M.; Khan, M.N.; Wang, S.; Li, Y.; Chen, Y.; Zhu, G. Alleviating sweetpotato salt tolerance through exogenous glutathione and melatonin: A profound mechanism for active oxygen detoxification and preservation of photosynthetic organs. *Chemosphere* **2024**, *350*, 141120. [CrossRef]
22. Powers, J.; Zhang, X.; Reyes, A.; Zavaliev, R.; Ochakovski, R.; Xu, S.L.; Dong, X.N. Next-generation mapping of the salicylic acid signaling hub and transcriptional cascade. *Mol. Plant* **2024**, *17*, 1558–1572. [CrossRef]
23. Tahjib-Ul-Arif, M.; Zahan, I.; Hossain, M.S.; Imran, S.; Hasanuzzaman, M.; Dawood, M.F.A.; Dawood, A.F.A.; Asaduzzaman, M.; Rhaman, M.S.; Souri, Z.; et al. Melatonin-mediated ionic homeostasis in plants: Mitigating nutrient deficiency and salinity stress. *Discov. Plants* **2025**, *2*, 143. [CrossRef]
24. Wang, J.; Yan, D.; Liu, R.; Wang, T.; Lian, Y.; Lu, Z.; Hong, Y.; Wang, Y.; Li, R. The Physiological and Molecular Mechanisms of Exogenous Melatonin Promote the Seed Germination of Maize (*Zea mays* L.) under Salt Stress. *Plants* **2024**, *13*, 2142. [CrossRef]
25. Hao, S.; Wang, Y.; Yan, Y.; Liu, Y.; Wang, J.; Chen, S. A Review on Plant Responses to Salt Stress and Their Mechanisms of Salt Resistance. *Horticulturae* **2021**, *7*, 132. [CrossRef]
26. Zhao, X.; Lin, S.; Yu, S.; Zhang, Y.; Su, L.; Geng, L.; Cheng, C.; Jiang, X. Exogenous calcium enhances the physiological status and photosynthetic capacity of rose under drought stress. *Hortic. Plant J.* **2024**, *10*, 853–865. [CrossRef]
27. *GB/T 3543.4-1995*; Rules for agricultural seed testing—Germination test. National Standard of the People's Republic of China; S.A.O.: Beijing, China, 1995.
28. Doerge, D.R.; Divi, R.L.; Churchwell, M.I. Identification of the colored guaiacol oxidation product produced by peroxidases. *Anal. Biochem.* **1997**, *250*, 10–17. [CrossRef]
29. Peskin, A.V.; Winterbourn, C.C. A microtiter plate assay for superoxide dismutase using a water-soluble tetrazolium salt (WST-1). *Clin. Chim. Acta* **2000**, *293*, 157–166. [CrossRef]
30. Johansson, L.H.; Borg, L.A. A spectrophotometric method for determination of catalase activity in small tissue samples. *Anal. Biochem.* **1988**, *174*, 331–336. [CrossRef]
31. Spitz, D.R.; Oberley, L.W. An assay for superoxide dismutaze activity in mammalian tissue homogenates. *Anal. Biochem.* **1989**, *179*, 8–18. [CrossRef]
32. Owens, C.W.; Belcher, R.V. A Colorimetric Micro-Method for the Determination of Glutathione. *Biochem. J.* **1965**, *94*, 705–711. [CrossRef]
33. Bodelon, O.G.; Blanch, M.; Sanchez-Ballesta, M.T.; Escribano, M.I.; Merodio, C. The effects of high CO_2 levels on anthocyanin composition, antioxidant activity and soluble sugar content of strawberries stored at low non-freezing temperature. *Food Chem.* **2010**, *122*, 673–678. [CrossRef]
34. Janshekar, H.; Brown, C.; Fiechter, A. Determination of biodegraded lignin by ultraviolet spectrophotometry. *Anal. Chim. Acta* **1981**, *130*, 81–91. [CrossRef]
35. Hayat, Q.; Hayat, S.; Irfan, M.; Ahmad, A. Effect of exogenous salicylic acid under changing environment: A review. *Environ. Exp. Bot.* **2010**, *68*, 14–25. [CrossRef]
36. Nawaz, M.A.; Huang, Y.; Bie, Z.; Ahmed, W.; Reiter, R.J.; Niu, M.; Hameed, S. Melatonin: Current Status and Future Perspectives in Plant Science. *Front. Plant Sci.* **2016**, *6*, 1230. [CrossRef]
37. Shaki, F.; Maboud, H.E.; Niknam, V. Effects of salicylic acid on hormonal cross talk, fatty acids profile, and ions homeostasis from salt-stressed safflower. *J. Plant Interact.* **2019**, *14*, 340–346. [CrossRef]
38. Imran, M.; Widemann, E.; Shafiq, S.; Bakhsh, A.; Chen, X.; Tang, X. Salicylic Acid and Melatonin Synergy Enhances Boron Toxicity Tolerance via AsA-GSH Cycle and Glyoxalase System Regulation in Fragrant Rice. *Metabolites* **2024**, *14*, 520. [CrossRef]
39. Sidek, N.; Nulit, R.; Yap, C.K.; Yong, C.S.Y.; Sekeli, R. In vitro development of salt tolerant *Malaysian indica* rice 'MARDI Siraj 297' and enhancement of salinity tolerance using salicylic acid. *Chil. J. Agric. Res.* **2024**, *84*, 3–14. [CrossRef]
40. Saharan, B.S.; Brar, B.; Duhan, J.S.; Kumar, R.; Marwaha, S.; Rajput, V.D.; Minkina, T. Molecular and Physiological Mechanisms to Mitigate Abiotic Stress Conditions in Plants. *Life-Basel* **2022**, *12*, 1634. [CrossRef]
41. Swain, R.; Sahoo, S.; Behera, M.; Rout, G.R. Instigating prevalent abiotic stress resilience in crop by exogenous application of phytohormones and nutrient. *Front. Plant Sci.* **2023**, *14*, 1104874. [CrossRef]
42. Vatanparast, M.; Merkel, L.; Amari, K. Exogenous Application of dsRNA in Plant Protection: Efficiency, Safety Concerns and Risk Assessment. *Int. J. Mol. Sci.* **2024**, *25*, 6530. [CrossRef]
43. Saitanis, C.J.; Agathokleous, E. Exogenous application of chemicals for protecting plants against ambient ozone pollution: What should come next? *Curr. Opin. Environ. Sci. Health* **2021**, *19*, 100215. [CrossRef] [PubMed]
44. Tu, Y.; Fu, L.; Wang, F.; Wu, D.; Shen, Q.; Zhang, G. GWAS and transcriptomic integrating analysis reveals key salt-responding genes controlling Na+ content in barley roots. *Plant Physiol. Biochem.* **2021**, *167*, 596–606. [CrossRef] [PubMed]

45. Ulhassan, Z.; Huang, Q.; Gill, R.A.; Ali, S.; Mwamba, T.M.; Ali, B.; Hina, F.; Zhou, W. Protective mechanisms of melatonin against selenium toxicity in *Brassica napus*: Insights into physiological traits, thiol biosynthesis and antioxidant machinery. *BMC Plant Biol.* **2019**, *19*, 1–16. [CrossRef] [PubMed]
46. Foyer, C.H.; Noctor, G. Redox homeostasis and antioxidant signaling: A metabolic interface between stress perception and physiological responses. *Plant Cell* **2005**, *17*, 1866–1875. [CrossRef]
47. Colombage, R.; Singh, M.B.; Bhalla, P.L. Melatonin and Abiotic Stress Tolerance in Crop Plants. *Int. J. Mol. Sci.* **2023**, *24*, 7447. [CrossRef]
48. Awan, S.A.; Khan, I.; Wang, Q.; Gao, J.; Tan, X.; Yang, F. Pre-treatment of melatonin enhances the seed germination responses and physiological mechanisms of soybean (*Glycine max* L.) under abiotic stresses. *Front. Plant Sci.* **2023**, *14*, 1149873. [CrossRef]
49. Yin, Y.; Yang, T.; Li, S.; Li, X.; Wang, W.; Fan, S. Transcriptomic analysis reveals that methyl jasmonate confers salt tolerance in alfalfa by regulating antioxidant activity and ion homeostasis. *Front. Plant Sci.* **2023**, *14*, 1258498. [CrossRef]
50. Razmi, N.; Ebadi, A.; Daneshian, J.; Jahanbakhsh, S. Salicylic acid induced changes on antioxidant capacity, pigments and grain yield of soybean genotypes in water deficit condition. *J. Plant Interact.* **2017**, *12*, 457–464. [CrossRef]
51. Wang, L.; Tanveer, M.; Wang, H.; Arnao, M.B. Melatonin as a key regulator in seed germination under abiotic stress. *J. Pineal Res.* **2024**, *76*, e12937. [CrossRef]
52. Bhattacharjee, A.; Debnath, S.; Sikdar, P.; Bhattacharya, K.; Chanu, N.R. Melatonin in Plants: Biosynthesis, Occurrence and Role in plants. In *Melatonin: Role in Plant Signaling, Growth and Stress Tolerance: Phytomelatonin in Normal and Challenging Environments*; Mukherjee, S., Corpas, F.J., Eds.; Springer International Publishing: Cham, Switzerland, 2023; Volume 4, pp. 29–44. [CrossRef]
53. Faize, L.; Faize, M. Functional Analogues of Salicylic Acid and Their Use in Crop Protection. *Agronomy* **2018**, *8*, 5. [CrossRef]
54. Zhao, D.; Luan, Y.; Xia, X.; Shi, W.; Tang, Y.; Tao, J. Lignin provides mechanical support to herbaceous peony (*Paeonia lactiflora* Pall.) stems. *Hortic. Res.* **2020**, *7*, 213. [CrossRef]
55. Yang, M.; Lu, K.; Zhao, F.-J.; Xie, W.; Ramakrishna, P.; Wang, G.; Du, Q.; Liang, L.; Sun, C.; Zhao, H.; et al. Genome-Wide Association Studies Reveal the Genetic Basis of Ionomic Variation in Rice. *Plant Cell* **2018**, *30*, 2720–2740. [CrossRef] [PubMed]
56. Sackey, O.K.; Feng, N.; Mohammed, Y.Z.; Dzou, C.F.; Zheng, D.; Zhao, L.; Shen, X. A comprehensive review on rice responses and tolerance to salt stress. *Front. Plant Sci.* **2025**, *16*, 1561280. [CrossRef]
57. Talaat, N.B.; Shawky, B.T. Synergistic Effects of Salicylic Acid and Melatonin on Modulating Ion Homeostasis in Salt-Stressed Wheat (*Triticum aestivum* L.) Plants by Enhancing Root H^+-Pump Activity. *Plants* **2022**, *11*, 416. [CrossRef] [PubMed]
58. Ahmadzai, A.S.; Hu, C.; Zhang, C.; Li, Y. Mechanisms of anthocyanin-mediated salt stress alleviation and cellular homeostasis in plants. *Plant Growth Regul.* **2025**, *105*, 655–673. [CrossRef]
59. Ali, M.; Malik, Z.; Abbasi, G.H.; Irfan, M.; Ahmad, S.; Ameen, M.; Ali, A.; Sohaib, M.; Rizwan, M.; Ali, S. Potential of melatonin in enhancing antioxidant defense system and yield of maize (*Zea mays* L.) hybrids under saline condition. *Sci. Hortic.* **2024**, *325*, 112665. [CrossRef]
60. Hanin, M.; Ebel, C.; Ngom, M.; Laplaze, L.; Masmoudi, K. New Insights on Plant Salt Tolerance Mechanisms and Their Potential Use for Breeding. *Front. Plant Sci.* **2016**, *7*, 1787. [CrossRef]
61. Justamante, M.S.; Larriba, E.; Luque, A.; Nicolás-Albujer, M.; Pérez-Pérez, J.M. A systematic review to identify target genes that modulate root system architecture in response to abiotic stress. *Sci. Rep.* **2025**, *15*, 13219. [CrossRef] [PubMed]
62. Rabeh, K.; Hnini, M.; Oubohssaine, M. A comprehensive review of transcription factor-mediated regulation of secondary metabolites in plants under environmental stress. *Stress Biol.* **2025**, *5*, 15. [CrossRef]
63. Munns, R.; Tester, M. Mechanisms of salinity tolerance. *Annu. Rev. Plant Biol.* **2008**, *59*, 651–681. [CrossRef]
64. Castro, C.B.; Whittock, L.D.; Whittock, S.P.; Leggett, G.; Koutoulis, A. DNA sequence and expression variation of hop (*Humulus lupulus*) valerophenone synthase (VPS), a key gene in bitter acid biosynthesis. *Ann. Bot.* **2008**, *102*, 265–273. [CrossRef]

current issues in molecular biology

MDPI

Article

Genome-Wide Characterization of the Heat Shock Transcription Factor Gene Family in *Begonia semperflorens* Reveals Promising Candidates for Heat Tolerance

Zhirou Liu [1], Nan Lin [2], Qirui Wang [1], Enkai Xu [1,*] and Kaiming Zhang [1,*]

[1] College of Landscape Architecture and Art, Henan Agricultural University, Zhengzhou 450002, China; liuzhirou@126.com (Z.L.)

[2] College of Life Sciences, Henan Agricultural University, Zhengzhou 450046, China

* Correspondence: xek1206@henau.edu.cn (E.X.); k.m.zhang@henau.edu.cn (K.Z.).; Tel.: +86-13673637626 (E.X.); +86-15039078696 (K.Z.)

Abstract: *Begonia semperflorens* (*B. semperflorens*) is a popular ornamental plant widely used in landscapes such as plazas and flower beds, and it is also commonly grown as a potted plant indoors. It is known for its adaptability to high temperatures, drought, and shade. Under heat-tolerant conditions, heat shock transcription factors (HSFs) are key transcriptional regulatory proteins that play crucial roles in cellular processes. Despite extensive studies on the *HSF* family in various species, there has been no specific analysis targeting *B. semperflorens*. In this study, we identified 37 members of the *BsHSF* gene family in *B. semperflorens* based on its genome scaffold, which are unevenly distributed across the genome. Phylogenetic analysis reveals that these 37 members can be divided into three subfamilies. Analysis of their physicochemical properties shows significant diversity among these proteins. Except for the BsHSFB7 protein located in the cytoplasm, all other BsHSF proteins were found to be nuclear-localized. A comparison of the amino acid sequences indicates that all BsHSF proteins contain a conserved DNA-binding domain structure. Analysis of the promoter cis-acting elements also suggests that *BsHSFs* may be associated with heat stress and plant secondary metabolism. We further investigated the duplication events of *BsHSF* genes and their collinearity with genes from other Begonia species. Finally, through real-time quantitative PCR, we examined the expression patterns of the 37 *BsHSFs* in different plant tissues (roots, stems, leaves, and flowers) and their expression levels under heat stress treatment. The results show that, except for *BsHSF29*, all *BsHSFs* were expressed in various tissues, with varying expression levels across tissues. Except for *BsHSF33* and *BsHSF34*, the expression levels of almost all *BsHSF* genes increased in response to heat treatment. In summary, these findings provide a better understanding of the role and regulatory mechanisms of HSFs in the heat stress response of *B. semperflorens* and lay the foundation for further exploration of the biological functions of *BsHSFs* in the stress responses of *B. semperflorens*.

Keywords: *Begonia semperflorens*; *HSF* gene family; bioinformatics; expression pattern analysis; transcription factor

1. Introduction

Begonia semperflorens (*B. semperflorens*) is a widely cultivated ornamental plant with a broad economic market [1]. It is popular due to its diverse flower colors, year-round blooming characteristics, and adaptability to the environment [2]. However, with the rise in global temperatures, its production and application are being challenged, which threatens its production and ornamental value [3]. Heat stress can disrupt its cellular homeostasis, affecting its photosynthesis, cell membrane integrity, and metabolic processes, ultimately impacting its growth and flowering [4]. To address this issue, it is crucial to understand the molecular mechanisms of its heat tolerance. Transcription factors, as the main regulators of gene networks, play a key role in regulating stress responses to various adverse environments [5].

The heat shock transcription (HSF) gene family is a key transcription factor family that responds to heat stress and plays an important role in thermotolerance [6–8]. Previous studies have indicated that *HSF* family genes contain several conserved domains [9]. There is a DNA-binding domain (DBD) at the terminus, which can recognize the promoter elements in heat-responsive genes. An adjacent oligomerization domain (OD or HR-A/B), primarily composed of hydrophobic heptapeptide repeats, has been found in all HSF family genes [10,11]. The number of *HSF* gene family members varies among different plants, with 19 in *Rosa chinensis* [12], 21 in *Betula platyphylla* [13], 49 in *Salvia rosmarinus* [14], and 24 in *Ammopiptanthus mongolicus* [15]. Overall, higher plants generally possess more *HSF* family genes than lower plants. This may be attributed to multiple whole-genome duplication (WGD) or whole-genome triplication (WGT) events in plants, especially in higher plants [16]. Members of the HSF gene family respond to a variety of stress conditions [17–19]. Previous research has shown that in maize [20], *Hypericum perforatum* [21], and *Eucommia ulmoides* [22], members of the HSF gene family exhibit robust responses to high-temperature stress. Therefore, studying the HSF gene family in *B. semperflorens* is of great significance.

In this study, we systematically identified 37 HSF genes from the genome of *B. semperflorens* and conducted a comprehensive analysis of this gene family using bioinformatics methods. Given that research on the HSF gene in *Arabidopsis thaliana* (*A. thaliana*) is relatively comprehensive, we constructed a phylogenetic tree using the HSF genes from *A. thaliana* and *B. semperflorens* to infer the functions of the HSF gene in *B. semperflorens*. We further explored the heat stress response of *BsHSF* genes and identified a set of candidate genes that potentially play roles in thermotolerance. These findings are crucial for determining the key regulatory factors of stress resistance in *B. semperflorens*. Our results contribute to a deeper understanding of the evolutionary relationships and functional characteristics of the *B. semperflorens HSF* gene family and provide valuable insights for future research on the roles of HSF proteins in stress responses and stress-resistant breeding programs.

2. Materials and Methods

2.1. Identification of Gene Family Members

The genome data of *A. thaliana* were obtained from the *A. thaliana* Genome Database TAIR (https://www.arabidopsis.org/, accessed on 1 April 2025) [23]. The protein sequences of the *HSF* family members of *A. thaliana* were used as seed sequences to perform a BLAST (2.13.0) alignment against the proteome file of *B. semperflorens*, with an E-value of 1×10^{-5}. Sequences with homology greater than 50% were selected, and duplicates were removed to obtain the candidate sequences. Subsequently, the HMM model of HSF proteins (PF00447) was downloaded from the Pfam database [24], and hmmsearch from HMMER 3.0 [25] was used to screen the HSF protein members of *B. semperflorens*, with an E-value of 1×10^{-10}, and the proteome file of *B. semperflorens* set as the protein database. The candidate protein members obtained by the two methods were compared, and the common sequences were selected while removing duplicates. Finally, the protein members were verified by their domain structure using the CD-search tool from NCBI(https://www.ncbi.nlm.nih.gov/cdd, accessed on 1 April 2025) [26], with an E-value of 1×10^{-5}. Sequences containing at least one complete HSF_DNA-bind domain were selected. The resulting protein members were identified as the HSF protein family members of *B. semperflorens*.

2.2. Localization and Numbering of BsHSFs

Based on the genome annotation file and the identified *BsHSF* gene family members, only scaffolds containing *BsHSF* genes were selected for display due to the lack of chromosome data. The localization of *BsHSF* genes was visualized using TBtools v.2.069 [27], and they were numbered according to their order on the scaffolds.

2.3. Physicochemical Properties

The physicochemical properties of the 37 BsHSF sequences were analyzed using ExPASy (https://www.expasy.org/, accessed on 1 April 2025) [28]. The subcellular localization of the proteins was predicted using WoLF PSORT (https://wolfpsort.hgc.jp/?utm_source=chatgpt.com, accessed on 1 April 2025) [29].

2.4. Phylogenetic Tree Construction

The phylogenetic tree of HSF proteins from *B. semperflorens* and *A. thaliana* was constructed using MEGA-X 11.0 software [30]. Sequence alignment was performed using ClustalW, and the neighbor-joining method was employed for tree construction. The bootstrap method was used for testing, with the bootstrap value set to 1000, while other parameters were kept at their default settings. Additionally, an intraspecific phylogenetic tree for the HSF proteins of *B. semperflorens* was constructed using MEGA-X 11.0 software. Sequence alignment was performed using MUSCLE v5.1, and the neighbor-joining method was employed for tree construction. The bootstrap method was used for testing, with the bootstrap value set to 1000, while other parameters were kept at their default settings. The phylogenetic trees were visually enhanced using iTOL (https://itol.embl.de/, accessed on 1 April 2025) [31] and Adobe Illustrator 2022 software.

2.5. Collinearity Analysis

The genome file and gene annotation file of *Begonia darthvaderiana* (*B. darthvaderiana*) were downloaded from NCBI. The interspecific collinearity analysis between *B. semperflorens* and *B. darthvaderiana* was conducted using the One step MCScanX tool in TBtools-II v2.210 software. In addition, the intraspecific collinearity analysis of Begonia was performed using the One step MCScanX tool in TBtools-II v2.210 software. The images were then refined using Adobe Illustrator 2022 software.

2.6. Gene Structure Analysis

The annotation information of the *BsHSF* genes was extracted using TBtools-II v2.210. The gene structure analysis was conducted via the GSDS website (http://gsds.cbi.pku.edu.cn/, accessed on 1 April 2025) [32]. The images were then adjusted using Adobe Illustrator 2022 software.

2.7. Conserved Motif and Domain Analysis

The conserved motif analysis of the BsHSF protein was performed using the MEME online tool (https://meme-suite.org/meme/, accessed on 1 April 2025) [33], with the maximum number of motifs set to 10 and other settings kept as default. The conserved domain analysis was conducted using the NCBI website, with an E-value of 1×10^{-5}. The analysis results of motifs and domains were visualized using TBtools-II v2.210 software.

2.8. Cis-Acting Element Analysis in the Promoter Region

The sequence of 2000 bp upstream of the *BsHSF* gene promoter was extracted using TBtools-II v2.210. The prediction of cis-acting elements in the promoter region was performed using the PlantCARE website (https://bioinformatics.psb.ugent.be/webtools/plantcare/html/, accessed on 2 April 2025) [34]. The prediction results were visualized using TBtools-II v2.210. The images were then refined using Adobe Illustrator 2022 software.

2.9. Plant Materials and Treatment

The required *B. semperflorens* '*Super Olympia*' plants were cultivated at the College of Landscape Architecture and Art, Henan Agricultural University. The photoperiod was set at 12 h of light and 12 h of darkness, with a light intensity of 200 μmol/m^2/s and a cultivation temperature of 25 °C. Plants were planted in sterile bottles and placed in a growth chamber. Healthy and uniformly sized *B. semperflorens* plants, aged 3 to 4 months and grown in sterile bottles, were selected for heat stress treatment. Tissues of roots, stems, leaves, and flowers (Figure 1) were collected before treatment for the detection of tissue-specific expression. Other cultivation conditions remained unchanged, while the cultivation temperature was set at 38 °C, slightly lower than its high-temperature semi-lethal temperature. Samples were collected from the treated plants at 0 h, 12 h, 24 h, 36 h, and 48 h after treatment. Each sample weighed 0.1 g and was placed into a 2 mL centrifuge tube, then quickly frozen in liquid nitrogen and stored in a −80 °C freezer for later use. Each treatment had three biological replicates, with each replicate randomly selecting different 9 *B. semperflorens* plants.

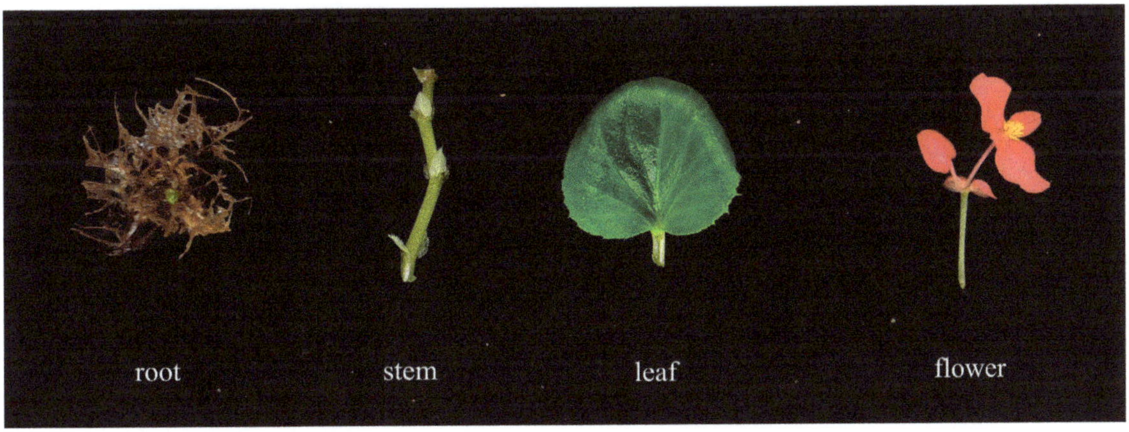

Figure 1. Sampling tissues of *B. semperflorens* for quantitative real-time PCR.

2.10. RNA Extraction and Real-Time Quantitative PCR Analysis

Total RNA was extracted from each tissue sample using the PureTotal RNA Extraction Kit (DP441) designed for polysaccharide–polyphenol plants. The RNA was then reverse-transcribed into cDNA using the HiScript III 1st Strand cDNA Synthesis Kit (Nanjing Vazyme Biotech Co., Ltd., Nanjing, China) with the gDNA Eraser. Primers for RT-qPCR were synthesized by Sangon Biotech, and all primer sequences are listed in Table S1. The *Bs18s* gene was used as the housekeeping gene [35]. The reaction system and program were configured according to the instructions of SYBR Premix Ex Taq™ II (TaKaRa BIO Inc., Beijing, China). The data were analyzed using the $2^{-\Delta\Delta CT}$ method. One-way analysis of variance (ANOVA) was performed using SPSS 22.0 software, followed by Duncan's multiple comparison test. GraphPad Prism 10.0 [36] was used for the graphical representation of the data.

3. Results

3.1. Identification and Physicochemical Characterization of the BsHSF Gene Family

After removing redundant sequences, we identified 37 HSF genes from the *B. semperflorens* genome using bioinformatics methods. These genes were named *BsHSF1* to *BsHSF37* based on their genomic locations. Table 1 lists the detailed information, including the physicochemical properties of the 37 BsHSF proteins. The lengths of the BsHSF proteins range from 81 amino acids (BsHSF7) to 521 amino acids (BsHSF17), with molecular weights varying from 9.62 kDa (BsHSF7) to 57.48 kDa (BsHSF17). The theoretical isoelectric points (pI) range from 4.63 (BsHSF11) to 9.86 (BsHSF7). Notably, with the exception of BsHSF7, BsHSF15, BsHSF3, BsHSF25, BsHSF2, BsHSF32, and BsHSF33, all BsHSF proteins are classified as acidic (pI < 7). The average hydrophilicity value of the BsHSF proteins is negative, indicating that they are predominantly hydrophilic. Moreover, the average instability index of these proteins is 56.7 (above the threshold of 40), suggesting that they are likely to be unstable. Subcellular localization prediction shows that, except for BsHSF7, which is localized in the cytoplasm, all other BpHSF proteins are located in the nucleus.

Table 1. Physicochemical properties of proteins encoded by the 37 *BsHSF* genes in *B. semperflorens*.

Rename	Gene ID	Number of Amino Acid	Molecular Weight	Theoretical pI	Instability Index	Aliphatic Index	Grand Average of Hydropathicity	Subcellular Localization
BsHSF1	g2919.t1	385	44,307.25	5.38	54.6	74.42	−0.794	Nuclear
BsHSF2	g4110.t1	246	28,643.14	7.69	49.15	66.59	−0.985	Nuclear
BsHSF3	g6017.t1	243	28,082.52	8.62	54.74	60.95	−1.03	Nuclear
BsHSF4	g7033.t2	458	52,759.41	5.15	66.23	68.76	−0.689	Nuclear
BsHSF5	g7541.t1	376	43,643.24	5.84	50.77	80.61	−0.67	Nuclear
BsHSF6	g7952.t1	390	44,297.31	5.43	54.34	73.74	−0.794	Nuclear
BsHSF7	g9408.t1	81	9621.97	9.86	32.63	68.64	−0.56	Cytoplasmic
BsHSF8	g10132.t1	346	39,200.81	4.8	58.47	76.91	−0.618	Nuclear
BsHSF9	g11656.t1	480	53,724.09	5.34	56.47	74.92	−0.614	Nuclear
BsHSF10	g11734.t1	310	35,367.96	5.76	61.04	69.13	−0.561	Nuclear
BsHSF11	g12866.t1	508	56,333.3	4.63	53.03	67.15	−0.726	Nuclear
BsHSF12	g12999.t1	322	35,388.58	5.05	59.62	71.74	−0.541	Nuclear
BsHSF13	g13272.t1	330	39,491.62	5.91	58.01	65.55	−0.898	Nuclear
BsHSF14	g13857.t1	335	39,352.21	5.61	53.66	64	−0.857	Nuclear
BsHSF15	g16136.t1	331	37,427.53	8.8	55.03	71.27	−0.501	Nuclear
BsHSF16	g16542.t1	474	53,737.31	5.21	64.51	70.97	−0.826	Nuclear
BsHSF17	g17820.t1	521	57,480.9	4.66	63.41	66.76	−0.643	Nuclear
BsHSF18	g17877.t2	333	36,687.07	4.97	55.75	78.47	−0.504	Nuclear
BsHSF19	g17997.t1	318	37,858.8	5.81	64.24	67.7	−0.846	Nuclear
BsHSF20	g18359.t1	480	54,441.75	5.87	62.63	73.52	−0.752	Nuclear
BsHSF21	g23282.t1	339	38,820.51	6.53	55.13	66.73	−0.6	Nuclear
BsHSF22	g25236.t1	399	45,983.14	4.74	48.14	75.04	−0.584	Nuclear
BsHSF23	g25942.t1	342	39,151.85	5.02	68.5	70.41	−0.714	Nuclear
BsHSF24	g26841.t1	343	39,072.9	6.73	55	73.03	−0.549	Nuclear
BsHSF25	g29399.t1	334	37,450.29	8.14	59.39	71.26	−0.578	Nuclear
BsHSF26	g30373.t1	520	57,020.55	5	60.02	67.87	−0.594	Nuclear
BsHSF27	g31168.t1	330	36,756.36	6.94	55.84	73.88	−0.534	Nuclear
BsHSF28	g31755.t1	344	40,447.32	5.42	54.09	64.88	−0.872	Nuclear
BsHSF29	g33546.t1	338	37,620.03	5.52	49.37	70.47	−0.693	Nuclear
BsHSF30	g34334.t1	312	35,559.17	5	73.79	68.4	−0.546	Nuclear
BsHSF31	g36487.t1	346	39,476.21	5.02	56.16	67.57	−0.804	Nuclear
BsHSF32	g36767.t1	281	31,424.19	7.59	43.38	61.42	−0.829	Nuclear
BsHSF33	g38782.t1	434	48,995.12	7.11	62.01	65.37	−0.614	Nuclear
BsHSF34	g38809.t1	318	36,390.19	5.67	60.89	68.71	−0.559	Nuclear
BsHSF35	g39393.t1	276	32,442.08	5.36	58.69	85.76	−0.592	Nuclear
BsHSF36	g40203.t1	326	37,186.68	6.46	54.7	67.61	−0.598	Nuclear
BsHSF37	g40444.t1	464	51,457.67	5.59	57.65	69.29	−0.486	Nuclear

3.2. Localization of BsHSF Genes

The localization analysis indicates that, despite the lack of chromosomal data, the distribution pattern of the 37 *BsHSF* genes is not uniform (Figure 2). Among the 71 scaffolds, only 22 contain *BsHSF* genes. The longest scaffold, ptg000035l, with a length of 24 MB, harbors only one *BsHSF* gene, *BsHSF25*. In contrast, three genes are found on scaffolds ptg000011l (19.7 MB), ptg000015l (6.7 MB), ptg000022l (6.5 MB), and ptg000036l (10.0 MB). Specifically, *BsHSF6*, *BsHSF7*, and *BsHSF8* are located on ptg000011l; *BsHSF11*, *BsHSF12*, and *BsHSF13* are on ptg000015l; *BsHSF17*, *BsHSF18*, and *BsHSF19* are on ptg000022l; and *BsHSF26*, *BsHSF27*, and *BsHSF28* are on ptg000036l.

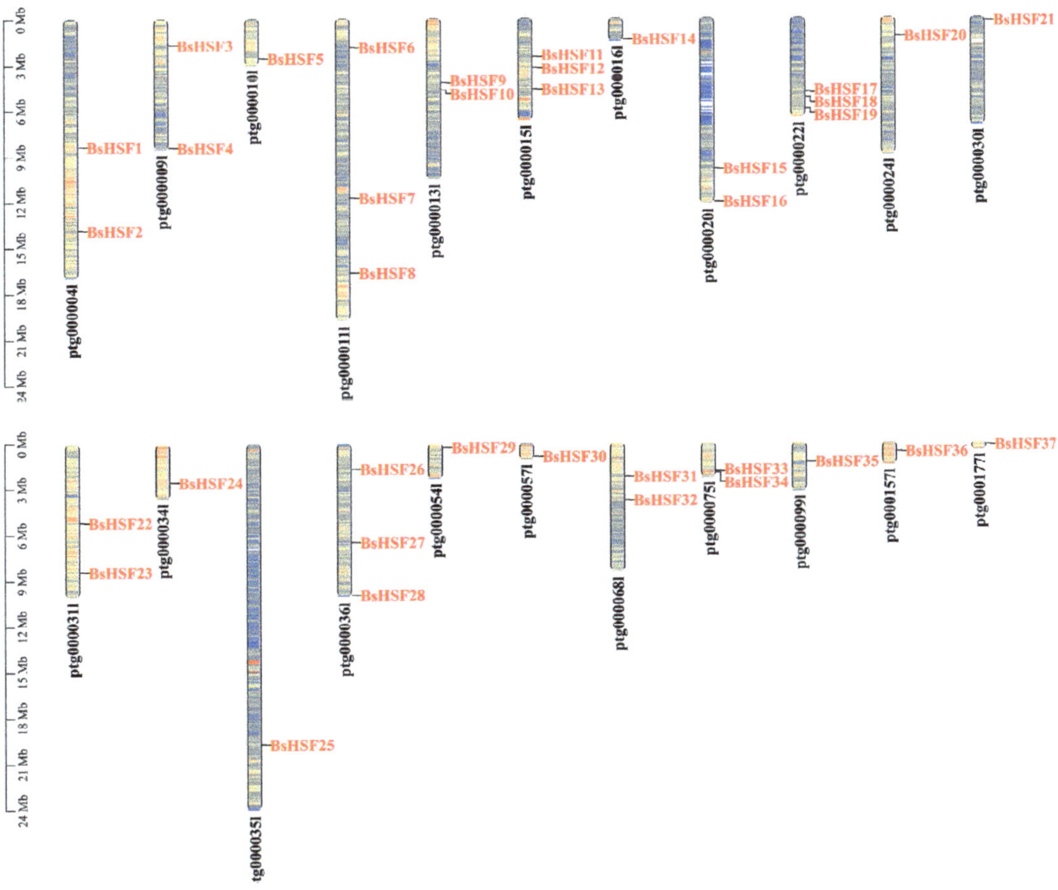

Figure 2. Localization of the *BsHSF* genes. Scale bars are in Mb, and scaffolds numbers are shown above the corresponding chromosomes.

3.3. Phylogenetic Analysis of BsHSF

A neighbor-joining phylogenetic tree was constructed to analyze the evolutionary relationships of *HSF* genes from *B. semperflorens* and *Arabidopsis thaliana* (Figure 3). This analysis included 61 protein sequences: 37 from *B. semperflorens* and 24 from *Arabidopsis thaliana*. Based on the classification system of the HSF protein family in the model plant *Arabidopsis thaliana*, the *BpHSF* genes were divided into three major groups corresponding to three subfamilies (A, B, and C). In *B. semperflorens*, subfamily A contains 20 BsHSF members, subfamily B contains 11 members, and subfamily C contains 6 members. In *Arabidopsis thaliana*, subfamily A contains 17 BsHSF members, subfamily B contains 5 members, and subfamily C contains 2 members. *B. semperflorens* has more members in each subfamily than *Arabidopsis thaliana*, with the greatest increase observed in subfamily C, where the number of members in *B. semperflorens* is three times that of *Arabidopsis thaliana*.

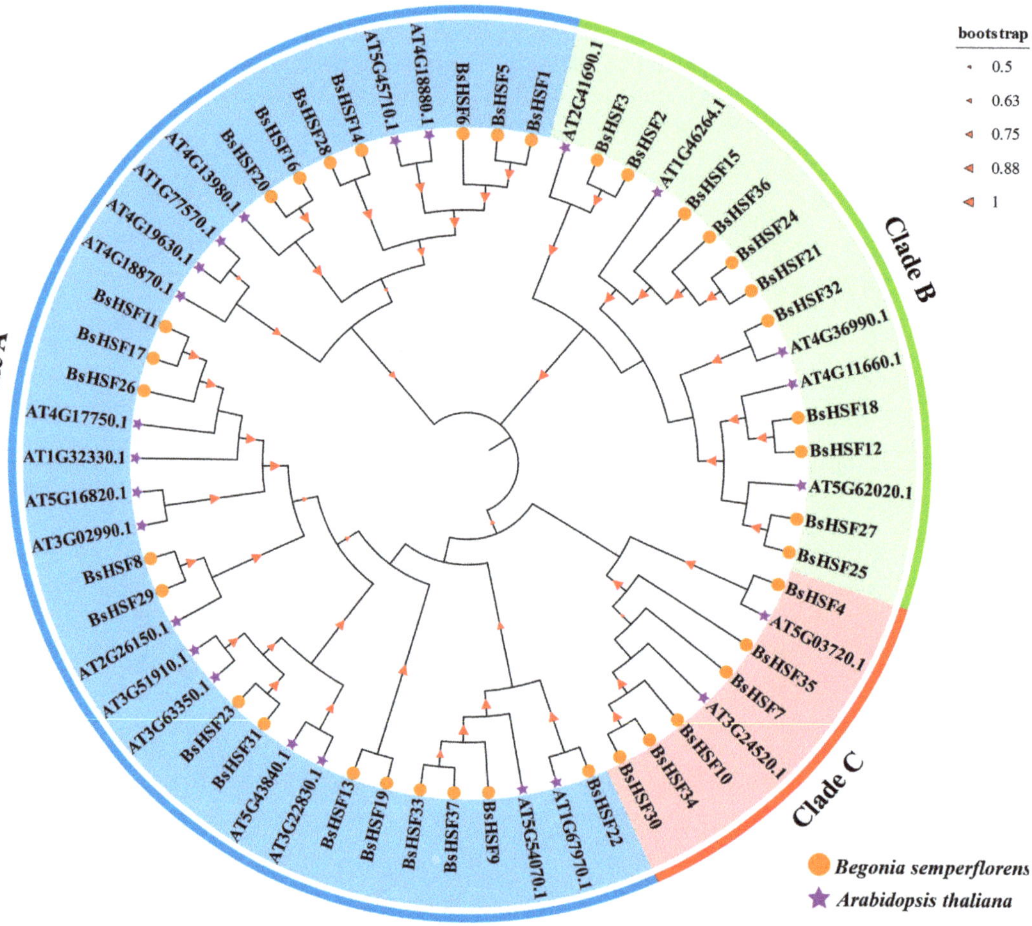

Figure 3. Phylogenetic analysis of *BsHSF* genes was conducted using HSF sequences from *B. semperflorens* (BsHSF) and Arabidopsis thaliana (AT). As a result, different HSF classes are distinguished in various colors, each representing a distinct class.

3.4. Protein Motifs and Gene Structure of BsHSF Genes

Analysis of gene structure and conserved motifs can provide deeper insights into gene function and evolution. Using the MEME tool, we identified 10 conserved motifs containing 20 to 50 amino acids in the BsHSF family, with variable distribution among different BsHSFs. The number of conserved motifs ranges from one to six, with BsHSF7 containing only one motif, while most BsHSFs contain 5–6 motifs (Figure 4). The differences in the number and distribution of motifs among family members suggest that different BsHSFs may have distinct biological functions. All BsHSF proteins contain Motif 1, indicating that Motif 1 is highly conserved in BsHSFs and may play an important role. Except for BsHSF7, all BsHSF proteins contain Motif 2 and Motif 3, while Motif 8 is only present in members of subfamily B. Except for BsHSF13 and BsHSF31, which have three exons, all other BsHSFs have two exons. This indicates that the evolution of BsHSFs is more conservative compared to other species.

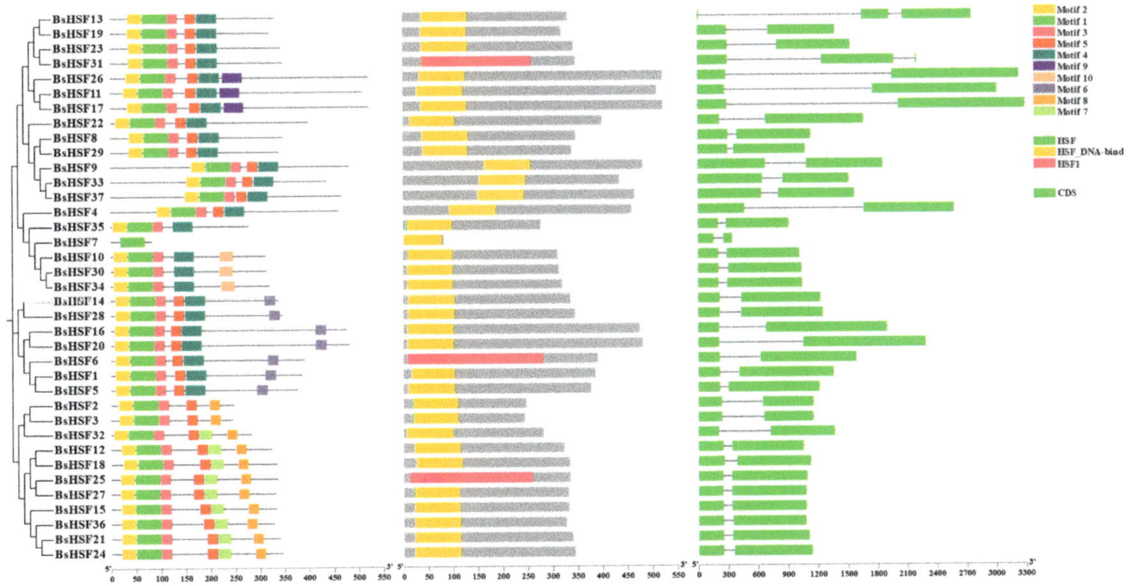

Figure 4. Gene structure and conserved motifs analysis of BsHSFs: phylogenetic relationship tree; conserved motifs of BsHSFs; SMART conserved HSF domain; gene structure of BsHSFs.

3.5. Analysis of Cis-Acting Elements in the Promoters of BsHSF Genes

Transcription factors play a crucial role in mediating plant responses to biotic and abiotic stresses by regulating various cis-regulatory elements in gene promoter sequences. To explore the potential biological functions and regulatory networks of *BsHSF* genes, we analyzed the cis-regulatory elements within the promoter regions of *BsHSF* genes (spanning 2000 bp upstream of the start codon) (Figure 5). A total of 598 cis-regulatory elements were identified in this analysis, with functional descriptions including abscisic acid responsiveness, anaerobic induction, auxin-responsive element, cell cycle regulation, circadian control, defense and stress responsiveness, differentiation of the palisade mesophyll cells, endosperm expression, flavonoid biosynthetic genes regulation, gibberellin-responsive element, low-temperature responsiveness, maximal elicitor-mediated activation, MeJA-responsiveness, meristem expression, the MYB binding site involved in drought-inducibility, phytochrome down-regulation expression, salicylic acid responsiveness, and zein metabolism regulation. Among these, the most abundant elements were abscisic acid responsiveness and MeJA-responsiveness, with 165 and 164 occurrences, respectively, indicating that *BsHSFs* may respond to abscisic acid and MeJA. The presence of other cis-acting elements also suggests that *BsHSFs* may be associated with abiotic stress responses and plant secondary metabolism.

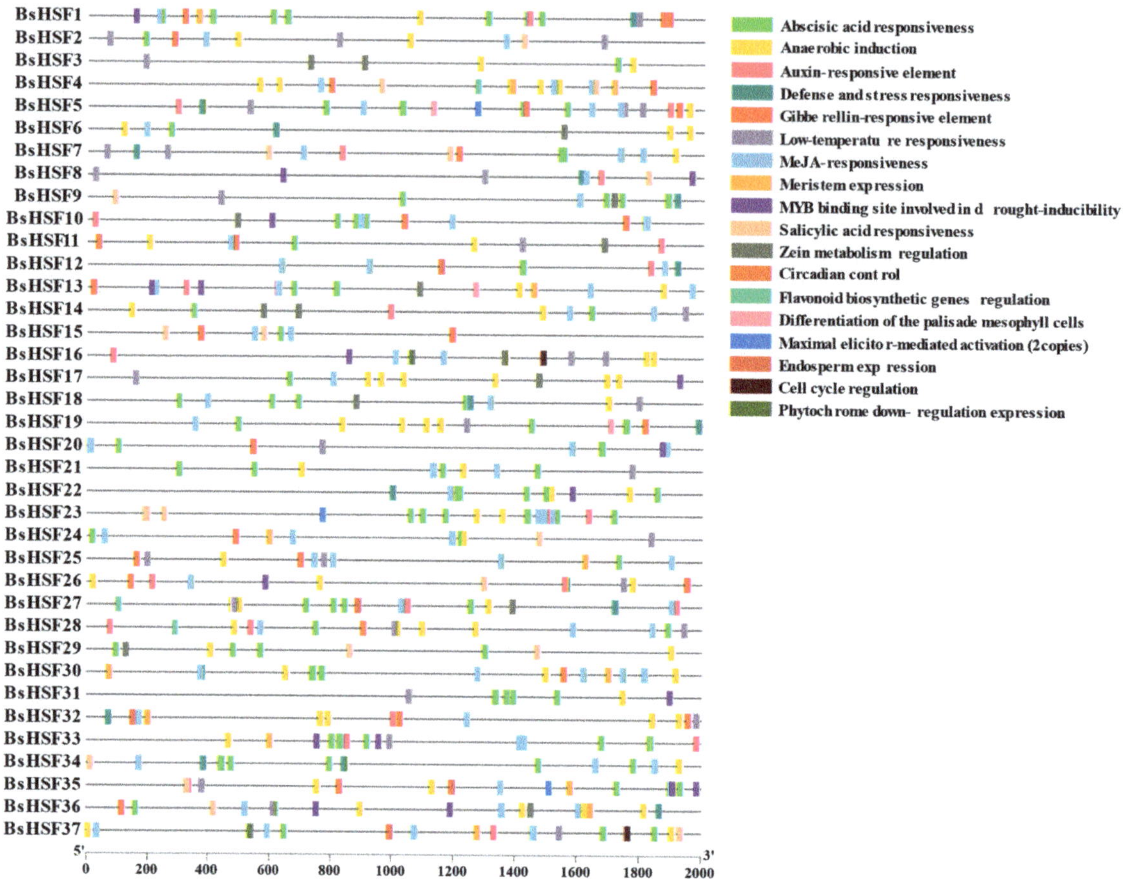

Figure 5. The distribution of cis-regulating elements in the promoter region of the HSF gene in *B. semperflorens*. Each promoter is located at the 2000 bp upstream region of BsHSF gene. The colored blocks represent 18 kinds of cis-elements.

3.6. Gene Duplication Events and Collinearity of BsHSF Genes

Gene duplication is crucial for the evolution of plant genomes. To identify duplication events of *BsHSF* genes, the gene pairs (GPs) were counted, and a homology analysis was conducted within the species (Figure 6A). The results show that 24 segmental duplications (SDs) occurred among the 34 *BsHSF* genes, with no tandem duplications detected. Only three *BsHSF* genes, namely, *BsHSF22*, *BsHSF7*, and *BsHSF4*, did not undergo SDs. The findings indicate that SDs play a key role in the expansion of *BsHSF* genes.

To better elucidate the evolutionary relationships of *B. semperflorens*, an interspecific collinearity map was constructed between *B. semperflorens* and its congener *B. darthvaderiana* (Figure 6B). It was found that 18 genes in *B. darthvaderiana* corresponded to 30 genes in *B. semperflorens*, with 47 gene pairs (GPs) identified between them, indicating strong collinearity. Moreover, most genes in *B. darthvaderiana* corresponded to more than one gene in *B. semperflorens*.

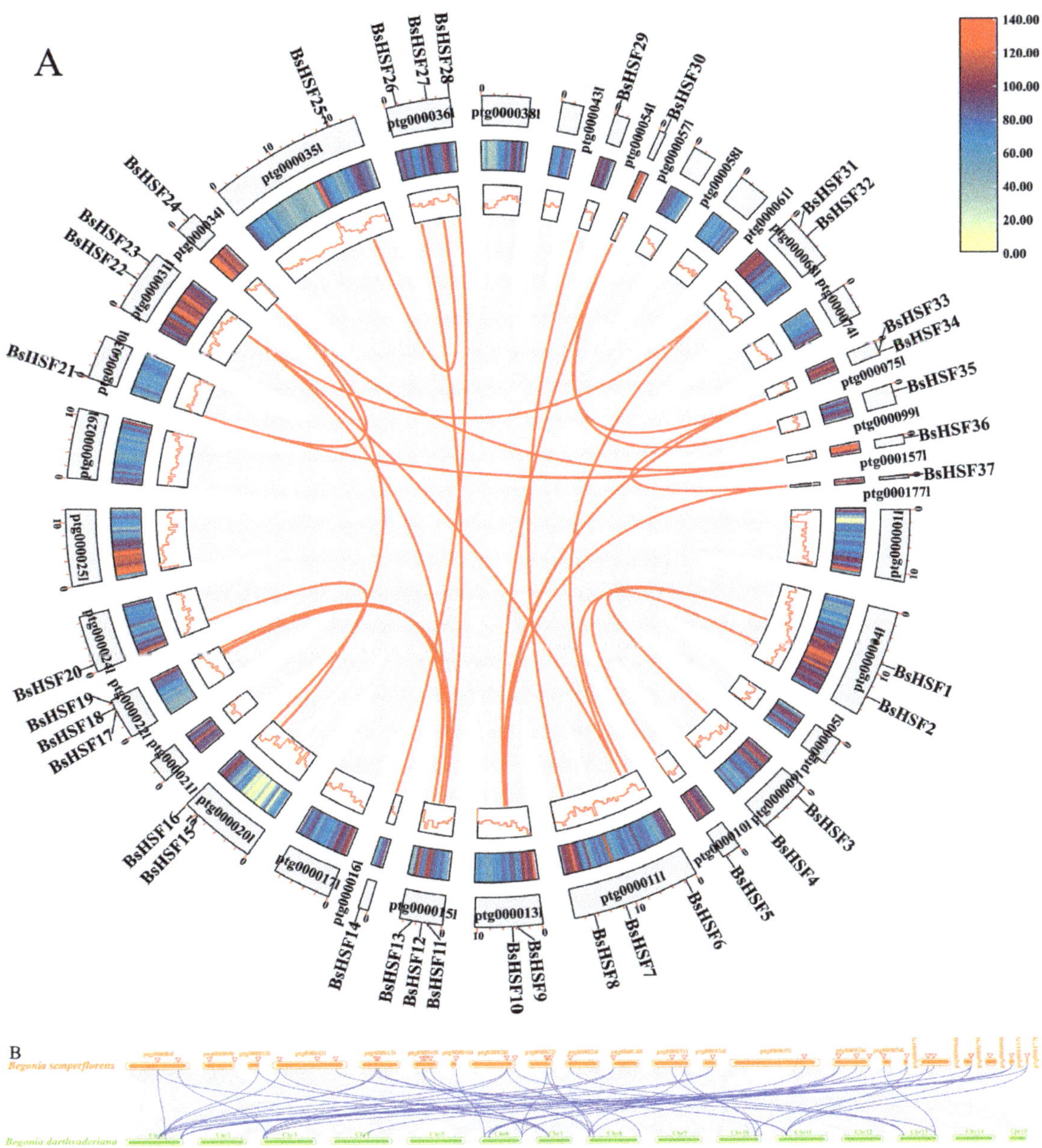

Figure 6. Collinearity of the *HSF* gene family in different species. (**A**) Collinearity of the HSF gene family in *B. semperflorens*; (**B**) Collinearity of the *HSF* gene family between *B. semperflorens* and *B. darthvaderiana*. The heatmaps and line plots illustrate the gene density within each chromosome. The grey curves in the background indicate all segmental duplications (SDs) within the *B. semperflorens* genome and between the genomes of *B. semperflorens* and *B. darthvaderiana*. The blue curves highlight the SDs involving *HSF* genes. Darker regions indicate a higher number of SDs and denser curves, while lighter regions indicate fewer SDs and sparser curves.

3.7. Expression Patterns of BsHSF Genes in Different Tissues

The expression of *BsHSF* genes in different tissues (root, stem, leaf, and flower) of *B. semperflorens* was detected using quantitative real-time PCR (Figure 7). Except for *BsHSF29*, which showed no detectable expression in root and leaf tissues, all other genes were expressed in each tissue. The expression levels of each gene varied among different tissues. In roots, higher expression levels were observed for *BsHSF5*, *BsHSF20*, *BsHSF22*, *BsHSF30*, *BsHSF32*, and *BsHSF35*. In stems, *BsHSF7*, *BsHSF20*, *BsHSF30*, *BsHSF33*, and *BsHSF35* exhibited higher expression levels. In leaves, the expression levels of the genes varied greatly, with *BsHSF30* showing a significantly higher expression than other genes. *BsHSF1*, *BsHSF7*, *BsHSF10*, *BsHSF11*, *BsHSF12*, *BsHSF20*, *BsHSF30*, and *BsHSF33* also had relatively high expression levels. In flowers, higher expression levels were detected for *BsHSF20*, *BsHSF22*, *BsHSF27*, *BsHSF30*, and *BsHSF33*.

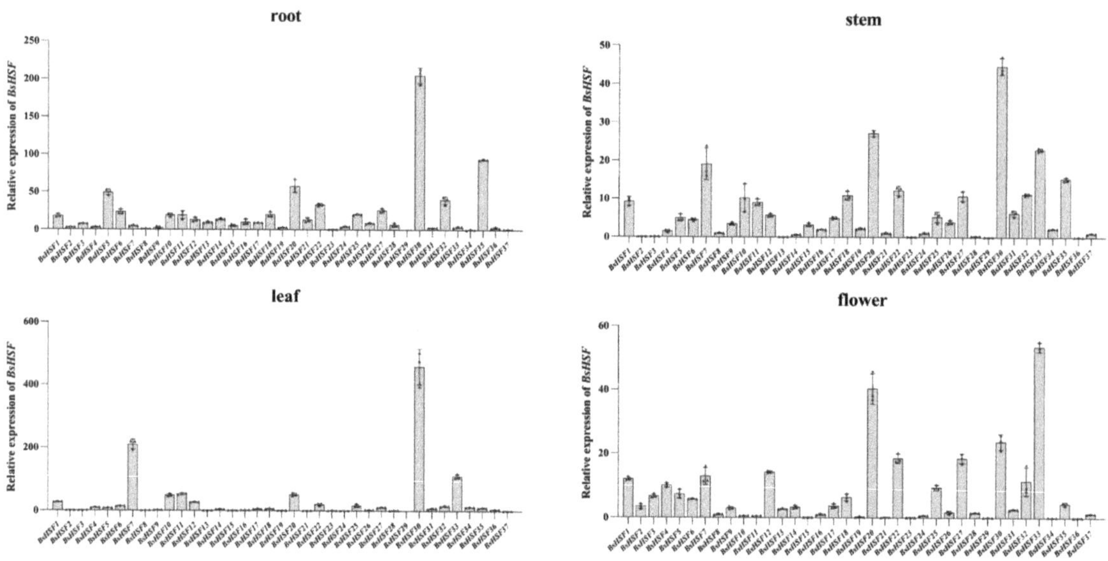

Figure 7. qRT-PCR analysis of *BsHSF* genes in four different tissues.

3.8. Expression Patterns of BsHSF Genes Under Heat Stress Treatment

To investigate the response patterns of *B. semperflorens* to heat stress, the expression of *BsHSF* genes after heat treatment was detected using quantitative real-time PCR (Figure 8). It was found that, overall, almost all *BsHSF* genes, except for *BsHSF33* and *BsHSF34*, showed increased expression levels in response to heat treatment. The expression levels of *BsHSF2*, *BsHSF3*, *BsHSF4*, *BsHSF19*, *BsHSF21*, *BsHSF23*, *BsHSF27*, and *BsHSF29* increased significantly in response to heat treatment. Notably, *BsHSF29*, which is almost undetectable or has extremely low expression under normal temperature conditions, showed an approximately 40,000-fold increase in expression after 48 h of heat treatment compared to 0 h. This indicates that *BsHSF29* strongly responds to heat treatment and may be associated with plant thermotolerance.

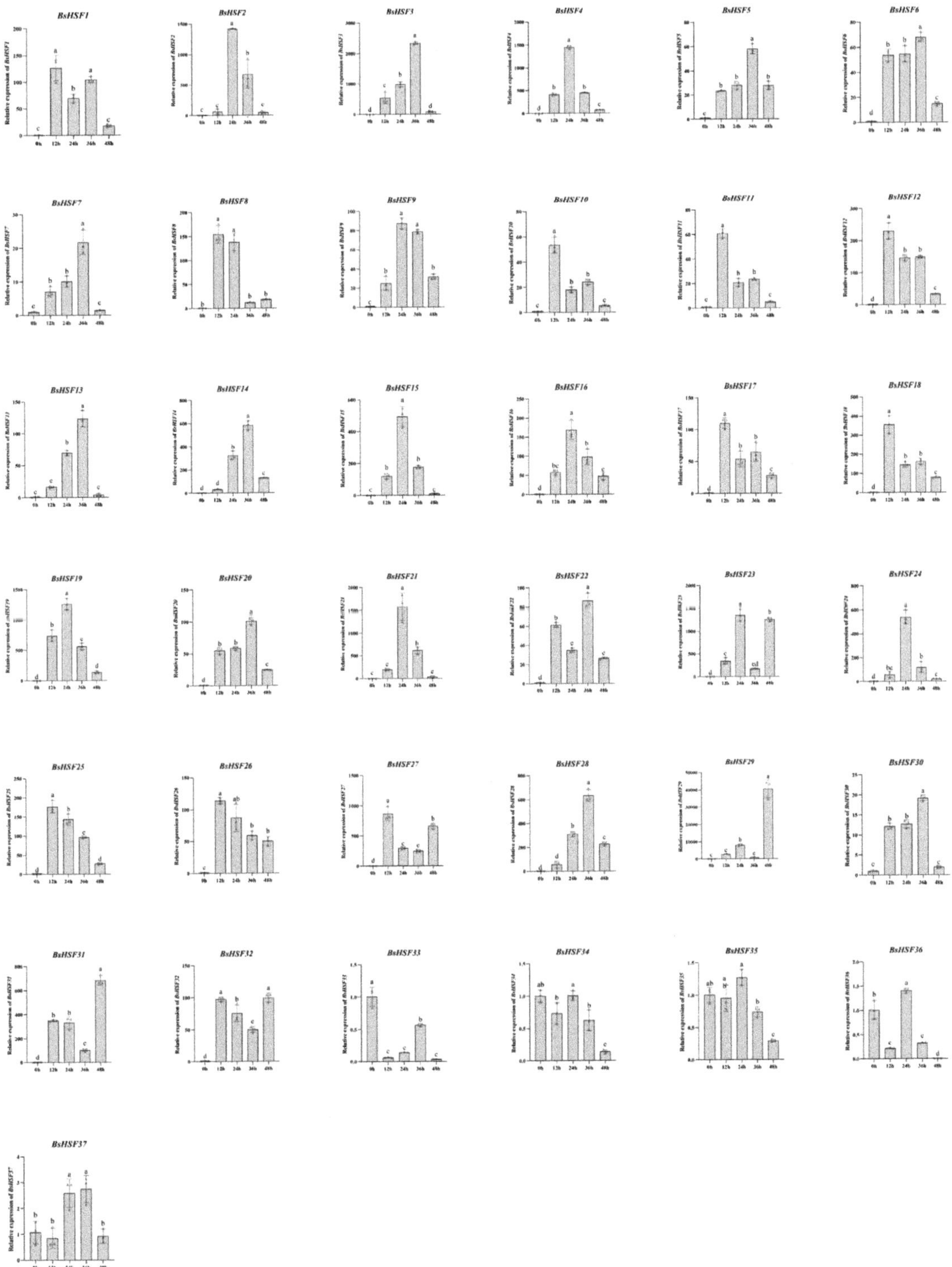

Figure 8. qRT-PCR analysis of BsHSF genes under heat stress. Different lowercase letters above the bars indicate statistically significant differences ($p < 0.05$) using one-way ANOVA.

4. Discussion

Begonia semperflorens is widely used for decorating various flower beds and squares due to its characteristic of blooming all year round. It has relatively strong resistance, with features of drought tolerance and high-temperature endurance [37]. Transcription factors play important regulatory roles in plant growth, development, and responses to stress, among which heat shock transcription factors are in a core position [8]. This study identified 37 *HSF* transcription factors, which is more than the number of *HSF* gene family members in *Arabidopsis thaliana* (21 genes), *Oryza sativa* (25 genes), *Solanum lycopersicum* (26 genes), *Brassica oleracea* (35 genes), and *Brassica rapa* (36 genes). Thus, we supposed that more *HSF* transcription factors may be related to the better heat tolerance of *B. semperflorens*. Higher plants have generated more members of the *HSF* gene family through multiple whole-genome duplication (WGD) or whole-genome triplication (WGT) events to adapt to high-temperature environments [38–40]. The distribution of these genes in the genome is not uniform, and the length of the segments is not correlated with the number of HSF genes distributed (Figure 2).

In angiosperms, the *HSF* gene family is divided into three subfamilies: A, B, and C (Figure 3). Based on gene structure and phylogenetic relationships, we classified the 37 BsHSF genes into subfamilies A, B, and C. Specifically, subfamily A contains 20 genes, subfamily B contains 11 genes, and subfamily C contains 6 genes. Although the total number is greater than that of the HSF gene family members in *Arabidopsis*, the main increase in members belongs to subfamilies B and C. The number of subfamily A members does not differ significantly. Such changes may also imply the functions of the *HSF* family genes in *B. semperflorens*. Most members of subfamily A have one or more AHA motifs at the C-terminus. These motifs act as transcriptional activators, functioning by binding to heat stress elements and playing a dominant role in the heat stress response [41–45]. Some members of subfamily B contain a tetrapeptide—LFGV—at their C-terminus, which may act as a repressive motif through interaction with an unknown co-repressor [46,47]. There is relatively less research on subfamily C. Studies on grass crops have found that there are multiple *HSFC2* members in rice and wheat. For example, *TaHSFC2a-B* in wheat has been reported to act as a trans-activator of heat shock proteins, thereby positively regulating thermotolerance [48]. In lilies, *LlHSFC2* is induced under heat shock conditions and exhibits the ability to repress the expression of heat protection genes. However, it can also interact with *HSFAs*, acting as a co-activator to activate their transcriptional activity and participate in the establishment of thermotolerance [49]. The increased number of subfamily C members in *B. semperflorens* may imply the existence of a unique heat stress resistance pathway.

Through amino acid sequence alignment analysis, it has been shown that all 37 BsHSF proteins contain the DNA-binding domain (DBD) conserved domain, which is consistent with studies in other plants. The structure of HSF family members is relatively conserved in plants, with most containing a DBD, an oligomeric domain (OD or HR-A/B, hydrophobic heptad repeat region), and a nuclear localization signal (NLS) [10]. In the analysis of the *HSF* promoter, a large number of ABA and MeJA response elements, as well as elements related to abiotic stress responses and plant secondary metabolism, were identified. Considering that transcription factors can exert transcriptional activation or repression by binding to cis-elements in the promoters of related genes, it is hypothesized that some functional genes may regulate gene expression by binding to certain cis-elements in the *BsHSF* promoters. This implies that *BsHSFs* may respond to ABA and MeJA and may be associated with abiotic stress responses and plant secondary metabolism. However, the mining of related genes and the specific regulatory mechanisms require further research in the future. Gene duplication and repetition play important roles in the evolution and expansion of gene families [50]. Studies have pointed out that the *AhHSF* gene family in peanuts has utilized gene duplication and repetition during the evolutionary process [51]. In *B. semperflorens*, a large number of homologous genes have been found in the HSF gene family, and compared with the *HSF* gene family members in the congener *B. darthvaderiana*, there has been an expansion. This indicates that there has been a significant amount of duplication and repetition of *HSF* genes in *B. semperflorens*, which may be related to the stronger resistance of *B. semperflorens*.

We can infer their primary regulatory regions by analyzing the expression patterns of the *HSF* gene family members in different tissues of *B. semperflorens*. We found that the member *BsHSF30*, which belongs to the C subfamily, is highly expressed in all tissues except flowers. This reveals that *BsHSF30* may play a role in plant growth and development. This aligns with previous studies indicating that the HSFC1 subclass, which is unique to dicots, may be involved in the developmental processes of plants [10]. Members of the HSF gene family have also been reported to be associated with root development. The *HSFB4* member of the B subfamily of the *A. thaliana* HSF gene family affects root development. *HSFB4* is expressed in various tissues, with the highest expression level in roots. Overexpression of the *HSFB4* gene in *A. thaliana* leads to rough root epidermis and promotes epidermal splitting, which, in turn, promotes the development of lateral roots [52]. Members of the B subfamily of the HSF gene family, including *BsHSF32*, *BsHSF21*, and *BsHSF18*, are highly expressed in roots. Among them, *BsHSF21* is highly expressed only in roots, while *BsHSF32* and *BsHSF18* also have relatively high expression in stems and flowers. It is speculated that these three genes may be related to the root development of *B. semperflorens*. Except for *BsHSF29*, which was not detected in root and leaf tissues, all other genes were expressed in all tissues. We also detected and analyzed the expression patterns of *BsHSFs* in *B. semperflorens* after heat stress treatment. Almost all *BsHSFs* showed increased expression levels in response to heat stress, with several being particularly notable, including *BsHSF2*, *BsHSF3*, *BsHSF4*, *BsHSF19*, *BsHSF21*, *BsHSF23*, *BsHSF27*, and *BsHSF29*. They are all highly likely to be associated with the heat stress resistance of *B. semperflorens*. Notably, the expression level of *BsHSF29* increased by approximately 40,000 times after 48 h of high-temperature treatment compared to 0 h. As previously mentioned, no expression of *BsHSF29* was detected in root and leaf tissues, suggesting that it may be a gene specifically induced by heat stress treatment. In *Solanum lycopersicum* and *A. thaliana*, the expression of some HSF genes is also inducible. Under normal temperature conditions, they are almost not expressed or their expression levels are maintained at an extremely low level. High temperatures can cause a dramatic increase in their expression levels [18,53]. In evolutionary analysis, *BsHSF29* clustered with *Arabidopsis AT2G26150.1*. Moreover, the overexpression of *Arabidopsis AtHsfA2* (*AT2G26150.1*) can enhance the tolerance of *Arabidopsis* to high-temperature stress [54], implying that *BsHSF29* is likely an important candidate gene for *B. semperflorens* to respond to high-temperature stress.

HSF is an important transcription factor that positively regulates the stress resistance of plants. In this study, based on bioinformatics analysis, we found that the expression levels of the *BsHSF* gene vary in different tissues and under heat stress, indicating that their functions in plants are not entirely the same. In the future, knockout and overexpression of the *BsHSF* gene are needed to further identify its molecular functions, which will provide new insights into the molecular regulatory pathways of the physiological mechanisms of *B. semperflorens*. Some members of the *BsHSF* gene family may be valuable candidate genes for the molecular breeding of *B. semperflorens* resistance.

Supplementary Materials: The following supporting information can be downloaded at: https://www.mdpi.com/article/10.3390/cimb47060398/s1.

Author Contributions: Conceptualization, Z.L.; data curation, Z.L.; formal analysis, Z.L.; funding acquisition, K.Z.; methodology, Z.L.; project administration, E.X.; supervision, E.X.; visualization, Z.L.; writing—original draft, Z.L.; writing—review and editing, Z.L., N.L., and Q.W. All authors have read and agreed to the published version of the manuscript.

Funding: This research was funded by the National Natural Science Foundation of China, grant number 32172622, and the College Students' Innovative Entrepreneurial Training Plan Program, grant number 202310466014.

Institutional Review Board Statement: Not applicable.

Informed Consent Statement: Not applicable.

Data Availability Statement: The original contributions presented in this study are included in the article/Supplementary Material. Further inquiries can be directed to the corresponding author(s).

Conflicts of Interest: The authors declare no conflicts of interest.

References

1. Tebbitt, M. *Begonias: Cultivation, Natural History, and Identification*; Timber Press: Portland, OR, USA, 2005.
2. Lim, T. *Begonia cucullata* var. *cucullata*. In *Edible Medicinal and Non-Medicinal Plants: Volume 7, Flowers*; Springer: Dordrecht, The Netherlands, 2014; pp. 551–555.
3. Qu, A.-L.; Ding, Y.-F.; Jiang, Q.; Zhu, C. Molecular mechanisms of the plant heat stress response. *Biochem. Biophys. Res. Commun.* **2013**, *432*, 203–207. [CrossRef] [PubMed]
4. Hasanuzzaman, M.; Nahar, K.; Alam, M.M.; Roychowdhury, R.; Fujita, M. Physiological, biochemical, and molecular mechanisms of heat stress tolerance in plants. *Int. J. Mol. Sci.* **2013**, *14*, 9643–9684. [CrossRef]
5. Koyama, T. Regulatory mechanisms of transcription factors in plant morphology and function. *Int. J. Mol. Sci.* **2023**, *24*, 7039. [CrossRef] [PubMed]
6. Ohama, N.; Kusakabe, K.; Mizoi, J.; Zhao, H.; Kidokoro, S.; Koizumi, S.; Takahashi, F.; Ishida, T.; Yanagisawa, S.; Shinozaki, K. The transcriptional cascade in the heat stress response of Arabidopsis is strictly regulated at the level of transcription factor expression. *Plant Cell* **2016**, *28*, 181–201. [CrossRef]
7. Panchuk, I.I.; Volkov, R.A.; Schöffl, F. Heat stress- and heat shock transcription factor-dependent expression and activity of ascorbate peroxidase in Arabidopsis. *Plant Physiol.* **2002**, *129*, 838–853. [CrossRef]
8. Guo, M.; Liu, J.-H.; Ma, X.; Luo, D.-X.; Gong, Z.-H.; Lu, M.-H. The plant heat stress transcription factors (HSFs): Structure, regulation, and function in response to abiotic stresses. *Front. Plant Sci.* **2016**, *7*, 114. [CrossRef]
9. Song, X.; Liu, G.; Duan, W.; Liu, T.; Huang, Z.; Ren, J.; Li, Y.; Hou, X. Genome-wide identification, classification and expression analysis of the heat shock transcription factor family in Chinese cabbage. *Mol. Genet. Genom.* **2014**, *289*, 541–551. [CrossRef] [PubMed]
10. Scharf, K.-D.; Berberich, T.; Ebersberger, I.; Nover, L. The plant heat stress transcription factor (Hsf) family: Structure, function and evolution. *Biochim. Biophys. Acta (BBA)-Gene Regul. Mech.* **2012**, *1819*, 104–119. [CrossRef]
11. Li, W.; Wan, X.-L.; Yu, J.-Y.; Wang, K.-L.; Zhang, J. Genome-wide identification, classification, and expression analysis of the Hsf gene family in carnation (*Dianthus caryophyllus*). *Int. J. Mol. Sci.* **2019**, *20*, 5233. [CrossRef]
12. Kang, Y.; Sun, P.; Yang, Y.; Li, M.; Wang, H.; Sun, X.; Jin, W. Genome-Wide Analysis of the Hsf Gene Family in Rosa chinensis and RcHsf17 Function in Thermotolerance. *Int. J. Mol. Sci.* **2025**, *26*, 287. [CrossRef]
13. Guo, S.; Chen, H.; Wu, H.; Xu, Z.; Yang, H.; Lin, Q.; Feng, H.; Zeng, Z.; Wang, S.; Liu, H.; et al. Genome-Wide Characterization of the Heat Shock Transcription Factor Gene Family in Betula platyphylla Reveals Promising Candidates for Heat Tolerance. *Int. J. Mol. Sci.* **2025**, *26*, 172. [CrossRef] [PubMed]
14. Cui, W.; Xu, Z.; Kong, Y.; Yang, L.; Dou, H.; Zhang, D.; Li, M.; Chen, Y.; Ding, S.; Yang, C.; et al. Genome-Wide Identification of the *Heat Shock Transcription Factor* Gene Family in Rosemary (*Salvia rosmarinus*). *Horticulturae* **2024**, *10*, 1250. [CrossRef]
15. Zhao, S.; Qing, J.; Yang, Z.; Tian, T.; Yan, Y.; Li, H.; Bai, Y.E. Genome-Wide Identification and Expression Analysis of the HSF Gene Family in Ammopiptanthus mongolicus. *Curr. Issues Mol. Biol.* **2024**, *46*, 11375–11393. [CrossRef]
16. Yu, T.; Bai, Y.; Liu, Z.; Wang, Z.; Yang, Q.; Wu, T.; Feng, S.; Zhang, Y.; Shen, S.; Li, Q. Large-scale analyses of heat shock transcription factors and database construction based on whole-genome genes in horticultural and representative plants. *Hortic. Res.* **2022**, *9*, uhac035. [CrossRef]
17. Huang, Y.-C.; Niu, C.-Y.; Yang, C.-R.; Jinn, T.-L. The heat stress factor HSFA6b connects ABA signaling and ABA-mediated heat responses. *Plant Physiol.* **2016**, *172*, 1182–1199. [CrossRef] [PubMed]

18. Nishizawa, A.; Yabuta, Y.; Yoshida, E.; Maruta, T.; Yoshimura, K.; Shigeoka, S. Arabidopsis heat shock transcription factor A2 as a key regulator in response to several types of environmental stress. *Plant J.* **2006**, *48*, 535–547. [CrossRef]
19. Guo, M.; Lu, J.-P.; Zhai, Y.-F.; Chai, W.-G.; Gong, Z.-H.; Lu, M.-H. Genome-wide analysis, expression profile of heat shock factor gene family (CaHsfs) and characterisation of CaHsfA2 in pepper (*Capsicum annuum* L.). *BMC Plant Biol.* **2015**, *15*, 151. [CrossRef]
20. Jiang, L.; Hu, W.; Qian, Y.; Ren, Q.; Zhang, J. Genome-wide identification, classification and expression analysis of the Hsf and Hsp70 gene families in maize. *Gene* **2021**, *770*, 145348. [CrossRef]
21. Zhou, L.; Yu, X.; Wang, D.; Li, L.; Zhou, W.; Zhang, Q.; Wang, X.; Ye, S.; Wang, Z. Genome-wide identification, classification and expression profile analysis of the HSF gene family in Hypericum perforatum. *PeerJ* **2021**, *9*, e11345. [CrossRef]
22. Shujuan, W.; Feng, X.; Guangan, W.; Chen, W. Identification of HSF Gene Family in *Eucommia ulmoides* and Its Expression Analysis under Adverse Stresses. *J. Henan Agric. Sci.* **2024**, *53*, 46–56. [CrossRef]
23. Swarbreck, D.; Wilks, C.; Lamesch, P.; Berardini, T.Z.; Garcia-Hernandez, M.; Foerster, H.; Li, D.; Meyer, T.; Muller, R.; Ploetz, L. The Arabidopsis Information Resource (TAIR): Gene structure and function annotation. *Nucleic Acids Res.* **2007**, *36*, D1009–D1014. [CrossRef] [PubMed]
24. Mistry, J.; Chuguransky, S.; Williams, L.; Qureshi, M.; Salazar, G.A.; Sonnhammer, E.L.; Tosatto, S.C.; Paladin, L.; Raj, S.; Richardson, L.J. Pfam: The protein families database in 2021. *Nucleic Acids Res.* **2021**, *49*, D412–D419. [CrossRef] [PubMed]
25. Finn, R.D.; Clements, J.; Eddy, S.R. HMMER web server: Interactive sequence similarity searching. *Nucleic Acids Res.* **2011**, *39*, W29–W37. [CrossRef] [PubMed]
26. Marchler-Bauer, A.; Bryant, S.H. CD-Search: Protein domain annotations on the fly. *Nucleic Acids Res.* **2004**, *32*, W327–W331. [CrossRef] [PubMed]
27. Chen, C.; Wu, Y.; Li, J.; Wang, X.; Zeng, Z.; Xu, J.; Liu, Y.; Feng, J.; Chen, H.; He, Y. TBtools-II: A "one for all, all for one" bioinformatics platform for biological big-data mining. *Mol. Plant* **2023**, *16*, 1733–1742. [CrossRef]
28. Artimo, P.; Jonnalagedda, M.; Arnold, K.; Baratin, D.; Csardi, G.; De Castro, E.; Duvaud, S.; Flegel, V.; Fortier, A.; Gasteiger, E. ExPASy: SIB bioinformatics resource portal. *Nucleic Acids Res.* **2012**, *40*, W597–W603. [CrossRef]
29. Horton, P.; Park, K.-J.; Obayashi, T.; Fujita, N.; Harada, H.; Adams-Collier, C.; Nakai, K. WoLF PSORT: Protein localization predictor. *Nucleic Acids Res.* **2007**, *35*, W585–W587. [CrossRef]
30. Kumar, S.; Stecher, G.; Li, M.; Knyaz, C.; Tamura, K. MEGA X: Molecular evolutionary genetics analysis across computing platforms. *Mol. Biol. Evol.* **2018**, *35*, 1547–1549. [CrossRef]
31. Letunic, I.; Bork, P. Interactive Tree Of Life (iTOL) v5: An online tool for phylogenetic tree display and annotation. *Nucleic Acids Res.* **2021**, *49*, W293–W296. [CrossRef]
32. Hu, B.; Jin, J.; Guo, A.-Y.; Zhang, H.; Luo, J.; Gao, G. GSDS 2.0: An upgraded gene feature visualization server. *Bioinformatics* **2015**, *31*, 1296–1297. [CrossRef]
33. Bailey, T.L.; Johnson, J.; Grant, C.E.; Noble, W.S. The MEME suite. *Nucleic Acids Res.* **2015**, *43*, W39–W49. [CrossRef] [PubMed]
34. Lescot, M.; Déhais, P.; Thijs, G.; Marchal, K.; Moreau, Y.; Van de Peer, Y.; Rouzé, P.; Rombauts, S. PlantCARE, a database of plant cis-acting regulatory elements and a portal to tools for in silico analysis of promoter sequences. *Nucleic Acids Res.* **2002**, *30*, 325–327. [CrossRef] [PubMed]
35. Li, Y.; Qu, Y.; Wang, Y.; Bai, X.; Tian, G.; Liu, Z.; Li, Y.; Zhang, K. Selection of suitable reference genes for qRT-PCR analysis of Begonia semperflorens under stress conditions. *Mol. Biol. Rep.* **2019**, *46*, 6027–6037. [CrossRef] [PubMed]
36. Swift, M.L. GraphPad prism, data analysis, and scientific graphing. *J. Chem. Inf. Comput. Sci.* **1997**, *37*, 411–412. [CrossRef]
37. Zhao, Z.; Liu, A.; Zhang, Y.; Yang, X.; Yang, S.; Zhao, K. Effects of Progressive Drought Stress on the Growth, Ornamental Values, and Physiological Properties of Begonia semperflorens. *Horticulturae* **2024**, *10*, 405. [CrossRef]
38. Sankoff, D.; Zheng, C. Whole genome duplication in plants: Implications for evolutionary analysis. In *Comparative Genomics: Methods and Protocols*; Humana Press: New York, NY, USA, 2018; pp. 291–315.
39. Godfree, R.C.; Marshall, D.J.; Young, A.G.; Miller, C.H.; Mathews, S. Empirical evidence of fixed and homeostatic patterns of polyploid advantage in a keystone grass exposed to drought and heat stress. *R. Soc. Open Sci.* **2017**, *4*, 170934. [CrossRef] [PubMed]
40. Song, X.; Nie, F.; Chen, W.; Ma, X.; Gong, K.; Yang, Q.; Wang, J.; Li, N.; Sun, P.; Pei, Q. Coriander Genomics Database: A genomic, transcriptomic, and metabolic database for coriander. *Hortic. Res.* **2020**, *7*, 55. [CrossRef]
41. Kotak, S.; Port, M.; Ganguli, A.; Bicker, F.; Von Koskull-Döring, P. Characterization of C-terminal domains of Arabidopsis heat stress transcription factors (Hsfs) and identification of a new signature combination of plant class A Hsfs with AHA and NES motifs essential for activator function and intracellular localization. *Plant J.* **2004**, *39*, 98–112. [CrossRef] [PubMed]
42. Zhang, L.; Li, Y.; Xing, D.; Gao, C. Characterization of mitochondrial dynamics and subcellular localization of ROS reveal that HsfA2 alleviates oxidative damage caused by heat stress in Arabidopsis. *J. Exp. Bot.* **2009**, *60*, 2073–2091. [CrossRef]
43. Liu, H.C.; Liao, H.T.; Charng, Y.Y. The role of class A1 heat shock factors (HSFA1s) in response to heat and other stresses in Arabidopsis. *Plant Cell Environ.* **2011**, *34*, 738–751. [CrossRef]

44. Nishizawa-Yokoi, A.; Nosaka, R.; Hayashi, H.; Tainaka, H.; Maruta, T.; Tamoi, M.; Ikeda, M.; Ohme-Takagi, M.; Yoshimura, K.; Yabuta, Y. HsfA1d and HsfA1e involved in the transcriptional regulation of HsfA2 function as key regulators for the Hsf signaling network in response to environmental stress. *Plant Cell Physiol.* **2011**, *52*, 933–945. [CrossRef]
45. Wu, Z.; Liang, J.; Wang, C.; Zhao, X.; Zhong, X.; Cao, X.; Li, G.; He, J.; Yi, M. Overexpression of lily HsfA3 s in Arabidopsis confers increased thermotolerance and salt sensitivity via alterations in proline catabolism. *J. Exp. Bot.* **2018**, *69*, 2005–2021. [CrossRef] [PubMed]
46. Ikeda, M.; Ohme-Takagi, M. A novel group of transcriptional repressors in Arabidopsis. *Plant Cell Physiol.* **2009**, *50*, 970–975. [CrossRef] [PubMed]
47. Ikeda, M.; Mitsuda, N.; Ohme-Takagi, M. Arabidopsis HsfB1 and HsfB2b act as repressors of the expression of heat-inducible Hsfs but positively regulate the acquired thermotolerance. *Plant Physiol.* **2011**, *157*, 1243–1254. [CrossRef] [PubMed]
48. Hu, X.J.; Chen, D.; Lynne Mclntyre, C.; Fernanda Dreccer, M.; Zhang, Z.B.; Drenth, J.; Kalaipandian, S.; Chang, H.; Xue, G.P. Heat shock factor C2a serves as a proactive mechanism for heat protection in developing grains in wheat via an ABA-mediated regulatory pathway. *Plant Cell Environ.* **2018**, *41*, 79–98. [CrossRef] [PubMed]
49. Wu, Z.; Li, T.; Ding, L.; Wang, C.; Teng, R.; Xu, S.; Cao, X.; Teng, N. Lily LlHSFC2 coordinates with HSFAs to balance heat stress response and improve thermotolerance. *New Phytol.* **2024**, *241*, 2124–2142. [CrossRef]
50. Bai, Y.; Meng, Y.; Huang, D.; Qi, Y.; Chen, M. Origin and evolutionary analysis of the plant-specific TIFY transcription factor family. *Genomics* **2011**, *98*, 128–136. [CrossRef]
51. Wang, Q.; Zhang, Z.; Guo, C.; Zhao, X.; Li, Z.; Mou, Y.; Sun, Q.; Wang, J.; Yuan, C.; Li, C. Hsf transcription factor gene family in peanut (*Arachis hypogaea* L.): Genome-wide characterization and expression analysis under drought and salt stresses. *Front. Plant Sci.* **2023**, *14*, 1214732. [CrossRef]
52. Begum, T.; Reuter, R.; Schöffl, F. Overexpression of AtHsfB4 induces specific effects on root development of Arabidopsis. *Mech. Dev.* **2013**, *130*, 54–60. [CrossRef]
53. Yu, H.-D.; Yang, X.-F.; Chen, S.-T.; Wang, Y.-T.; Li, J.-K.; Shen, Q.; Liu, X.-L.; Guo, F.-Q. Downregulation of chloroplast RPS1 negatively modulates nuclear heat-responsive expression of HsfA2 and its target genes in Arabidopsis. *PLoS Genet.* **2012**, *8*, e1002669. [CrossRef]
54. Li, C.; Chen, Q.; Gao, X.; Qi, B.; Chen, N.; Xu, S.; Chen, J.; Wang, X. AtHsfA2 modulates expression of stress responsive genes and enhances tolerance to heat and oxidative stress in Arabidopsis. *Sci. China Ser. C Life Sci.* **2005**, *48*, 540–550. [CrossRef] [PubMed]

current issues in molecular biology

MDPI

Article

Synergistic Effect of *Serratia fonticola* and *Pseudomonas koreensis* on Mitigating Salt Stress in *Cucumis sativus* L.

Sajid Ali [1,*,†], Murtaza Khan [2,†] and Yong-Sun Moon [1,*]

1 Department of Horticulture and Life Science, Yeungnam University, Gyeongsan 38541, Republic of Korea
2 Agriculture and Life Sciences Research Institute, Kangwon National University, Chuncheon 24341, Republic of Korea
* Correspondence: sajidbioali@gmail.com (S.A.); hangulmys@ynu.ac.kr (Y.-S.M.)
† These authors contributed equally to this work.

Abstract: Beneficial microbes enhance plant growth and development, even under stressful conditions. *Serratia fonticola* (S1T1) and *Pseudomonas koreensis* (S4T10) are two multi-trait plant growth-promoting rhizobacteria (PGPRs) that are resistant to saline conditions. This study evaluated the synergistic effect of these PGPRs on mitigating salinity stress (200 mM) in *Cucumis sativus*. Presently, the synergistic effect of both strains enhances the plant growth-promoting attributes of cucumber, and the growth parameters were significantly higher than those of uninoculated plants. The PGPR-treated plants revealed a significantly higher biomass and improved chlorophyll content. The inoculation of S1T1 and S4T10 and the synergistic effect of both promoted 23, 24, and 28% increases, respectively, in the fresh biomass and 16, 19.8, and 24% increases, respectively, in the dry biomass. Similarly, S1T1 and S4T10 and their synergistic effects led to 16.5, 28.4, and 38% increases, respectively, in the water potential and 18, 22, and 28% decreases, respectively, in abscisic acid (ABA). A reduction in the electrolytic leakage (EL) was additional proof of successful PGPR activities. Similarly, a decrease in the antioxidant levels, including those of malondialdehyde (21–30%), hydrogen peroxide (19–38%), and superoxide anions (24–34%), was observed, alongside an increase in antioxidant enzymes such as catalase (22–29%) and superoxide dismutase (17–27%). Additionally, the synergistic inoculation of the PGPRs enhanced the NaCl stress tolerance by upregulating the expression of the ion transporter genes *HKT1* (1–2-fold), *NHX* (1–3-fold), and *SOS1* (2–4-fold). Conclusively, the synergistic effect of the multi-trait PGPRs significantly enhances *C. sativus* L. growth under salt stress.

Keywords: abiotic stress; biofertilizers; cucumber; consortia; PGPR; salinity; stress-related gene

1. Introduction

Abiotic stress factors, such as salinity, drought, extreme temperature, heavy metals, and waterlogging, have become more frequent and severe because of global climate change, resulting in reduced crop growth and productivity [1,2]. According to estimates, abiotic stress is responsible for the loss of more than 50% of agricultural yields, and salt stress alone results in a 1–2% annual loss in arable land [3,4]. Salinity is a major problem in agricultural soil because of its adverse impact on different crops worldwide [1,5]. The increase in NaCl ions (Na^+ and Cl^-) in the soil causes salinity, where various factors including irrigation with saline water, the use of chemical fertilizers, and high evaporation contribute to salt accumulation and negatively impact different crop plants [6]. Salt stress can trigger several different unwanted responses in plants, lead to metabolic toxicity and reactive oxygen species formation, reduce photosynthesis, and affect the nutrient uptake capability of plants [7]. Moreover, an increase in the salt content causes ionic stress conditions that constrain plants from absorbing adequate amounts of nutrients and water from within close vicinity of the plant's rhizosphere [8]. In other words, salt stress is a combination of osmotic, ionic, and oxidative stresses that lowers the water potential of leaves and tissues and creates ionic imbalance and osmotic stress [9].

Environmentally friendly strategies, such as applying plant growth-promoting rhizobacteria (PGPRs), are essential for tackling such stressful conditions and enhancing crop yields in the future [2]. Various direct and indirect processes have shown that PGPRs develop symbiotic relationships with plants and support plant growth under stressful conditions [10]. In nature, plant and soil bacteria continue to interact closely, where the bacteria are typically mutualistic and beneficial to plants. A plant root system produces exudates that significantly affect the growth of rhizospheric bacteria. Thus, the interactions between rhizobacteria and plants have been studied extensively [11], and communication between interacting partners involves physiological and molecular mechanisms, even under abiotic stress conditions. Rhizobacteria relieve stressful conditions using different mechanisms; some are known as direct mechanisms, while others are indirect mechanisms [10]. The functional roles of a plethora of multi-trait PGPR strains have been reported in crop nutrition and health, while the mitigation of abiotic stress conditions has been reported from various environmental conditions. Subsequently, some PGPR strains belonging to *Arthrobacter*, *Azobacetr*, *Azospirillium*, *Bacillus*, *Burkholderia*, *Enterobacter*, *Flavobacterium*, and *Pseudomonas putida* are used as inoculants to improve plant growth attributes under such stressful conditions [12–14].

The studies of Yuan et al. [15] and Gholizadeh et al. [16] reported that *HKT1*, *NHX*, and *SOS1* transporters work together to regulate ion balance and enhance salinity tolerance in plants by reducing Na^+ toxicity while maintaining K^+ levels. *HKT1* retrieves Na^+ from the xylem, limiting its accumulation in aerial tissues. *NHX* transporters, located on the vacuolar membrane, sequester Na^+ into vacuoles, mitigating cytoplasmic toxicity and preserving osmotic stability. *SOS1*, a plasma membrane Na^+/H^+ antiporter, expels Na^+ from root cells into the rhizosphere to prevent excess accumulation, with its function regulated by calcium-binding proteins [15,16]. The use and efficacy of a single PGPR with one or more than one plant growth-promoting characteristics have been reported in different crops. However, the synergistic effect of PGPRs with multiple traits has received less attention. Using numerous PGPRs rather than a single bacteria in producing biofertilizers may be preferred because it synergistically affects nutrient mobilization and boosts efficacy, stability, and uniformity when utilized in different crops [14]. Our previous study reported on the multi-trait plant growth-promoting activities of both strains (*S. fonticola* (S1T1) and *P. koreensis* (S4T10)) individually and deposited the sequences in the NCBI database (MZ612851 and MZ612855). The study also revealed different activities of *S. fonticola* (S1T1) and *P. koreensis* (S4T10) including siderophore production, phosphate solubilization, and the production of indole-acetic acid (IAA), and the aminocyclopropane-1-carboxylic acid (ACC) deaminase enzyme [17]. These PGPRs can be used as biofertilizers, and their mode of application in the rhizosphere and phyllosphere needs to be investigated further in different crop plants.

Cucumis sativus L., commonly known as cucumber, is a plant species in the Cucurbitaceae family. Additionally, *C. sativus* L. is widely cultivated and serves as a versatile ingredient in various everyday meals. In the Republic of Korea, cucumber fruit is a substantial part of everyday meals consumed as part of traditional Kimchi recipes [18]. Different forms of fertilizers (including chemicals) are used to enhance the quality of the soil in which they grow. However, the applications of chemicals contaminate the soil, decrease soil fertility, and adversely affect groundwater quality [19]. Conversely, applying biofertilizers improves plant health and sustains soil fertility and the environment [20]. This study evaluated the synergistic effects of newly isolated *S. fonticola* (S1T1) and *P. koreensis* (S4T10) on cucumber plants under normal and NaCl stress conditions. Parameters such as electrolyte leakage (EL), relative water content (RWC), antioxidant activity, and pattern of gene expression in the salt-related genes under both normal and NaCl stress conditions were investigated to determine the role of these PGPRs on salt stress.

2. Materials and Methods

2.1. PGPR Strain Cultures

The isolation, characterization, and identification of *S. fonticola* (S1T1) and *P. koreensis* (S4T10) from plant species in Pohang Beach have previously been reported [17]. The cultures of these PGPRs were kept at 4 °C in about equal proportion of nutrient broth and glycerol (40%) for long-term storage. PGPR strains were added to the nutrient-rich Luria–Bertani (LB) broth medium (MB cell, Seoul, South Korea). This medium offers vital elements necessary for bacterial growth, including vitamins, minerals, and amino acids. The strains were cultured in the broth at 28 °C for 24 h to prepare the inoculum.

2.2. Cucumber Seed Sterilization

The detailed procedure described by Ofek et al. [21] was used to sterilize the surfaces of the cucumber seeds. Briefly, the seeds were surface-sterilized by soaking and gentle shaking in 3% sodium hypochlorite for 90 s (Chungnam, Republic of Korea). The seeds were then placed into 70% ethanol (Ulsan, South Korea) for 90 s before being cleaned with sterile deionized water [21]. The sterile filter paper was used to dry the seed surface.

2.3. Evaluation of the Range of Cucumber Seedling Tolerance to Salinity Stress

Surface-sterilized seeds were sown for germination in a 50-cell seed planting tray (one seedling per pot) containing 25 g of sterilized horticultural soil. Eight days later, uniform-sized seedlings were selected for transplantation. The selected seedlings were transplanted into 10 cm diameter pots (one seedling per pot) containing 210 g of soil and cultivated in a growth chamber in a natural dark and light photoperiod at 20 °C ± 2 °C with 55% relative humidity. The seedlings were watered to saturation thrice weekly; ten were considered per treatment. Different NaCl concentrations (0 mM, 100 mM, 200 mM, 300 mM, and 400 mM) were used to evaluate the range of plant tolerance to NaCl stress.

2.4. Salt Stress Application, PGPR Interactions, and NaCl Stress

The experimental design used eight distinct sets of plants to assess selected PGPRs as inoculum under NaCl stress. The treatments were as follows: (i) control (plants without PGPRs and salt stress); (ii) S1T1-treated; (iii) S4T10-treated; (iv) both S1T1- and S4T10-treated; (v) salt-stressed only; (vi) S1T1 and NaCl stress; (vii) S4T10 and NaCl stress; (viii) S1T1 + S4T10 + NaCl stress. Autoclaved soil (210 g) was removed from each pot (12 × 10 cm), and the seedlings were grown for three consecutive weeks in a greenhouse (28.2 °C); seedlings were watered using either distilled water or NaCl solution (DUKSAN, Ansan, Republic of Korea). Previously, the NaCl tolerance-related growth of the PGPRs was examined in LB medium at various NaCl concentrations (0, 1, 2, and 3%), and growth was measured at OD_{600} (Supplementary Figure S1). The selected PGPRs were used for a subsequent experiment to assess any influence on the cucumber plants. According to the recommended protocols, 40 mL of the bacterial isolate (10^8 CFU/mL) was applied twice to the autoclaved soil [22]. The bacterial suspension was applied twice before and after transplantation, with an interval of 48 h between applications. Subsequently, the plants were grown in the greenhouse at 28 ± 2 °C for five weeks, and salt stress was applied from week one to week five. A salt (NaCl) solution was used for high salt stress (200 mM), and three replicates of each treatment were prepared. Cucumber seedlings were subjected to high salt stress for one week to evaluate the effects of salinity on growth and development. Specifically, a 200 mM NaCl solution was utilized since this concentration is known to lead to significant osmotic and ionic stresses in plants [23,24]. The experiment focused on five-week-old seedlings since this represents a crucial developmental stage, during which the root systems and leaf structure are established. The seedlings were closely monitored for changes in various growth parameters, including plant height, overall biomass, and total chlorophyll content, and evaluated by using a SPAD meter (SPAD-502-Minolta Ltd, Osaka, Japan). When plants were harvested, shoot and root lengths and fresh and dried weights were measured. The dry weights following all treatments were calculated after the plants were dried in an oven at 70 °C for 48 h.

2.5. Determination of Leaf Relative Water Content (LRWC)

For the evaluation of LRWC, the Barrs and Weatherley [25] method was adopted with a few minor modifications, including the imbibition duration (24 h), to calculate the leaf water potential percentage (%). Identical leaf varieties were removed from the plants in each treatment and weighed (leaf fresh weight, LFW). To calculate the turgid fresh weight of leaves (TFWL) after calculating the LFW, in a sealed container, the leaves were submerged in the distilled water and left to float for 24 h. Then, leaves were blotted dry using blotting paper before being weighed. The leaves were heated to 70 °C in an oven for 24 h to achieve the dry biomass of the leaf (DWL). Subsequently, to determine the LRWC, the values of the LFW, TW, and DWL were applied to the following formula (Equation (1)):

$$\mathrm{LRWC}(\%) = (\frac{\mathrm{LFW} - \mathrm{DWL}}{\mathrm{TFWL} - \mathrm{DWL}}) \times 100 \quad (1)$$

2.6. Measuring the Electrolyte Leakage (EL)

The protocols reported by Dellagi et al. [26] were adopted to measure the electrolyte leakage with a few minor modifications. Here, a total of 200 g of fresh leaves samples was used in this study. The samples, each measuring 5 mm in length, were placed in test tubes containing 10 mL of distilled deionized water to assess the extent of EL. All tubes were sealed in a water bath heated up to 32 °C. An electrical conductivity meter was used to measure the initial electrical conductivity of the medium (EC1) (HURIBA Twin Cond B-173, Japan) after 120 min. The final electrical conductivity (EC2) of the sample was measured after autoclaving at 121 °C for 20 min and cooling to 25 °C. The quantity of electrolyte leakage was calculated using the following formula (Equation (2)):

$$\text{Percentage } (\%)\mathrm{E.L} = (\frac{\mathrm{EC1}}{\mathrm{EC2}}) \times 100 \quad (2)$$

2.7. Evaluation of Oxidative Stress Markers

Antioxidant activity analysis was conducted by pulverizing 400 mg of freeze-dried leaf tissues using a cold mortar and pestle before homogenizing the tissue in phosphate buffer (50 mM, pH 7.5; DUKSAN, Ansan, Republic of Korea) with polyvinylpyrrolidone (PVP) 1.0% (*w*/*v*), EDTA (0.1 mM), and Triton X-100 0.5% (*w*/*v*) [27]. Superoxide anions (SOAs) were precisely measured at a wavelength of 580 nm using the previously described method [28,29]. Likewise, the H_2O_2 levels were measured following various treatments using the previously described methods [30]. Lipid peroxidation (LPO) values were determined at a wavelength of 532 nm using the comprehensive method reported by Srivastava et al. [31].

2.8. Enzymatic Antioxidant Activity

The catalase (CAT) activity was evaluated using the methodology outlined by Radhakrishnan and Lee [32], and the resulting absorbance was measured at 240 nm. Spectrophotometer measurements were taken at 560 nm using the detailed technique for determining superoxide dismutase (SOD) activities [33].

2.9. Estimation of Endogenous Abscisic Acid (ABA)

The endogenous ABA content was extracted and measured using a previously described detailed approach [34]. The extracts were dried, methylated with diazomethane, and analyzed using GC/MS-SIM (6890N GC and 5973 mass selective detector, Agilent Technologies Inc. Palo Alto, CA, USA). Me-ABA and Me-[2H6]-ABA were quantified based on ion responses (m/e 162, 190 and m/e 166, 194, respectively) using Lab-Base (ThermoQuest, Manchester, UK) data system software.

2.10. Gene Expression Analysis Using qRT-PCR

Gene expression was analyzed through quantitative real-time PCR (qRT-PCR) following a previously described method [35]. RNA was extracted from treated cucumber leaf samples using TRIzol solution (Invitrogen, Carlsbad, CA, USA). Reverse transcription was carried out using the DiaStar™ RT kit (SolGent, Daejeon, Republic of Korea) in an Eco™ real-time PCR thermocycler. The qRT-PCR reactions included cDNA, 2X Real-Time PCR Master Mix with SYBR Green I (BioFACT™, Daejeon, Republic of Korea), and gene-specific primers (Table 1). β-actin was used as a reference gene for normalization, and expression levels in control plants were used to assess the impact of treatments on gene expression.

Table 1. Primer sequences of *SOS1*, *HKT1*, and *NHX1*.

Symbol of Gene	Forward Primer	Reverse Primer
SOS1	5′-AGGAAGGTTCAAAGCCTAGTG-3′	5′-CATGAGTAAATGTGGGGTGCA-3′
HKT1	5′-GGACCTCTCCACCTTGTCGT-3′	5′-CTGCTACCGTTTGTTTGTCACTCT-3′
NHX1	5′-TGCTTTTGCCACCCTTTCA-3′	5′-TTCCAACCAGAACCAATCCC-3′

2.11. Statistical Analysis

All data were collected in triplicate to ensure reliability and statistical robustness. To analyze the differences among treatment means, Duncan's Multiple Range Test (DMRT) was conducted using SAS version 9.2 software, which allows for multiple pairwise comparisons while controlling Type I error rates. The statistical analysis facilitated the identification of significant variations among treatments. Following the statistical evaluation, the results were visualized using GraphPad Prism software (Version 6.01, San Diego, CA, USA), which enabled the creation of high-quality graphical representations for better interpretation and comparison of the data.

3. Results

3.1. Effect of PGPRs (S. fonticola and P. koreensis) on Plant Growth Under Salt Stress

NaCl stress adversely affected the growth and development of cucumber plants; however, this effect was alleviated following inoculation with multi-trait PGPRs. Plants inoculated with PGPRs demonstrated improved growth compared to the uninoculated control plants (Figure 1).

Figure 1. Effects of PGPR (*S. fonticola* (S1T1) and *P. koreensis* (S4T10)) inoculation on cucumber plants under NaCl stress. The figure presents the phenotypic response of cucumber plants subjected to different treatments, including control (untreated), inoculation with *S. fonticola* (S1T1), *P. koreensis* (S4T10), and their combined application (S1T1+S4T10), both in the absence and presence of NaCl-induced salinity stress.

The inoculation of the PGPRs revealed a prominent effect on the growth of the cucumber plants and significantly mitigated the harmful effects of salinity stress. The results showed that the root and shoot lengths of the cucumber plants were greatly improved by applying PGPRs compared to the untreated control plants. In contrast to S1T1, the rhizobacterium *P. koreensis* (S4T10) treatment significantly promoted plant growth during NaCl-induced stress (Table 2). The inoculation with PGPRs such as S1T1 and S4T10 demonstrated prominent effects on the cucumber plants and significantly augmented the fresh biomass by 23.06% and 24.34%, respectively. By contrast, the synergistic effect of inoculating with both PGPRs revealed a 28.45% increase in the fresh biomass of the plants under normal control conditions. Similarly, the shoot length under *S. fonticola* (S1T1) and *P. koreensis* (S4T10) were 16.92% and 21.08% higher, respectively, compared to the control plants under normal conditions. In comparison, the consortia of *S. fonticola* (S1T1) and *P. koreensis* (S4T10) revealed a more than 20% increase in the shoot length of the inoculated plants under salinity conditions (Table 2).

The PGPR inoculations significantly alleviated the adverse effects of salt stress and enhanced the growth attributes of the cucumber plants. In comparison to the untreated control plants, those treated with *P. koreensis* (S4T10) and *S. fonticola* (S1T1) demonstrated a significant increase in the fresh and dry weight, shoot and root length, and chlorophyll content. This suggests that the application of these bacterial strains positively influenced plant growth-promoting attributes, including the chlorophyll content, potentially enhancing the photosynthetic efficiency and overall plant health. These results showed that the cucumber plants under salt stress had considerably smaller shoots (39.8%) and reduced fresh biomass (36.4%) than the control plants. Compared to the untreated control plants, the plants treated with *P. koreensis* (S4T10) and *S. fonticola* (S1T1) exhibited significantly higher chlorophyll contents. Moreover, the plants treated with *S. fonticola* (S1T1) and *P. koreensis* (S4T10) under NaCl stress had 24.56% and 25.52% higher chlorophyll contents, respectively. Thus, the combined application of S1T1 and S4T10 under salinity stress revealed improved plant chlorophyll contents. The treatment with *P. koreensis* (S4T10) and S1T1 also significantly improved the biomass (fresh and dry). The synergistic effect of inoculating with both PGPRs significantly enhanced the plant growth-promoting attributes under stressful conditions (Table 2; Figure 1). Furthermore, applying *P. koreensis* (S4T10) and *S. fonticola* (S1T1) significantly alleviated salt stress compared to in the untreated control plants.

Table 2. Impact of NaCl stress on plant growth with and without *S. fonticola* (S1T1) and *P. koreensis* (S4T10). The data presented show the mean values from three replicates. Different letters indicate significant differences according to DMRT analysis.

Treatment	Shoot Fresh Weight (gm/plant)	Shoot Dry Weight (gm/plant)	Shoot Length/(cm)	Root Length/(cm)	Chlorophyll Content/(SPAD)
Control	23.35 ± 1.32 ^{b}	5.98 ± 1.18 ^{b}	22.41 ± 1.11 ^{b}	21.27 + 0.14 ^{b}	40.11 ± 1.20 ^{b}
S1T1-Treated	27.62 ± 0.24 ^{a}	7.12 ± 0.35 ^{a}	26.86 ± 0.23 ^{a}	27.81 ± 1.81 ^{a}	44.08 ± 1.05 ^{a}
S4T10-Treated	28.15 ± 1.21 ^{a}	7.41 ± 0.17 ^{a}	26.90 ± 0.22 ^{a}	27.08 ± 0.35 ^{a}	44.62 ± 0.12 ^{a}
S1T1+ S4T10 Consortia-Treated	29.17 ± 0.11 ^{a}	7.60 ± 1.15 ^{a}	28.21 ± 0.41 ^{a}	30.67 ± 0.56 ^{a}	44.98 ± 0.46 ^{a}
NaCl (200mM) Stress	14.83 ± 1.41 ^{e}	3.68 ± 0.22 ^{e}	13.47 ± 1.12 ^{e}	15.34 ± 0.41 ^{e}	29.23 ± 0.22 ^{e}
S1T1+ NaCl Stress	18.25 ± 0.64 ^{d}	4.28 ± 0.56 ^{d}	15.75 ± 0.32 ^{d}	18.17 ± 0.11 ^{d}	36.41 ± 0.48 ^{d}
S4T10 + NaCl Stress	18.44 ± 0.52 ^{d}	4.41 ± 0.47 ^{d}	16.31 ± 0.65 ^{d}	19.05 ± 0.52 ^{d}	36.69 ± 0.34 ^{d}
Consortia (S1T1+ S4T10) + NaCl Stress	19.05 ± 1.54 ^{c}	4.58 ± 0.64 ^{c}	17.04 ± 0.41 ^{c}	20.13 ± 0.55 ^{c}	37.05 ± 0.61 ^{d}

3.2. PGPRs' Role in Leaf Water Potential Under NaCl Stress

The capability and capacity of a plant to store water are significantly affected by salt stress. The non-inoculated control cucumber plants exhibited an inferior leaf water status to those treated with the PGPRs. The combined application of *P. koreensis* (S4T10) and *S. fonticola* (S1T1) promoted a significant increase in the LRWC under salinity conditions (Figure 2). Compared to the *P. koreensis* (S4T10) and *S. fonticola* (S1T1)-inoculated cucumber plants, the LRWC of the control plants was 12.6% and 11.4% lower, respectively. Conversely, NaCl stress had a more distinct effect on the uninoculated control plants. The uninoculated control plants had a considerably lower water potential than the *P. koreensis* (S4T10)- and *S. fonticola* (S1T1)-inoculated plants under salt stress. When *P. koreensis* (S4T10) and *S. fonticola* (S1T1) were applied synergistically to the cucumber plants under salt stress, the resulting leaf water potential was increased by 38.0%. In contrast, the *S. fonticola* (S1T1)- and *P. koreensis* (S4T10)-treated plants under NaCl stress revealed LRWCs of 16.5% and 28.4%, respectively.

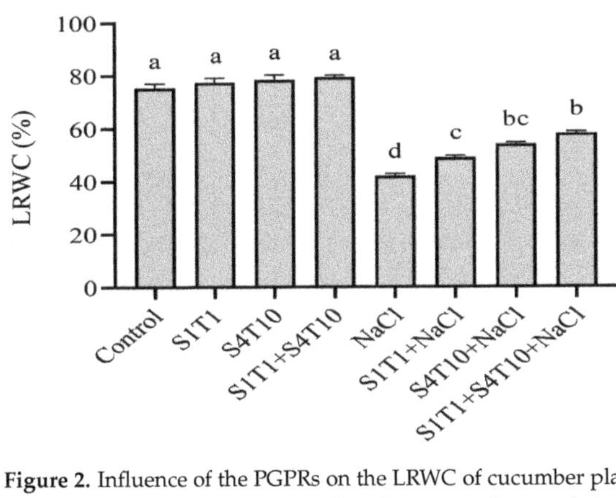

Figure 2. Influence of the PGPRs on the LRWC of cucumber plants under both normal and salt stress conditions. The data are presented as the mean values and standard error bars from three replicates. Different letters on the mean bars indicate significant differences according to DMRT analysis.

3.3. Effects of PGPRs on Electrolyte Leakage, MDA, H_2O_2, and SOAs Under NaCl Stress

The EL, MDA, H_2O_2, and superoxide anion contents were assessed to determine the impact of the PGPRs and NaCl stress on the membrane integrity of the plant. The findings indicated a notable increase in electrolyte leakage, MDA, H_2O_2, and superoxide anion levels under NaCl-induced stress (Figure 3). The plants under salt stress revealed more electrolyte leakage from the leaf tissues than the control (treated only with water) plants. Comparatively, applying S1T1 and S4T10 individually and their combined application under NaCl stress reduced electrolyte leakage up to 19, 27, and 31%, respectively (Figure 3A). Salt stress adversely affected the membranes of these plants compared to the plants treated with PGPRs. The analysis of the MDA contents revealed the extent of lipid peroxidation (LPO); higher MDA levels were observed in the cucumber plants treated with NaCl compared to those in the normal control plants. In contrast, the individual and synergistic inoculation of the PGPRs decreased the MDA levels of the cucumber plants by 21, 23, and 30%, respectively, under salinity conditions (Figure 3B).

Likewise, the ROS content was evaluated in the cucumber plants subjected to NaCl stress. Salt stress enhances the production of ROS, such as H_2O_2 and superoxide anions (SOAs). Moreover, the H_2O_2 level increased in response to NaCl stress. However, the H_2O_2 production in the cucumber plants was reduced significantly following the application of the PGPRs: 19% after the *S. fonticola* treatment and 22% following the *P. koreensis* treatment (Figure 3C). A similar pattern was observed in the superoxide anion (SOA) levels of the cucumber plants exposed to NaCl stress (Figure 3D). Indeed, SOA production was reduced significantly in the plants inoculated with *S.fonticola* (24%) and *P. koreensis* (27%) compared to the plants under NaCl stress conditions.

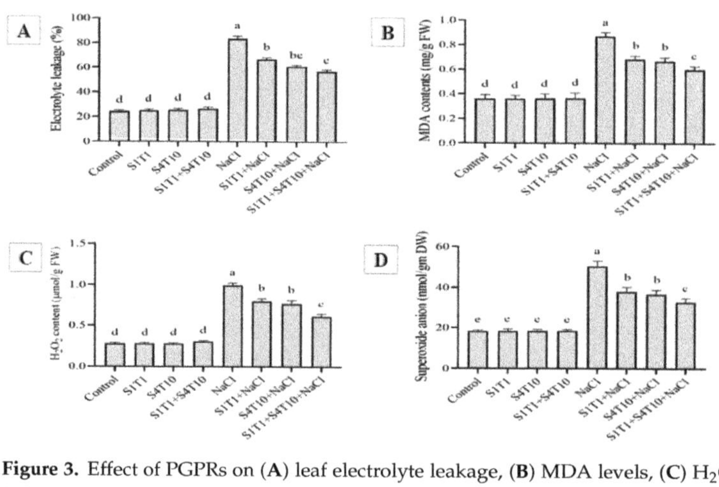

Figure 3. Effect of PGPRs on (**A**) leaf electrolyte leakage, (**B**) MDA levels, (**C**) H_2O_2 content, and (**D**) superoxide anion levels in cucumber plants under both normal and NaCl stress conditions. Data represent the mean ± standard error from three replicates. Different letters on the bars denote statistically significant differences based on DMRT analysis.

3.4. Effects of PGPRs on Antioxidant Enzymes Under NaCl Stress

Plants activate antioxidants to regulate ROS generation and mitigate the detrimental effects of external stressors, including NaCl stress. Higher salt concentrations affect both soil properties and microbial processes, including respiration and enzyme activity. The enzymatic activity of SOD and CAT in the cucumber plants under normal and NaCl stress conditions, with and without PGPRs, was assessed (Figure 4). PGPR application under NaCl stress enhanced the SOD and CAT activity compared to that in the plants subjected to NaCl stress alone. The inoculation of *S. fonticola* (S1T1) and *P. koreensis* (S4T10) increased CAT biosynthesis by 22% and 23%, respectively. Furthermore, the combined application of the PGPRs further enhanced CAT production to 29% under salt stress conditions (Figure 4A). Likewise, the cucumber plants showed a similar pattern for SOD under NaCl stress conditions. Under salt stress conditions, the PGPRs enhanced SOD biosynthesis, whereby S1T1 produced a 17% increase and S4T10 produced an approximately 21% increase in the biosynthesis of SOD. On the other hand, the synergistic application of the PGPRs revealed a 27% increase in the activity of SOD (Figure 4B). Therefore, the levels of the analyzed antioxidant enzymes increased in the cucumber plants upon the inoculation with the multi-trait PGPRs.

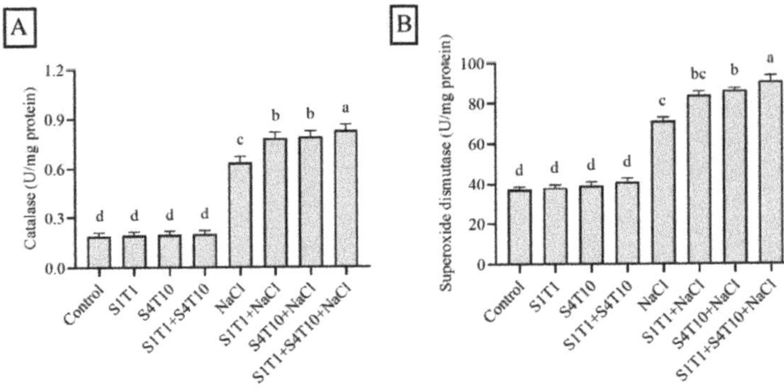

Figure 4. Effect of PGPRs on (**A**) catalase activity and (**B**) superoxide dismutase. The data are presented as the mean and standard error bars from three replicates. Different letters on the mean bars indicate significant differences according to DMRT analysis.

3.5. Regulation of Endogenous ABA by PGPRs Under NaCl Stress

ABA is essential in modulating plant growth and performance, particularly in response to abiotic stressors, such as NaCl stress. Furthermore, PGPRs have been reported to promote prominent effects on regulating ABA in plants under stressful conditions. Therefore, the impact of PGPRs on the ABA content in the cucumber plants was assessed under NaCl stress conditions. Based on the research results, the plants under stress had significantly higher ABA concentrations than those in the control plants. Conversely, the plants treated with PGPRs under NaCl stress exhibited a modulated ABA content. As a result, under NaCl stress settings, the levels of ABA produced increased exponentially, in contrast to the plants treated with *S. fonticola* (S1T1) and *P. koreensis* (S4T10), which showed decreases of 18% and 22%, respectively. By contrast, the combined application of these PGPRs demonstrated a 28% decrease in the ABA content under salt stress conditions (Figure 5).

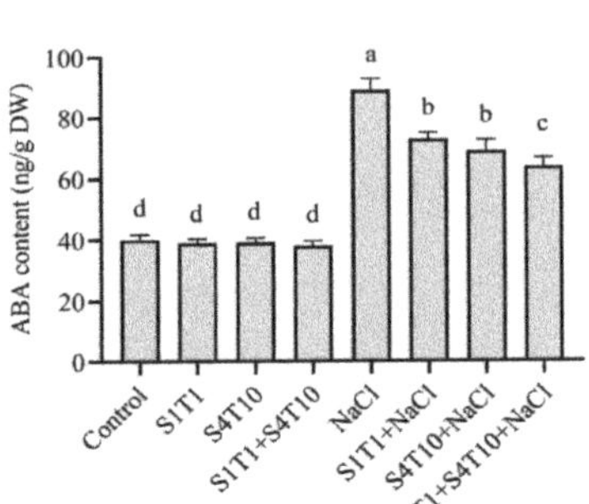

Figure 5. Effects of PGPRs on abscisic acid (ABA) regulation under NaCl stress. Different letters on the mean bars indicate significant differences according to DMRT analysis.

3.6. Expression of NaCl Stress-Related Genes and Effects of PGPRs

Plants upregulate the expression of candidate genes under stressful conditions. Recent studies have reported that salt-responsive genes are modulated to combat the unwanted effects of salinity. Therefore, sodium transporters play a crucial role in plant defense. These transporters can be either antiporters or symporters; the former remove sodium ions from the root cell and redistribute them throughout the body to reduce the adverse effects of salt stress and restore water homeostasis. The expressions of three putative genes (*HKT1*, *SOS1*, and *NHX*) were examined to increase the systematic knowledge of the potential mechanism through which cucumber plants activate NaCl stress tolerance.

HKT1, *NHX*, and SOS1 were upregulated in the cucumber plants under NaCl stress. On the other hand, transcript accumulation of the *HKT1*, *NHX*, and SOS1 genes was augmented in the cucumber plants inoculated with PGPRs. The expression of *HKT1* was enhanced by 1-fold under *S. fonticola* (S1T1) and 1.3-fold under *P. koreensis* (S4T10) compared to that in the NaCl-stressed control plants. Meanwhile, the combined application of S1T1 and S4T10 promoted a 1.7-fold increase in the expression of *HKT1* (Figure 6A). Thus, applying *S. fonticola* (S1T1) and *P. koreensis* (S4T10) to cucumber plants under NaCl stress can induce adaptability, which results in a significant increase in *HKT1*.

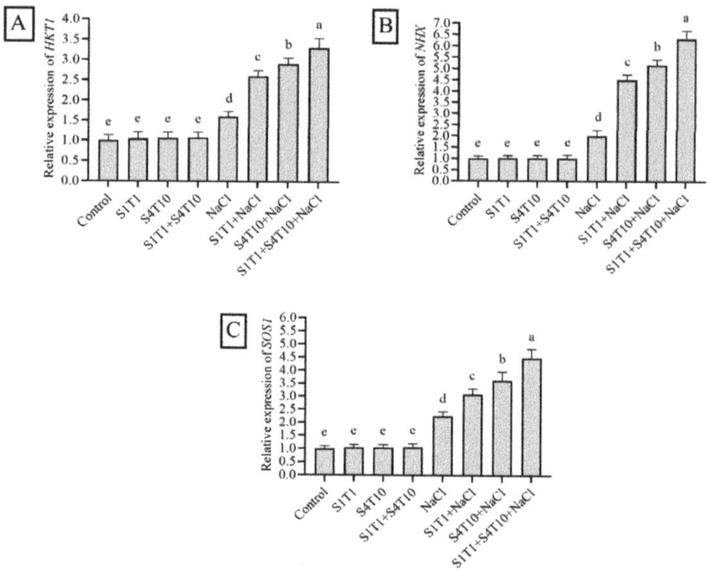

Figure 6. Effects of PGPRs on gene expression under normal and salt stress conditions. Leaf samples were collected from plants across all treatment groups to analyze gene expression. The relative expression levels of (**A**) HKT1, (**B**) NHX, and (**C**) SOS1 genes under normal and salt stress conditions were quantified using β-actin as the reference gene. Data are presented as mean ± standard error from three replicates, with different letters on the bars indicating statistically significant differences based on DMRT analysis.

Similarly, the *NHX* gene expression increased significantly by 0.9-fold, 1.4-fold, and 2.3-fold upon the inoculation with S1T1 and S4T10 and their synergistic application, respectively, compared to the salt stress control plants (Figure 6B). The same pattern was observed for the *SOS1* gene, whereby a significant increase in expression was observed in the cucumber plants treated with the PGPRs (Figure 6C). The application *S. fonticola* (S1T1) and *P. koreensis* (S4T10) and their synergistic effect on cucumber plants improved NaCl resistance by increasing *SOS1* expression by 2.5-fold, 3.17-fold, and 4.34-fold, respectively, compared to in the uninoculated cucumber plants under salt stress (Figure 6C).

4. Discussion

Major environmental stress factors adversely affecting plant development and agricultural productivity include salinity, drought, heavy metals, extreme temperature, and ultraviolet (UV) radiation [36]. Among these factors, salt stress alone affects approximately 11% of the world's irrigated land, which can result in a loss of more than 50% of crop yields in species sensitive to salt [4,37]. Salinity stress inhibits plant growth by increasing sodium uptake, increasing the amount of endogenous ABA, generating increased and toxic amounts of ROS, and decreasing the photosynthetic rate and potassium uptake [38]. Thus, salt stress has detrimental effects on the morphological, biochemical, and molecular aspects of plants.

The use of PGPRs can have significant environmental impacts, particularly on soil health and microbial diversity. PGPRs enhance soil fertility by improving the nutrient availability, increasing organic matter decomposition, and promoting plant–microbe interactions. They contribute to soil structure stabilization and increase water retention, reducing erosion risks [39]. Microbial inoculation offers a sustainable and cost-effective alternative to chemical amendments, genetic modifications, and advanced irrigation techniques, offering significant ecological benefits. Unlike chemical fertilizers, which contribute to soil degradation and environmental pollution, microbial inoculants enhance nutrient cycling, improve soil structure, and promote long-term fertility while minimizing synthetic input dependency [40]. Compared to genetic modifications, which require extensive regulatory approval and pose potential ecological risks, microbial inoculation naturally enhances plant resilience without altering genetic integrity. Additionally, while advanced irrigation strategies optimize water use, microbial inoculants improve drought tolerance by enhancing root development and soil water retention, thereby reducing irrigation demands. This approach supports sustainable agricultural practices by maintaining soil biodiversity, minimizing environmental impacts, and ensuring long-term productivity [40,41].

Several approaches have been used to mitigate the adverse impacts of abiotic stressors, including salt stress, and to enhance crop plant growth and productivity [20], including using various chemicals, producing genetically modified crops, and modern agricultural methods and irrigation systems. Ultimately, the current agricultural system and crop production greatly benefit from the contributions of these approaches. However, natural ecosystems have consequently suffered, since most fertilizers and chemicals end up in fields where they are wasted or kill beneficial microbes. Notably, various microbial plant biostimulants have recently been used to improve crop yields, enhance abiotic stress tolerance, and improve nutrient utilization [42]. The findings in this paper offer insight into improving the approach of applying PGPRs to combat NaCl stress. The rhizobacteria *S. fonticola* and *P. koreensis* have synergistic efficiency in stimulating plant growth-promoting features to induce salinity stress tolerance.

In the current investigation, the cucumber plant growth and development were efficiently supported by the selected PGPRs (*S. fonticola* and *P. koreensis*), which also reduced the detrimental effects of NaCl stress. Nadeem et al. [43] also reported that PGPRs rescued cucumber plants following exposure to salt stress. Furthermore, Sapre et al. [44] reported that the application of PGPRs reduced the toxic effects of salt stress on pea plants, and the inoculation of rhizobacteria enhanced the parameters of pea seedling development under salinity stress. Furthermore, the inoculation of rhizobacteria decreased salt stress by altering physiological, biochemical, and molecular factors. In addition, the PGPR-inoculated plants displayed reduced electrolyte leakage and H_2O_2 concentrations under NaCl stress compared to those of the uninoculated seedlings.

Salt stress leads to water deficit conditions in plant cells, affecting plant growth and development. The first observable sign of salt stress is a decrease in the RWC; hence, assessing the RWC is a good method for evaluating the water condition of the plant [45]. Under salt stress conditions, applying *S. fonticola* (S1T1) and *P. koreensis* (S4T10) enhanced the RWC compared to that of control plants. The current findings are consistent with those of Singh et al. [46], in that PGPRs assist the plants under salt stress. Abiotic stresses, including salt stress, induce electrolyte leakage and increase the MDA, H_2O_2, and superoxide anion contents [47]. Conversely, applying PGPRs reduces the production of these oxidative stress markers [18]. Similarly, salt concentrations that are too high interfere with microbial processes, including respiration and the activity of soil enzymes, and affect the physicochemical characteristics of the soil. PGPR application increases the activities of antioxidant enzymes, including CAT and SOD [48]. The findings of this study are also in line with those of these previous studies, and although salt stress increased the production of oxidative stress markers, the inoculation with *S. fonticola* and *P. koreensis*, particularly via the root zone, lowered their production. These multi-trait PGPRs significantly enhance cucumber growth and resilience under NaCl stress through several key mechanisms. Firstly, the PGPRs improve nutrient acquisition by solubilizing essential minerals, such as phosphorus, producing siderophores and ACC deaminase, thus supporting root development and overall growth in saline conditions [49]. Secondly, the PGPRs produce plant hormones, such as IAA, which promote root elongation and cell division, helping cucumber plants absorb more water and nutrients despite the inhibitory effects of salt [50,51].

ABA is an essential hormone that is crucial for growth and stress adaptation. Plants possess an improved ability to withstand abiotic stresses after activating physiological changes that improve survival and preserve water [52,53]. To reduce water loss during water shortage conditions, the plant can also control the opening and closing of stomata—microscopic pores on leaves. Furthermore, ABA affects cell division and elongation, which normally slows growth under stress to conserve resources. ABA also affects the development of roots, assisting them in adjusting to changes in the availability of nutrients and water. Abiotic stress conditions cause an increase in the levels of ABA in plants. Intriguingly, these results showed that ABA production was considerably reduced in the presence of the PGPRs compared to in the control plants under stress conditions. Although some studies suggested that adding bacteria to plant roots and leaves could boost the degree of ABA accumulation, the effect may vary depending on the microbe and plant species [54,55].

The coordinated action of transporters including *HKT1*, *NHX*, and *SOS1* enables plants to maintain ion homeostasis and improve their salinity tolerance by minimizing Na^+ toxicity while preserving essential K^+ levels. *HKT1*, *NHX*, and *SOS1* are key transporters involved in salinity stress tolerance by regulating Na^+ exclusion, vacuolar compartmentalization, and ion homeostasis in plants. *HKT1* (typically, a mediator of relatively selective Na^+ transport) functions as a Na^+ transporter that facilitates Na^+ retrieval from the xylem, preventing its excessive accumulation in aerial tissues. *NHX* transporters, primarily localized in the vacuolar membrane, mediate Na^+ sequestration into vacuoles, reducing cytoplasmic Na^+ toxicity and maintaining cellular osmotic balance. *SOS1*, a plasma membrane Na^+/H^+ antiporter, actively extrudes Na^+ from root cells into the rhizosphere, preventing Na^+ overload in plant tissues and modulation by calcium-binding proteins [15]. These transporters allow plants to maintain ion homeostasis and improve salt tolerance. Several PGPRs increase the ability of a plant to withstand salt stress [16,56]. By regulating ion transporter gene expression, PGPRs improve the resistance of plants to salt stress. Plants that grow in salty environments experience difficulties, such as ionic imbalances and elevated osmotic stress levels, which can hinder their development. By controlling the expression of particular ion transporter genes, which are critical for preserving ion homeostasis and controlling the absorption and distribution of vital nutrients, PGPRs help resolve these problems [16]. Moreover, PGPRs assist plants in more effectively managing excess salt by modifying the expression of these genes, ultimately reducing the negative effects on plants and maintaining development and yield in difficult environmental conditions.

Salt stress tolerance in plants is conferred via various stress-related genes involved in signal transduction, ion transporters, transcription regulation, and metabolic pathways. The current study also found that applying *S. fonticola* (S1T1) and *P. koreensis* (S4T10) significantly increased the expression of *HKT1*, *NHX*, and *SOS1* in response to salt stress. Furthermore, the study by Liu et al. [57] reported that the inoculation of *Bacillus amyloliquefaciens* enhanced the expression of *HKT1* in the shoots of *Arabidopsis thaliana*. The enhancement in the growth attributes under salinity can be connected with the improved expression of the aforementioned genes under the inoculation of the PGPRs. Conclusively, applying multi-trait PGPRs could be an efficient way to confer NaCl stress resistance in crop plants. In the present study, the synergistic application of S1T1 and S4T10 mitigated NaCl (200 mM) stress in cucumber plants. The experimental data showed that the multi-trait PGPRs *S. fonticola* (S1T1) and *P. koreensis* (S4T10) played a pivotal role in the recovery of cucumber plants by activating antioxidants (SOD and CAT), modulating the phytohormone (ABA) levels, maintaining the RWC, and regulating ion transporter genes (*HKT1*, *NHX*, and *SOS1*). Therefore, the current study validates using a consortium of PGPRs to produce microbial plant biostimulants for the increased growth and quality yield of crop plants growing under saline stress conditions. Applying PGPRs in cucumber cultivation can promote growth and stress tolerance in saline areas, thereby encouraging sustainable agricultural methods. Moreover, future studies must be designed to include dose-dependent trials to evaluate the efficacy of PGPRs across different stress intensities, strain-specific quantification techniques, and confirm the survival and interaction dynamics of more than one strain in soil.

5. Conclusions

Overall, *Cucumis sativus* L. yielded prominent benefits from synergistic inoculation with two multi-trait PGPRs, *S. fonticola* (S1T1) and *P. koreensis* (S4T10). This study demonstrates that combining these PGPRs increases plant stress tolerance and improves growth indices, including the biomass and chlorophyll content. The observed improvements in the fresh and dry biomass, water potential, and a decrease in ABA and electrolyte leakage demonstrate the efficiency of these PGPRs in reducing salinity stress. The ability of these PGPRs to increase plant resistance is further supported by the downregulation of oxidative stress indicators and the overexpression of ion transporter genes. Thus, the synergistic inoculation of these PGPRs appears to be a viable method for enhancing the growth and tolerance of plants under saline conditions.

Supplementary Materials: The following supporting information can be downloaded at: https://www.mdpi.com/article/10.3390/cimb47030194/s1.

Author Contributions: S.A. and Y.-S.M. conceived and designed the research. S.A. and M.K. conducted experiments. Y.-S.M. contributed reagents and analytical tools. S.A., M.K. and Y.-S.M. analyzed data. S.A., Y.-S.M. and M.K. wrote the manuscript. All authors have read and agreed to the published version of the manuscript.

Funding: This research received no external funding.

Institutional Review Board Statement: This article does not contain any studies with animals performed by any of the authors.

Informed Consent Statement: Not applicable.

Data Availability Statement: Datasets generated during the current study are available from the corresponding author on reasonable request.

Conflicts of Interest: All authors declare that there are no conflicts of interest.

References

1. Singh, A. Soil salinity: A global threat to sustainable development. *Soil Use Manag.* **2022**, *38*, 39–67. [CrossRef]
2. de Andrade, L.A.; Santos, C.H.B.; Frezarin, E.T.; Sales, L.R.; Rigobelo, E.C. Plant growth-promoting rhizobacteria for sustainable agricultural production. *Microorganisms* **2023**, *11*, 1088. [CrossRef]
3. Etesami, H.; Maheshwari, D.K. Use of plant growth promoting rhizobacteria (PGPRs) with multiple plant growth promoting traits in stress agriculture: Action mechanisms and future prospects. *Ecotoxicol. Environ. Saf.* **2018**, *156*, 225–246. [CrossRef] [PubMed]
4. Sangiorgio, D.; Cellini, A.; Donati, I.; Pastore, C.; Onofrietti, C.; Spinelli, F.J.A. Facing climate change: Application of microbial biostimulants to mitigate stress in horticultural crops. *Agronomy* **2020**, *10*, 794. [CrossRef]
5. Truşcă, M.; Gâdea, Ş.; Vidican, R.; Stoian, V.; Vâtcă, A.; Balint, C.; Stoian, V.A.; Horvat, M.; Vâtcă, S. Exploring the Research Challenges and Perspectives in Ecophysiology of Plants Affected by Salinity Stress. *Agriculture* **2023**, *13*, 734. [CrossRef]
6. Machado, R.M.A.; Serralheiro, R.P. Soil salinity: Effect on vegetable crop growth. Management practices to prevent and mitigate soil salinization. *Horticulturae* **2017**, *3*, 30. [CrossRef]
7. Zhao, C.; Zhang, H.; Song, C.; Zhu, J.-K.; Shabala, S. Mechanisms of plant responses and adaptation to soil salinity. *Innovation* **2020**, *1*, 100017. [CrossRef]
8. Ismail, A.; Takeda, S.; Nick, P. Life and death under salt stress: Same players, different timing? *J. Exp. Bot.* **2014**, *65*, 2963–2979. [CrossRef]
9. Shah, A.N.; Tanveer, M.; Abbas, A.; Fahad, S.; Baloch, M.S.; Ahmad, M.I.; Saud, S.; Song, Y. Targeting salt stress coping mechanisms for stress tolerance in Brassica: A research perspective. *Plant Physiol. Biochem.* **2021**, *158*, 53–64. [CrossRef]
10. Giannelli, G.; Potestio, S.; Visioli, G. The contribution of PGPR in salt stress tolerance in crops: Unravelling the molecular mechanisms of cross-talk between plant and bacteria. *Plants* **2023**, *12*, 2197. [CrossRef]
11. Glick, B.R. The enhancement of plant growth by free-living bacteria. *Can. J. Microbiol.* **1995**, *41*, 109–117. [CrossRef]
12. Samain, E.; Ernenwein, C.; Aussenac, T.; Selim, S. Effective and durable systemic wheat-induced resistance by a plant-growth-promoting rhizobacteria consortium of Paenibacillus sp. strain B2 and *Arthrobacter* spp. strain AA against Zymoseptoria tritici and drought stress. *Physiol. Mol. Plant Pathol.* **2022**, *119*, 101830. [CrossRef]
13. Sivasakthi, S.; Usharani, G.; Saranraj, P. Biocontrol potentiality of plant growth promoting bacteria (PGPR)-Pseudomonas fluorescens and Bacillus subtilis: A review. *Afr. J. Agric. Res.* **2014**, *9*, 1265–1277.
14. Martínez, O.A.; Jorquera, M.A.; Crowley, D.E.; de la Luz Mora, M. Influence of nitrogen fertilisation on pasture culturable rhizobacteria occurrence and the role of environmental factors on their potential PGPR activities. *Biol. Fertil. Soils* **2011**, *47*, 875–885. [CrossRef]
15. Yuan, J.; Cao, H.; Qin, W.; Yang, S.; Zhang, D.; Zhu, L.; Song, H.; Zhang, Q. Genomic and Modern Biotechnological Strategies for Enhancing Salt Tolerance in Crops. *New Crops* **2024**, *2*, 100057. [CrossRef]
16. Gholizadeh, F.; Mirmazloum, I.; Janda, T. Genome-wide identification of HKT gene family in wheat (*Triticum aestivum* L.): Insights from the expression of multiple genes (HKT, SOS, TVP and NHX) under salt stress. *Plant Stress* **2024**, *13*, 100539. [CrossRef]
17. Moon, Y.-S.; Ali, S. Isolation and identification of multi-trait plant growth–promoting rhizobacteria from coastal sand dune plant species of Pohang beach. *Folia Microbiol.* **2022**, *67*, 523–533. [CrossRef] [PubMed]
18. Moon, Y.-S.; Khan, M.; Khan, M.A.; Ali, S. Ameliorative symbiosis of Serratia fonticola (S1T1) under salt stress condition enhance growth-promoting attributes of *Cucumis sativus* L. *Symbiosis* **2023**, *89*, 283–297. [CrossRef]
19. Rashmi, I.; Roy, T.; Kartika, K.S.; Pal, R.; Coumar, V.; Kala, S.; Shinoji, K.C. Organic and inorganic fertilizer contaminants in agriculture: Impact on soil and water resources. In *Contaminants in Agriculture: Sources, Impacts and Management*; Springer: Cham, Switzerland, 2020; pp. 3–41.
20. Bittencourt, P.P.; Alves, A.F.; Ferreira, M.B.; da Silva Irineu, L.E.S.; Pinto, V.B.; Olivares, F.L. Mechanisms and applications of bacterial inoculants in plant drought stress tolerance. *Microorganisms* **2023**, *11*, 502. [CrossRef]
21. Ofek, M.; Hadar, Y.; Minz, D. Colonization of cucumber seeds by bacteria during germination. *Environ. Microbiol.* **2011**, *13*, 2794–2807. [CrossRef]

22. Jhuma, T.A.; Rafeya, J.; Sultana, S.; Rahman, M.T.; Karim, M.M. Isolation of endophytic salt-tolerant plant growth-promoting rhizobacteria from Oryza sativa and evaluation of their plant growth-promoting traits under salinity stress condition. *Front. Sustain. Food Syst.* **2021**, *5*, 687531. [CrossRef]
23. Liu, D.; Dong, S.; Bo, K.; Miao, H.; Li, C.; Zhang, Y.; Zhang, S.; Gu, X. Identification of QTLs controlling salt tolerance in cucumber (*Cucumis sativus* L.) seedlings. *Plants* **2021**, *10*, 85. [CrossRef] [PubMed]
24. Li, L.; Du, L.; Cao, Q.; Yang, Z.; Liu, Y.; Yang, H.; Duan, X.; Meng, Z. Salt tolerance evaluation of cucumber germplasm under sodium chloride stress. *Plants* **2023**, *12*, 2927. [CrossRef]
25. Barrs, H.D.; Weatherley, P.E. A re-examination of the relative turgidity technique for estimating water deficits in leaves. *Aust. J. Biol. Sci.* **1962**, *15*, 413–428. [CrossRef]
26. Dellagi, A.; Brisset, M.-N.; Paulin, J.-P.; Expert, D. Dual role of desferrioxamine in Erwinia amylovora pathogenicity. *Mol. Plant-Microbe Interact.* **1998**, *11*, 734–742. [CrossRef]
27. Khan, M.A.; Hamayun, M.; Asaf, S.; Khan, M.; Yun, B.-W.; Kang, S.-M.; Lee, I.-J. Rhizospheric *Bacillus* spp. rescues plant growth under salinity stress via regulating gene expression, endogenous hormones, and antioxidant system of *Oryza sativa* L. *Front. Plant Sci.* **2021**, *12*, 665590. [CrossRef]
28. Gajewska, E.; Skłodowska, M. Effect of nickel on ROS content and antioxidative enzyme activities in wheat leaves. *Biometals* **2007**, *20*, 27–36. [CrossRef]
29. Khan, M.A.; Ullah, I.; Waqas, M.; Hamayun, M.; Khan, A.L.; Asaf, S.; Kang, S.-M.; Kim, K.-M.; Jan, R.; Lee, I.-J. Halo-tolerant rhizospheric Arthrobacter woluwensis AK1 mitigates salt stress and induces physio-hormonal changes and expression of GmST1 and GmLAX3 in soybean. *Symbiosis* **2019**, *77*, 9–21. [CrossRef]
30. Park, H.-S.; Kazerooni, E.A.; Kang, S.-M.; Al-Sadi, A.M.; Lee, I.-J. Melatonin enhances the tolerance and recovery mechanisms in *Brassica juncea* (L.) Czern. under saline conditions. *Front. Plant Sci.* **2021**, *12*, 593717. [CrossRef]
31. Srivastava, S.; Sinha, P.; Sharma, Y.K. Status of photosynthetic pigments, lipid peroxidation and anti-oxidative enzymes in Vigna mungo in presence of arsenic. *J. Plant Nutr.* **2017**, *40*, 298–306. [CrossRef]
32. Radhakrishnan, R.; Lee, I.-J. Spermine promotes acclimation to osmotic stress by modifying antioxidant, abscisic acid, and jasmonic acid signals in soybean. *J. Plant Growth Regul.* **2013**, *32*, 22–30. [CrossRef]
33. Marklund, S.; Marklund, G. Involvement of the superoxide anion radical in the autoxidation of pyrogallol and a convenient assay for superoxide dismutase. *Eur. J. Biochem.* **1974**, *47*, 469–474. [CrossRef] [PubMed]
34. Qi, Q.; Rose, P.A.; Abrams, G.D.; Taylor, D.C.; Abrams, S.R.; Cutler, A.J. (+)-Abscisic acid metabolism, 3-ketoacyl-coenzyme A synthase gene expression, and very-long-chain monounsaturated fatty acid biosynthesis in *Brassica napus* embryos. *Plant Physiol.* **1998**, *117*, 979–987. [CrossRef] [PubMed]
35. Shahid, M.; Cai, G.; Zu, F.; Zhao, Q.; Qasim, M.U.; Hong, Y.; Fan, C.; Zhou, Y. Comparative transcriptome analysis of developing seeds and silique wall reveals dynamic transcription networks for effective oil production in *Brassica napus* L. *Int. J. Mol. Sci.* **2019**, *20*, 1982. [CrossRef] [PubMed]
36. Niyoifasha, C.J.; Borena, B.M.; Ukob, I.T.; Minh, P.N.; Al Azzawi, T.N.I.; Imran, M.; Ali, S.; Inthavong, A.; Mun, B.-G.; Lee, I.-J. Alleviation of Hg-, Cr-, Cu-, and Zn-Induced Heavy Metals Stress by Exogenous Sodium Nitroprusside in Rice Plants. *Plants* **2023**, *12*, 1299. [CrossRef] [PubMed]
37. Zörb, C.; Geilfus, C.M.; Dietz, K.J. Salinity and crop yield. *Plant Biol.* **2019**, *21*, 31–38. [CrossRef]
38. Ansari, M.; Shekari, F.; Mohammadi, M.H.; Juhos, K.; Végvári, G.; Biró, B. Salt-tolerant plant growth-promoting bacteria enhanced salinity tolerance of salt-tolerant alfalfa (*Medicago sativa* L.) cultivars at high salinity. *Acta Physiol. Plant.* **2019**, *41*, 195. [CrossRef]
39. Vocciante, M.; Grifoni, M.; Fusini, D.; Petruzzelli, G.; Franchi, E. The role of plant growth-promoting rhizobacteria (PGPR) in mitigating plant's environmental stresses. *Appl. Sci.* **2022**, *12*, 1231. [CrossRef]
40. Shah, A.; Nazari, M.; Antar, M.; Msimbira, L.A.; Naamala, J.; Lyu, D.; Rabileh, M.; Zajonc, J.; Smith, D.L. PGPR in agriculture: A sustainable approach to increasing climate change resilience. *Front. Sustain. Food Syst.* **2021**, *5*, 667546. [CrossRef]
41. Mohanty, P.; Singh, P.K.; Chakraborty, D.; Mishra, S.; Pattnaik, R. Insight into the role of PGPR in sustainable agriculture and environment. *Front. Sustain. Food Syst.* **2021**, *5*, 667150. [CrossRef]
42. Ali, S.; Moon, Y.-S.; Hamayun, M.; Khan, M.A.; Bibi, K.; Lee, I.-J. Pragmatic role of microbial plant biostimulants in abiotic stress relief in crop plants. *J. Plant Interact.* **2022**, *17*, 705–718. [CrossRef]
43. Nadeem, S.M.; Ahmad, M.; Naveed, M.; Imran, M.; Zahir, Z.A.; Crowley, D.E. Relationship between in vitro characterization and comparative efficacy of plant growth-promoting rhizobacteria for improving cucumber salt tolerance. *Arch. Microbiol.* **2016**, *198*, 379–387. [CrossRef] [PubMed]
44. Sapre, S.; Gontia-Mishra, I.; Tiwari, S. Plant growth-promoting rhizobacteria ameliorates salinity stress in pea (*Pisum sativum*). *J. Plant Growth Regul.* **2022**, *41*, 647–656. [CrossRef]
45. Shabaan, M.; Asghar, H.N.; Zahir, Z.A.; Zhang, X.; Sardar, M.F.; Li, H. Salt-tolerant PGPR confer salt tolerance to maize through enhanced soil biological health, enzymatic activities, nutrient uptake and antioxidant defense. *Front. Microbiol.* **2022**, *13*, 901865. [CrossRef]

46. Singh, R.; Soni, S.K.; Patel, R.P.; Kalra, A. Technology for improving essential oil yield of *Ocimum basilicum* L.(sweet basil) by application of bioinoculant colonized seeds under organic field conditions. *Ind. Crops Prod.* **2013**, *45*, 335–342. [CrossRef]
47. Ilyas, N.; Mazhar, R.; Yasmin, H.; Khan, W.; Iqbal, S.; Enshasy, H.E.; Dailin, D.J. Rhizobacteria isolated from saline soil induce systemic tolerance in wheat (*Triticum aestivum* L.) against salinity stress. *Agronomy* **2020**, *10*, 989. [CrossRef]
48. Yasmeen, R.; Shaheed Siddiqui, Z. Physiological responses of crop plants against Trichoderma harzianum in saline environment. *Acta Bot. Croat.* **2017**, *76*, 154–162. [CrossRef]
49. Bennis, M.; Kaddouri, K.; Badaoui, B.; Bouhnik, O.; Chaddad, Z.; Perez-Tapia, V.; Lamin, H.; Alami, S.; Lamrabet, M.; Abdelmoumen, H. Plant growth promoting activities of Pseudomonas sp. and Enterobacter sp. isolated from the rhizosphere of Vachellia gummifera in Morocco. *FEMS Microbiol. Ecol.* **2023**, *99*, fiad114. [CrossRef]
50. Sumreen, S.; Sharif, M.; Tariq Sultan, D.M.; Khan, A. Effect of isolated plant growth promoting rhizobacteria on growth and nutrients uptake by maize in acidic and alkaline soil conditions. *Pak. J. Bot* **2025**, *57*, 37–46. [CrossRef]
51. Yilmaz, H. Enhancements in morphology, biochemicals, nutrients, and L-Dopa in Faba bean through plant growth promoting rhizobacteria and arbuscular mycorrhizal Fungi. *Sci. Rep.* **2025**, *15*, 7390. [CrossRef]
52. Zareen, S.; Ali, A.; Yun, D.-J. Significance of ABA Biosynthesis in Plant Adaptation to Drought Stress. *J. Plant Biol.* **2024**, *67*, 175–184. [CrossRef]
53. Zulfiqar, B.; Raza, M.A.S.; Saleem, M.F.; Ali, B.; Aslam, M.U.; Al-Ghamdi, A.A.; Elshikh, M.S.; Hassan, M.U.; Toleikienė, M.; Ahmed, J. Abscisic acid improves drought resilience, growth, physio-biochemical and quality attributes in wheat (*Triticum aestivum* L.) at critical growth stages. *Sci. Rep.* **2024**, *14*, 20411. [CrossRef] [PubMed]
54. Herrera-Medina, M.J.; Steinkellner, S.; Vierheilig, H.; Ocampo Bote, J.A.; Garcia Garrido, J. Abscisic acid determines arbuscule development and functionality in the tomato arbuscular mycorrhiza. *New Phytol.* **2007**, *175*, 554–564. [CrossRef] [PubMed]
55. Evelin, H.; Kapoor, R.; Giri, B. Arbuscular mycorrhizal fungi in alleviation of salt stress: A review. *Ann. Bot.* **2009**, *104*, 1263–1280. [CrossRef]
56. Roy, S.; Chakraborty, A.P.; Chakraborty, R. Understanding the potential of root microbiome influencing salt-tolerance in plants and mechanisms involved at the transcriptional and translational level. *Physiol. Plant.* **2021**, *173*, 1657–1681. [CrossRef]
57. Liu, S.; Tian, Y.; Jia, M.; Lu, X.; Yue, L.; Zhao, X.; Jin, W.; Wang, Y.; Zhang, Y.; Xie, Z. Induction of salt tolerance in Arabidopsis thaliana by volatiles from Bacillus amyloliquefaciens FZB42 via the jasmonic acid signaling pathway. *Front. Microbiol.* **2020**, *11*, 562934. [CrossRef]

current issues in molecular biology

MDPI

Review

Transgenerational Memory of Phenotypic Traits in Plants: Epigenetic Regulation of Growth, Hormonal Balance, and Stress Adaptation

Erna Karalija [1], Saida Ibragić [2], Sabina Dahija [1] and Dunja Šamec [3,*]

[1] Department of Biology, Faculty of Science, University of Sarajevo, Zmaja od Bosne 35, 71000 Sarajevo, Bosnia and Herzegovina; erna.k@pmf.unsa.ba (E.K.); sabina.dahija@pmf.unsa.ba (S.D.)
[2] Department of Chemistry, Faculty of Science, University of Sarajevo, Zmaja od Bosne 33, 71000 Sarajevo, Bosnia and Herzegovina; saida.i@pmf.unsa.ba
[3] Department of Food Technology, University North, Trg Dr. Žarka Dolinara 1, 48 000 Koprivnica, Croatia
* Correspondence: dsamec@unin.hr

Abstract: Plants exhibit remarkable adaptability to environmental stresses, with epigenetic modifications playing a key role in stress memory and adaptation. This review explores how epigenetic mechanisms influence hormonal regulation in plants, shaping growth, development, and stress responses. Specifically, we focus on the roles of DNA methylation, histone modifications, and small RNAs in modulating auxin, abscisic acid (ABA), gibberellin (GA), and jasmonic acid (JA) pathways. These pathways influence the plant's ability to cope with abiotic and biotic stresses and can be inherited by progeny, enhancing stress resilience across generations. By understanding the epigenetic regulation of these hormones, we aim to provide insights into how epigenetic priming can be harnessed in crop improvement to address the challenges posed by climate change.

Keywords: crop improvement; epigenetic priming; epigenetic inheritance; hormonal regulation; plant resilience; stress memory; transgenerational adaptation

1. Introduction

Environmental stresses such as drought, salinity, extreme temperatures, and pathogen attacks have become increasingly problematic for global agricultural productivity, exacerbated by climate change. In response, plants have evolved diverse physiological, biochemical, and molecular mechanisms to cope with such stresses. One remarkable mechanism that has emerged is transgenerational stress memory, where plants "remember" past stress events and pass this memory to their progeny, enhancing their ability to adapt to recurring environmental challenges [1,2]. Plants, as sessile organisms, have evolved sophisticated mechanisms to "remember" environmental stresses, enabling them to respond more effectively upon re-exposure. This phenomenon, known as stress memory, manifests in two main forms: somatic and transgenerational memory [3,4].

Somatic stress memory refers to the retention of stress-induced epigenetic modifications—such as DNA methylation, histone modifications, and the accumulation of small RNAs—within an individual plant's lifespan. These changes allow for a faster or more robust transcriptional response during subsequent stress events. For example, in *Arabidopsis thaliana*, drought-induced memory genes such as RD29A retain permissive chromatin states (e.g., H3K4me3) and poised RNA Polymerase II during recovery, enabling rapid re-induction upon renewed drought stress [4,5].

https://doi.org/10.3390/cimb47060404

In contrast, transgenerational stress memory involves the inheritance of epigenetic marks that escape meiotic resetting and are transmitted to progeny. These heritable modifications include DNA methylation maintained by *MET1* or *DRM2*, histone marks such as H3K27me3, and the activity of RNA-directed DNA methylation (RdDM) pathways. Such mechanisms have been shown to confer improved stress resilience in offspring, as seen in wheat and rice exposed to drought or salt stress, where progeny exhibit altered expression of ABA- and auxin-related genes [6,7].

While the genetic basis of stress tolerance has been extensively studied, the role of epigenetic mechanisms in plant stress responses remains an evolving field. Epigenetic modifications, such as DNA methylation, histone modifications, and small RNA regulation, do not alter the DNA sequence but influence gene expression, thereby modifying phenotypic traits related to stress adaptation [3]. These epigenetic alterations significantly impact plant growth, development, and responses to environmental stressors, particularly through the regulation of plant hormones such as abscisic acids (ABAs), auxins, gibberellins (GAs), and jasmonic acids (JAs), which play central roles in stress signaling and resilience (Figure 1).

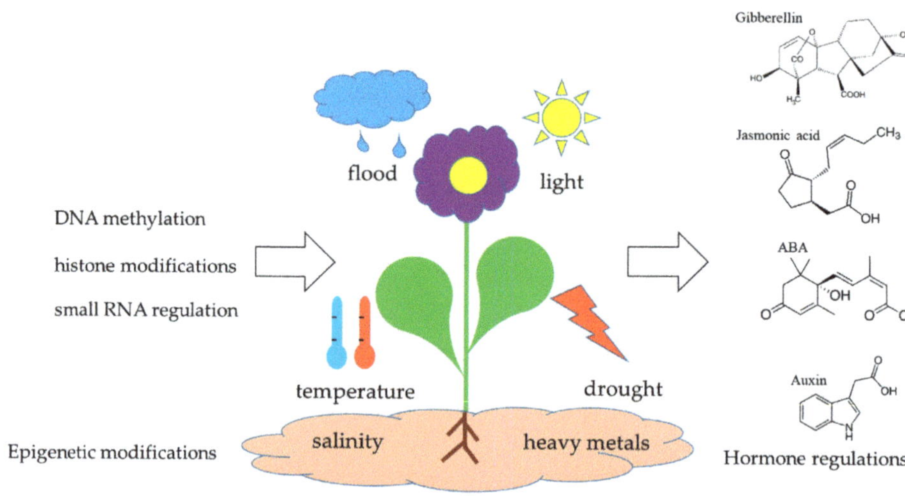

Figure 1. Epigenetic alterations significantly impact plant growth, development, and responses to environmental stressors, particularly through the regulation of plant hormones such as abscisic acids (ABAs), auxins, gibberellins (GAs), and jasmonic acids (JAs), which play central roles in stress signaling and resilience.

Despite the increasing body of research on plant stress memory, significant gaps remain in understanding the mechanisms by which epigenetic modifications regulate hormonal pathways, particularly in the context of transgenerational memory. Specifically, how these epigenetic modifications influence progeny fitness, how stable or reversible these modifications are under fluctuating conditions, and how the different epigenetic mechanisms interact to regulate hormonal responses are not well understood. Furthermore, the long-term implications of epigenetic priming for crop improvement under field conditions remain an open area for investigation.

This review aims to address these gaps by exploring how epigenetic regulation of hormonal pathways influences stress adaptation and resilience. We will explore the role of DNA methylation, histone modifications, and small RNA regulation in shaping hormonal balance in plants and examine how these modifications can be inherited, enhancing stress tolerance across generations. We will also discuss the potential applications of epigenetic priming in crop improvement, with a focus on future directions that could enhance our understanding of epigenetic memory in plants.

2. Epigenetic Mechanisms Governing Hormonal Regulation

Epigenetic modifications, particularly DNA methylation, have emerged as central regulatory players that govern the dynamic transcription of hormone-related genes in plants. Through changes in cytosine methylation at promoters and gene bodies, DNA methylation orchestrates hormonal biosynthesis, signaling, and downstream responses, contributing to the fine-tuning of plant growth, development, and environmental plasticity [8,9].

2.1. DNA Methylation Regulates Hormonal Synthesis and Signaling: A Crucial Epigenetic-Hormonal Interface in Plant Development and Stress Adaptation

DNA methylation influences hormone levels by directly targeting genes involved in their biosynthetic pathways. In soybeans, experimental demethylation via 5-azacytidine (5-azaC) significantly altered the concentrations of ABA, auxin (IAA), and ethylene, indicating that the methylation state modulates hormone biosynthesis from the early stages of seedling development [10]. The upregulation of *DRM2* and downregulation of *ROS1* in roots under hypomethylating conditions suggests methylation-dependent control of hormone-related transcriptional cascades.

In *Capsicum annuum*, ripening-associated hypomethylation was found to induce ABA biosynthesis genes while repressing auxin and gibberellin biosynthesis, suggesting a role of DNA methylation in the hormonal reprogramming that governs the ripening process [11]. The expression of *CaDML2-like* (a demethylase) was induced by exogenous ABA, and conversely, IAA promoted transcription of *CaMET1-like1*, revealing feedback between methylation machinery and hormonal signals.

Methylation of DNA impacts the transcriptional regulation of entire metabolic cascades by altering chromatin accessibility and the binding affinity of transcription factors to hormone biosynthesis genes. For instance, DNA hypomethylation enhances growth-sustaining hormonal pathways under cadmium stress in *Arabidopsis thaliana*, particularly auxin, cytokinin, and gibberellin pathways, in the methylation-defective ddc mutant [12].

Recent evidence demonstrates that plant hormones can regulate DNA methylation by altering the expression of methyltransferases and demethylases. Using GUS reporter lines in *Arabidopsis*, Bennett et al. [13] showed that phytohormones such as ABA, auxin, ethylene, SA, cytokinin, and GA dynamically alter promoter activity of key genes involved in both de novo methylation (*DRM1/2/3*) and demethylation (*ROS1, DME, DML2, DML3*). These hormone-induced shifts in the expression of methylation-related genes were confirmed by RT-qPCR, substantiating the feedback loop wherein hormonal signals influence the epigenetic landscape [13].

For example:

- Auxin decreased *DRM2* and *DML3* expression in roots and shoots, while slightly upregulating *CMT3*.
- ABA upregulated *VIM1* (a *MET1* cofactor for CG methylation) in roots and shoots but suppressed *DRM3* and *DML3*.
- GA treatment enhanced the expression of *DML2* and *DML3*, suggesting active demethylation supports GA-mediated growth.
- SA broadly suppressed *MET1*, *VIM1*, *DRM2*, and demethylases like *ROS1*, reducing methylation turnover.

The hormonal regulation of methylation genes is highly tissue- and stage-specific. For instance, in reproductive tissues, *DRM2* and *DME* are co-expressed in petals, anthers, and stigma, suggesting coordinated methylation–demethylation dynamics are essential for flower development [13]. During seed development, *DRM2*, *MET1*, and *DME* were highly active in both the endosperm and embryo, reflecting their role in shaping hormone-responsive transcription during embryogenesis [13].

Methylation profiles in root meristems and shoot apices align with the high activity of hormone signaling and biosynthesis. The activation of demethylases in mature root tissues versus methyltransferases in root tips indicates developmental reprogramming influenced by hormonal flux and epigenetic remodeling [13].

The ability of DNA methylation to modulate hormone levels enables plants to prioritize either growth or defense responses based on environmental cues. For instance, SA, typically associated with biotic stress, represses genes for both methylation and demethylation in roots and older leaves, possibly stabilizing a defense-associated gene expression state [13]. In contrast, gibberellin promotes the expression of demethylases, aligning with growth activation and stress recovery [13]. These findings suggest that DNA methylation acts not only as a molecular memory of hormonal responses but also as a regulatory integrator, ensuring a dynamic balance between plasticity and stability in plant hormone networks.

2.2. Histone Modifications in Hormonal Signaling: Epigenetic Regulation of Plant Development and Stress Responses

Histone modifications are pivotal chromatin-level epigenetic mechanisms that regulate transcriptional outcomes in response to developmental and environmental cues. These modifications—primarily acetylation, methylation, and phosphorylation—impact chromatin accessibility and consequently the expression of hormone-related genes. A growing body of research highlights the intricate crosstalk between histone modifications and hormonal pathways in plants, positioning histone dynamics as critical regulators of hormonal biosynthesis, signaling, and response coordination during key stages such as germination, ripening, and stress adaptation [14,15].

Histone acetylation, catalyzed by histone acetyltransferases (HATs), typically correlates with transcriptional activation by loosening chromatin structure, whereas histone deacetylases (HDACs) remove acetyl groups and restore repression [14]. Histone methylation is more precise: tri-methylation of H3K4 and H3K36 is associated with active transcription, while H3K27me3 and H3K9me2 generally mediate gene silencing [16]. These marks are dynamically altered in response to internal cues such as hormone levels or external stressors like salinity and pathogens [17].

During seed germination, the repression of maturation-phase genes and activation of vegetative genes is controlled by Polycomb Repressive Complexes PRC1 and PRC2 through H3K27me3-mediated silencing. These complexes are recruited by transcriptional repressors such as VAL1 and VAL2 to suppress LAFL genes (*LEC1*, *ABI3*, *FUS3*, *LEC2*)—central to ABA signaling—thereby promoting GA-mediated germination [15,18]. PICKLE (PKL), a chromatin remodeling factor, also plays a role by antagonizing LAFL activity, further facilitating hormonal transition [19].

In *Arabidopsis*, histone acetylation is dynamically modulated during seedling development to control the expression of GA biosynthetic genes, whereas HDACs such as *HDA6* and *HDA19* repress ABA-responsive genes during seedling growth [14,15,18]. Likewise, in tomatoes, *HDA3* and *HDA1* repress ripening by inhibiting ethylene biosynthesis and perception genes until developmental cues remove repression via histone acetylation [20,21].

Epigenetic control over hormone biosynthesis genes is well established. In climacteric fruits such as tomatoes and bananas, H3K27me3 represses key ripening genes including ACO and ACS, which encode enzymes for ethylene biosynthesis. Demethylation of H3K27me3 at these loci coincides with ethylene bursts and the onset of ripening [20,21]. Additionally, ABA biosynthesis genes (NCED) show activation through increased H3K4me3 levels, reflecting permissive chromatin under stress or ripening cues [16,21].

In sugar signaling contexts, which integrate with hormonal pathways, chromatin remodelers and histone acetylation facilitate the activation of hormone-sensitive genes. HXK1, a glucose sensor, recruits histone-modifying complexes to modulate auxin and cytokinin signaling genes under sugar-rich conditions [22].

Under abiotic stresses like salinity or drought, histone modifications provide plants with transcriptional plasticity. In response to salt stress, increased H3 acetylation (e.g., H3K9ac, H3K27ac) at promoters of stress-responsive genes correlates with ABA accumulation and activation of signaling pathways [16]. Similarly, methylation of H3K4 at genes involved in JA and SA signaling has been documented during pathogen response, reflecting pathogen-triggered immunity (PTI) activation [17].

Moreover, specific histone methyltransferases (HMTs) and demethylases (e.g., JMJ family proteins) respond to hormonal signals and ROS cues, modulating the histone code to control stress hormone pathways [16,18]. PRC2 and HDA19 repression of ABA-inducible transcription factors under non-stress conditions ensure homeostasis, while their derepression during stress allows rapid hormonal responses [14,16].

Histone modifications frequently work in tandem with DNA 14 and miRNAs, especially in developmental reprogramming events. For example, sugar- and hormone-responsive genes often exhibit coordinated changes in DNA methylation and histone acetylation/methylation, forming a multi-layered regulatory landscape [22]. In seed germination, ROS-induced chromatin remodeling leads to increased H3 acetylation and reduced H3K27me3, thereby promoting GA biosynthesis and ABA degradation [18].

2.3. Small RNAs in Hormonal Pathways: Post-Transcriptional Regulation of Plant Development and Immunity

Small RNAs (sRNAs), including microRNAs (miRNAs) and small interfering RNAs (siRNAs), are critical components of gene regulatory networks in plants. By guiding post-transcriptional gene silencing, they fine-tune gene expression with spatial and temporal precision. In recent years, sRNAs have been linked to the regulation of hormonal biosynthesis and signaling pathways, positioning them as central players in the coordination of growth, development, and stress responses [8].

Plant miRNAs typically function by guiding the RNA-induced silencing complex (RISC) to complementary mRNA targets, resulting in transcript cleavage or translational repression. Many hormone-responsive genes—including transcription factors and receptors—are direct targets of miRNAs [23,24]. For example, the miR160 and miR167 families regulate auxin signaling by targeting ARF10/16/17 and ARF6/8, respectively, which encode auxin response factors [23]. Similarly, miR159 and miR319 modulate GA and JA signaling through MYB and TCP transcription factors [25,26].

In the ABA pathway, miR168 regulates ARGONAUTE1 (AGO1), a key effector of the miRNA pathway itself, forming a feedback loop between sRNA biogenesis and ABA sensitivity [27]. miR393, which targets auxin receptors TIR1 and AFB2, is induced by abiotic stress and serves as a rapid switch that attenuates auxin signaling, thereby enhancing stress tolerance [23,26].

sRNAs play essential roles in regulating hormone-driven developmental transitions. The miR156–SPL module, for instance, is involved in the juvenile-to-adult phase transition, flowering time, and inflorescence development through interactions with GA and ABA pathways [23,25]. In rice, miR5488 and miR399 are implicated in thermosensitive male sterility, partly through influencing lignin biosynthesis and flavonoid metabolism—processes tightly linked to hormone action [25].

During seed germination, a finely tuned interplay between ABA and GA determines dormancy and activation. Small RNAs modulate this balance by targeting key hormone metabolism genes. For example, miR159-mediated repression of MYB33/MYB101 transcription factors reduces ABA sensitivity, promoting germination [27].

sRNAs also act as mediators of hormone-regulated defense responses. The miR393–TIR1 module enhances SA signaling and suppresses auxin signaling, creating a hormonal environment favoring defense overgrowth [24,26]. Viral infections further trigger sRNA expression shifts that impact hormonal crosstalk; for instance, induction of miR171 and miR168 affects ABA, auxin, and JA signaling, which viruses may exploit to enhance replication and systemic spread [24,28].

In *Arabidopsis*, SA upregulates miR472, which suppresses resistance (R) genes, fine-tuning the balance between immune activation and growth [26]. JA- and ET-pathways are also targeted by sRNAs such as miR319, adjusting defense output based on the invading pathogen [29].

Stress exposure can lead to transgenerational reprogramming of sRNA expression, impacting hormone-related gene expression in offspring. In durum wheat, water-deficit stress experienced by parental plants triggered changes in miRNAs regulating ABA and cytokinin pathways, enhancing progeny resilience [6].

These effects align with stress-induced epigenetic memory, where sRNAs help stabilize hormone-related transcription patterns across generations, offering potential targets for breeding stress-resilient crops [6,27].

3. Hormonal Balance in Progeny: Epigenetic Effects on Fitness

The capacity of plants to transmit environmental experience to their progeny has fundamentally reshaped our understanding of plant adaptation and fitness. Far beyond traditional Mendelian inheritance, transgenerational epigenetic mechanisms—most notably DNA methylation, histone modifications, and small RNAs—can modulate hormonal balance in offspring, influencing developmental trajectories and ecological success.

Plants subjected to abiotic stressors such as drought or salinity often exhibit persistent changes in hormonal regulation, which are partially inherited by their descendants. This transmission of "stress memory" is frequently mediated by epigenetic modifications that alter the expression of hormone biosynthesis or signaling genes [7]. For example, enhanced ABA signaling, a hallmark of stress priming, can persist in progeny due to maintained chromatin states or stable DNA methylation patterns established during parental stress exposure [12].

The maintenance of such epigenetic marks during reproductive development—especially in germline precursor cells—enables the inheritance of primed states. These modifications modulate hormone homeostasis in developing seeds, influencing dormancy, germination, and early growth responses to environmental conditions.

Maternal plants exposed to stress conditions often alter the hormonal content of seeds, either through direct provisioning or epigenetic programming of hormone metabolism. For example, altered auxin or gibberellin levels in seeds derived from stressed parents have been associated with delayed germination or enhanced stress tolerance [30].

This effect can be further shaped by environmental interactions, such as beneficial symbioses. Arbuscular mycorrhizal (AM) symbiosis can alleviate drought stress in *Taraxacum brevicorniculatum* but also can trigger transgenerational phenotypic plasticity, including changes in root architecture and resource-use traits in progeny [31]. These changes were closely linked to hormonal pathways and suggest that epigenetic mechanisms control the balance between stress-responsiveness and growth-promoting hormones across generations.

Although epigenetic memory can confer enhanced resilience, it can also result in hormonal imbalances that compromise growth under favorable conditions. For instance, a primed ABA state may increase survival during drought but impair seedling vigor if water is abundant [32]. Thus, the persistence of epigenetic marks regulating hormonal crosstalk must be finely tuned to environmental predictability.

Importantly, fitness advantages conveyed by epigenetic-hormonal modifications tend to be most pronounced during early developmental stages—especially in seedling establishment and root growth—where hormonal regulation plays a dominant role [31]. In perennial plants, partial maintenance of the epigenetic landscape across seasons may contribute to long-term adaptations, such as adjusted hormone sensitivity and seasonal phenology [7].

Plants exhibit dynamic control over the inheritance of epigenetic marks. While some marks are faithfully transmitted through meiosis, others are reset during gametogenesis, a process influenced by both hormonal signals and developmental context [12]. This controlled resetting allows flexibility: it preserves beneficial adaptations while avoiding maladaptive stress legacies.

Recent studies suggest that selective retention of epigenetic states regulating hormone biosynthesis and perception allows for a refined strategy—transmitting only those hormonal cues that increase progeny fitness under anticipated environmental scenarios [30,32].

3.1. *Auxin and Root–Shoot Allocation: Epigenetic Regulation of Growth Plasticity*

Plant developmental plasticity, especially in root–shoot allocation, is a key adaptive mechanism for optimizing resource acquisition in dynamic environments. Central to this plasticity is the plant hormone auxin, which orchestrates differential growth patterns and organ development through tightly regulated spatial gradients. Emerging evidence has revealed that epigenetic regulation—including chromatin remodeling, histone modifications, DNA methylation, and non-coding RNAs—modulates auxin biosynthesis, transport, and signaling to fine-tune root–shoot balance under varying environmental and physiological contexts.

Auxin distribution is critical in shaping root system architecture (RSA) and shoot development. Local auxin maxima, particularly in the root apical meristem, govern lateral root initiation and elongation, while systemic auxin gradients integrate signals from shoot-derived photosynthates, nutrient availability, and stress conditions [33]. Mutations in YUCCA, TAA1, and PIN gene families, which regulate auxin biosynthesis and transport, consistently disrupt root development and alter root-to-shoot ratios in both model and crop species such as rice, maize, and pea [33].

Moreover, auxin interacts with other hormones such as cytokinins, ABA, and gibberellins to modulate organ growth in a context-dependent manner. For example, heat stress triggers root–shoot signaling involving auxin and ABA, with the auxin pool in roots contributing to meristem maintenance and stress adaptation [34]. Crosstalk with cytokinin and ethylene further refines RSA to optimize growth under nutrient stress or thermal extremes.

Environmental cues frequently alter auxin-related gene expression through epigenetic modifications. Under phosphate and sulfur starvation, DNA methylation patterns in YUCCA and PIN promoters are reorganized to induce local auxin biosynthesis and redirect root growth to favorable zones [23]. Simultaneously, histone acetylation and H3K4me3 marks activate auxin-responsive genes in the root tip, enhancing adaptive root elongation and branching [34].

The Target of Rapamycin (TOR) kinase pathway integrates nutrient availability and sugar signaling with auxin responses. Under sulfur deficiency, autophagy is upregulated in shoots while root TOR activity is preserved, facilitating carbon allocation and supporting auxin-induced meristem activity [23]. This balance is critical for root growth under nutrient stress and directly links energy sensing with hormone-regulated developmental plasticity.

The TOR–auxin axis also intersects with chromatin-level regulation. TOR activity modulates the expression of WUSCHEL and E2Fa in a tissue-specific manner, thereby controlling meristem size and activity in a way that responds both to internal metabolic status and external environmental signals [23].

Sulfate limitation, for instance, downregulates TOR activity in shoots but maintains it in roots via glucose-TOR signaling, preserving auxin-mediated root apical meristem activity. This organ-specific control enables enhanced root growth at the expense of shoots, a classical example of an increased root-to-shoot ratio during stress [23].

Chromatin remodeling under thermal and salt stress also regulates auxin responses by modulating access to auxin transporters and signaling genes. This epigenetic plasticity contributes to maintaining functional RSA under hostile conditions, such as salinity and high temperature, where meristem function must be preserved despite compromised cellular metabolism [34,35].

Root–shoot plasticity is not merely a physiological adjustment but an evolved developmental strategy. Plants dynamically regulate root branching, elongation, and shoot architecture to optimize carbon–nutrient balance, especially under stress or soil constraints [36,37]. Genetic diversity in wheat, for instance, has revealed alleles contributing to enhanced plasticity under sodic and acidic soils—traits associated with epigenetically regulated auxin and ABA pathways [37].

Modern breeding strategies increasingly target these epigenetically controlled traits to enhance resilience in cereal crops. Epigenomic profiling, coupled with high-resolution phenotyping of RSA traits, is enabling the identification of candidate regulators and epialleles that confer beneficial root–shoot allocation patterns [35].

3.2. *Abscisic Acid and Stress Tolerance: Epigenetic Memory of Drought Response*

Abscisic acid is a central hormonal mediator of plant responses to drought, functioning as both a rapid-response signal and a key regulator of long-term acclimation. Recent research has revealed that epigenetic mechanisms—including DNA methylation, histone modifications, and chromatin remodeling—play essential roles in establishing and maintaining stress memory, enabling plants to mount enhanced responses to recurrent drought episodes. These insights not only deepen our understanding of plant resilience but also offer promising avenues for crop improvement in the face of climate change.

Under drought stress, increased ABA accumulation promotes stomatal closure, osmotic adjustment, and activation of ABA-responsive transcription factors (TFs) such as AREB/ABF and DREB families, which regulate gene networks related to dehydration tolerance [38]. Many of these ABA-inducible genes exhibit transcriptional memory, characterized by sustained expression or faster re-induction upon subsequent drought events.

Stress-induced changes in chromatin architecture—particularly DNA methylation and histone modifications—are essential for encoding drought memory. Repeated drought exposure leads to hypomethylation of drought-responsive gene promoters (e.g., RD29A, DREB2A) and enrichment of permissive histone marks such as H3K4me3 and H3K9ac, which maintain these loci in a "poised" transcriptional state [38,39].

In *Arabidopsis*, memory genes remain in a transcriptionally competent state during recovery phases between drought episodes, with paused RNA Polymerase II ready to resume transcription rapidly during renewed stress [39]. The stability of these marks during mitotic divisions enables somatic stress memory, ensuring that already differentiated tissues retain enhanced drought responsiveness.

Beyond somatic memory, drought-induced epigenetic changes can be transmitted to progeny, resulting in transgenerational stress memory. This has been observed in several crop species, including citrus, where grafting with buds from drought-experienced plants improved stress tolerance and photosynthetic efficiency in new individuals [40]. Such effects are often genotype-dependent, with specific rootstock–scion combinations better preserving epigenetic imprints related to ABA regulation.

Studies in wheat, rice, and poplar have shown that specific DNA methylation patterns and histone modifications established during drought can persist in germline cells, altering ABA-related gene expression in the next generation [25,41]. This heritability may occur via partial escape from meiotic resetting, allowing stress memory to bypass epigenetic reprogramming checkpoints [39,42].

Small RNAs, particularly siRNAs and miRNAs, modulate ABA signaling through post-transcriptional silencing of target genes involved in hormone biosynthesis and signal transduction. Moreover, the RNA-directed DNA methylation (RdDM) pathway reinforces transcriptional silencing and can contribute to stable memory marks [43]. These RNA-mediated pathways are integral in reinforcing chromatin states associated with ABA-responsive loci during stress priming and recovery.

Harnessing epigenetic memory of ABA-mediated drought tolerance offers great potential for sustainable agriculture. Priming strategies, such as drought pre-treatment or seed priming, induce beneficial epigenetic states that can be retained across developmental stages or even inherited, thereby enhancing tolerance without genetic modification [44]. The combination of natural variation, epigenome editing, and epigenetic breeding is an emerging frontier to exploit ABA-associated memory for climate-resilient crop varieties.

3.3. Gibberellins and Growth Plasticity: Epigenetic Regulation of Developmental Timing

Gibberellins are hormones that regulate a wide range of developmental processes, including seed germination, stem elongation, and flowering time. The regulation of GA signaling is particularly important for plant growth plasticity under stress, as it allows plants to adjust their growth patterns to optimize survival.

Epigenetic regulation of GA signaling is mediated by DNA methylation and histone modifications. DNA methylation of genes involved in GA biosynthesis or signaling, such as GA20ox, can influence seed germination and stem elongation under stress conditions [45]. For example, in response to heat stress, plants may adjust the timing of germination or elongate stems to avoid adverse conditions, with these changes being regulated by epigenetic modifications in GA-related genes.

In progeny, epigenetic modifications in GA pathways can have long-term effects on developmental timing. Transgenerational inheritance of epigenetic marks that regulate GA signaling can lead to offspring with altered growth patterns, such as delayed flowering or increased stem elongation, which may confer survival advantages under fluctuating environmental conditions [46]. This form of epigenetic regulation of GA signaling provides an example of how plant development can be tailored to specific environmental challenges, ensuring that offspring are better equipped to cope with future stressors.

3.4. Jasmonic Acid and Defense Responses: Epigenetic Regulation of Herbivore Resistance

Jasmonic acid is a key hormone involved in plant defense responses, particularly against herbivores and pathogens. Epigenetic regulation of JA signaling pathways has important implications for the defense responses of progeny, as it enables plants to "remember" prior herbivore attacks and enhance their defense mechanisms in subsequent generations.

DNA methylation and histone modifications regulate the expression of JA-responsive genes, which control defense-related proteins such as proteinase inhibitors and defensins. Epigenetic modifications at the promoters of these genes can enhance their expression in response to herbivory [47]. For example, DNA methylation of genes involved in the biosynthesis of JA or the expression of JA receptors can result in increased resistance to herbivores and pathogens. These epigenetic marks can be inherited, allowing progeny to mount a stronger defense response without the need for an initial herbivore attack.

Small RNA regulation also plays a role in the inheritance of JA-related defense responses. Small RNAs, particularly siRNAs, can target and silence genes involved in JA biosynthesis or receptor expression, providing a mechanism for transgenerational memory of herbivore attacks [48]. By inheriting these small RNA-mediated marks, progeny is able to respond more effectively to herbivore pressure, ensuring increased fitness in environments where biotic stresses are common.

4. Transgenerational Epigenetic Inheritance of Hormonal Balance: Implications for Crop Improvement

The inheritance of epigenetic modifications in hormonal pathways has significant implications for crop breeding and the development of stress-resilient varieties. Through epigenetic imprinting, plants can pass down enhanced stress responses to their progeny, enabling them to cope with similar environmental conditions in the future. This transgenerational adaptation provides a form of "stress priming", where progeny is better prepared to face environmental challenges such as drought, salinity, and pathogen attacks.

Understanding how epigenetic modifications influence hormonal pathways allows for the potential manipulation of these pathways to enhance crop resilience. By selecting crops with favorable epigenetic marks that promote stress tolerance, we can improve yield stability under adverse environmental conditions. Additionally, the use of epigenetic priming to activate stress-responsive pathways in crops can lead to enhanced productivity, particularly in regions affected by climate change [47].

Potential for Crop Improvement Through Epigenetic Priming

The ability to manipulate epigenetic pathways to improve crop resilience has significant implications for agriculture. Epigenetic priming involves the pre-conditioning of plants through environmental cues or controlled stress exposure, leading to beneficial epigenetic modifications that enhance stress tolerance in future generations. This process can be particularly useful in crop breeding, as it offers a way to enhance resilience without altering the plant's genetic makeup.

Recent studies have demonstrated that epigenetic priming can trigger long-lasting modifications in chromatin and DNA methylation patterns, particularly in genes involved in ABA, JA, and auxin pathways. For instance, in *Oryza sativa*, mild drought stress led to hypomethylation of promoters of ABA biosynthesis genes (e.g., *OsNCED*), and these changes persisted across at least two generations, resulting in enhanced drought avoidance and increased water use efficiency [49]. Similarly, in *Zea mays*, transgenerational memory of heat and drought stress was linked to differential expression of small RNAs targeting stress-responsive transcription factors, contributing to improved seedling vigor and root development in the progeny.

In *Zea mays* (maize), stress priming under controlled heat and drought conditions resulted in stable transgenerational upregulation of miRNAs such as miR168 and miR399, which regulate genes involved in hormone signaling and oxidative stress detoxification. These miRNAs were retained in seedlings derived from stressed parents, contributing to enhanced germination, seedling vigor, and stress preparedness, even in the absence of direct stress exposure [6].

Similarly, in *Brassica napus*, salt-primed parental plants produced seeds with altered expression of histone methyltransferases (e.g., *SUVH4*), resulting in differential H3K9me2 marks on stress-inducible genes in progeny. These chromatin changes correlated with increased expression of ion transporters like *NHX1*, promoting salt exclusion and maintaining ion homeostasis under salinity.

Moreover, priming with beneficial microbes, such as plant growth-promoting rhizobacteria (PGPR) or arbuscular mycorrhizal fungi (AMF), has been shown to trigger epigenetic changes in stress-related hormone pathways. For example, AMF-inoculated taraxacum plants exposed to drought exhibited not only improved stress tolerance in the parental generation but also transgenerational shifts in auxin-to-ABA balance and modifications in root architecture in their offspring, mediated by changes in DNA methylation [50].

By exposing plants to mild drought conditions, it is possible to induce epigenetic modifications in stress-responsive genes such as those involved in ABA biosynthesis. These modifications can be passed down to progeny, resulting in improved drought tolerance in future generations. Additionally, epigenetic priming can enhance the plant's ability to respond more rapidly and effectively to subsequent drought events, reducing water usage and improving crop productivity in water-limited regions.

Similarly, exposing plants to mild pathogen stress can induce epigenetic modifications in genes involved in the JA signaling pathway, which is crucial for plant defense responses. These modifications can be inherited by progeny, providing them with enhanced resistance to pathogens such as fungi, bacteria, or viruses. Epigenetic priming for pathogen resistance can improve crop health and yield, reducing the need for chemical pesticides and fostering more sustainable agricultural practices.

From a breeding perspective, epigenetic priming provides a complementary tool to traditional genetic selection. Marker-assisted selection targeting stable epialleles—heritable epigenetic variants affecting traits such as root architecture, flowering time, or pathogen resistance—is gaining momentum. Moreover, emerging technologies such as CRISPR/dCas9-based epigenome editing now allow targeted activation or silencing of stress-regulatory genes without changing the underlying DNA sequence. High-throughput epigenomic profiling platforms are also being integrated into pre-breeding programs to identify elite lines with desirable epigenetic marks [7,12].

In parallel, advances in epigenome editing technologies, such as CRISPR/dCas9 fused to DNA methyltransferases (e.g., DNMT3A) or demethylases (e.g., TET1), now enable targeted modifications of DNA methylation patterns at specific loci without altering nucleotide sequences [7]. These tools have the potential to precisely modulate gene expression, for instance by demethylating promoters of ABA or JA biosynthesis genes to enhance drought or pathogen resistance.

High-throughput epigenomic profiling platforms, including whole-genome bisulfite sequencing (WGBS) and chromatin immunoprecipitation sequencing (ChIP-seq), are increasingly incorporated into pre-breeding pipelines to map variation in methylation and histone modification landscapes. This facilitates the screening of natural epigenetic diversity across landraces, wild relatives, and cultivated varieties to identify epigenotypes with superior stress responses under real-world field conditions [12].

These advances offer a strategic framework for integrating epigenetic knowledge into breeding programs, enabling the development of crop varieties that are not only genetically robust but also epigenetically primed for resilience, particularly in the face of climate change. This epigenome-informed breeding approach provides a sustainable and innovative alternative to conventional improvement strategies, especially for traits where gene expression plasticity plays a central role [7].

5. Challenges and Future Directions

While the potential for epigenetic priming in crop improvement is promising, there are several challenges that need to be addressed. One key issue is understanding the stability and reversibility of epigenetic marks. Although some modifications are stable and can be passed on to progeny, others may be reversible or reset under favorable conditions. Future research should focus on identifying which epigenetic marks are most stable and beneficial for long-term stress tolerance and how these marks can be reliably inherited in crop species.

Another challenge is the genetic diversity of crops. While epigenetic priming offers a mechanism for enhancing stress tolerance, it must be integrated with traditional breeding practices to ensure that genetic diversity is maintained. Over-reliance on epigenetic priming could limit the adaptive potential of crops if genetic variation is reduced.

Precision breeding techniques, such as CRISPR/Cas9, offer the potential to directly edit epigenetic marks associated with stress tolerance. However, the long-term impacts of such interventions on crop performance, yield, and biodiversity must be carefully considered before widespread implementation.

6. Conclusions

The epigenetic regulation of hormonal pathways plays a crucial role in shaping plant responses to environmental stresses. These modifications not only influence growth and stress tolerance but also provide progeny with an adaptive advantage through transgenerational memory. By understanding the mechanisms that regulate these hormonal pathways, we can develop strategies to enhance crop resilience and productivity, particularly in the face of climate change. Future research will be essential to unravel the complexities of epigenetic inheritance and its role in improving plant resilience through epigenetic priming.

Author Contributions: Conceptualization, E.K. and D.Š.; data curation, S.D. and S.I.; writing—original draft preparation, E.K.; writing—review and editing, D.Š., S.D. and S.I.; visualization, E.K. and D.Š. All authors have read and agreed to the published version of the manuscript.

Funding: This research received no external funding.

Conflicts of Interest: The authors declare no conflicts of interest.

Abbreviations

The following abbreviations are used in this manuscript:

ABA	Abscisic acid
AM	Arbuscular mycorrhizal
ACS	1-aminocyclopropane-1-carboxylate synthase
ACO	1-aminocyclopropane-1-carboxylate oxidase
DME	DEMETER DNA glycosylase
DRM	Domains rearranged methyltransferase
GA	Gibberellin
GUS	β-glucuronidase
HAT	Histone acetyltransferase
HDAC	Histone deacetylase
HMT	Histone methyltransferase
JA	Jasmonic acid
LAFL	LEC1, ABI3, FUS3, and LEC2 gene network (seed maturation regulators)
miRNA	MicroRNA
PRC1/PRC2	Polycomb repressive complex 1/2
PTI	Pathogen-triggered immunity
RdDM	RNA-directed DNA methylation
RISC	RNA-induced silencing complex
RNA Pol II	RNA polymerase II
ROS	Reactive oxygen species
RSA	Root system architecture
SA	Salicylic acid

siRNA	Small interfering RNA
sRNA	Small RNA
TOR	Target of rapamycin

References

1. Boyko, A.; Kovalchuk, I. Transgenerational response to stress in *Arabidopsis thaliana*. *Plant Signal. Behav.* **2010**, *5*, 995–998. [CrossRef]
2. Espinas, N.A.; Saze, H.; Saijo, Y. Epigenetic control of defense signaling and priming in plants. *Front. Plant Sci.* **2016**, *7*, 1201. [CrossRef]
3. Kinoshita, T.; Seki, M. Epigenetic memory for stress response and adaptation in plants. *Plant Cell Physiol.* **2014**, *55*, 1859–1863. [CrossRef]
4. Lämke, J.; Bäurle, I. Epigenetic and chromatin-based mechanisms in environmental stress adaptation and stress memory in plants. *Genome Biol.* **2017**, *18*, 124. [CrossRef] [PubMed]
5. Ding, Y.; Fromm, M.; Avramova, Z. Multiple exposures to drought 'train' transcriptional responses in *Arabidopsis*. *Nat. Commun.* **2012**, *3*, 740. [CrossRef]
6. Liu, H.; Able, A.J.; Able, J.A. Small RNAs and their targets are associated with the transgenerational effects of water-deficit stress in durum wheat. *Sci. Rep.* **2021**, *11*, 3613. [CrossRef] [PubMed]
7. Gallusci, P.; Agius, D.R.; Moschou, P.N.; Dobránszki, J.; Kaiserli, E.; Martinelli, F. Deep inside the epigenetic memories of stressed plants. *Trends Plant Sci.* **2023**, *28*, 142–153. [CrossRef] [PubMed]
8. Ibragić, S.; Dahija, S.; Karalija, E. The Good, the Bad, and the Epigenetic: Stress-Induced Metabolite Regulation and Transgenerational Effects. *Epigenomes* **2025**, *9*, 10. [CrossRef]
9. Kaya, C.; Ashraf, M. Foliar Fertilization: A Potential Strategy for Improving Plant Salt Tolerance. *Crit. Rev. Plant Sci.* **2024**, *43*, 94–115. [CrossRef]
10. Coelho, F.S.; Miranda, S.S.; Moraes, J.L.; Hemerly, A.S.; Ballesteros, H.G.F.; Santa-Catarina, C.; dos Santos, R.C.; de Almeida, F.A.; Silveira, V.; Macedo, A.; et al. DNA methylation impacts soybean early development by modulating hormones and metabolic pathways. *Physiol. Plant.* **2024**, *176*, e14492. [CrossRef]
11. Xiao, K.; Chen, J.; He, Q.; Wang, Y.; Shen, H.; Sun, L. DNA methylation is involved in the regulation of pepper fruit ripening and interacts with phytohormones. *J. Exp. Bot.* **2020**, *71*, 1928–1942. [CrossRef]
12. Hemenway, C.A.; Gehring, M. Epigenetic regulation during plant development and the capacity for epigenetic memory. *Annu. Rev. Plant Biol.* **2023**, *74*, 87–109. [CrossRef] [PubMed]
13. Bennett, M.; Cleaves, K.; Hewezi, T. Expression patterns of DNA methylation and demethylation genes during plant development and in response to phytohormones. *Int. J. Mol. Sci.* **2021**, *22*, 9681. [CrossRef] [PubMed]
14. Kumar, V.; Thakur, J.K.; Prasad, M. Histone acetylation dynamics regulating plant development and stress responses. *Cell. Mol. Life Sci.* **2021**, *78*, 4467–4486. [CrossRef]
15. Sato, H.; Yamane, H. Histone modifications affecting plant dormancy and dormancy release: Common regulatory effects on hormone metabolism. *J. Exp. Bot.* **2024**, *75*, 6142–6158. [CrossRef]
16. Yung, W.-S.; Li, M.-W.; Sze, C.-C.; Wang, Q.; Lam, H.-M. Histone modifications and chromatin remodeling in plants in response to salt stress. *Physiol. Plant.* **2021**, *173*, 1495–1513. [CrossRef] [PubMed]
17. Kang, H.; Fan, T.; Wu, J.; Zhu, Y.; Shen, W.-H. Histone modification and chromatin remodeling in plant response to pathogens. *Front. Plant Sci.* **2022**, *13*, 986940 . [CrossRef]
18. Wang, Y.; Sun, X.; Peng, J.; Li, F.; Ali, F.; Wang, Z. Regulation of seed germination: ROS, epigenetic, and hormonal aspects. *J. Adv. Res.* **2024**, *71*, 107–125. [CrossRef]
19. Smolikova, G.; Strygina, K.; Krylova, E.; Leonova, T.; Frolov, A.; Khlestkina, E.; Medvedev, S. Transition from seeds to seedlings: Hormonal and epigenetic aspects. *Plants* **2021**, *10*, 1884. [CrossRef]
20. Ji, Y.; Xu, M.; Wang, A. Recent advances in the regulation of climacteric fruit ripening: Hormone, transcription factor and epigenetic modifications. *Front. Agric. Sci. Eng.* **2021**, *8*, 314–334. [CrossRef]
21. Li, X.; Wang, X.; Zhang, Y.; Zhang, A.; You, C.-X. Regulation of fleshy fruit ripening: From transcription factors to epigenetic modifications. *Hortic. Res.* **2022**, *9*, uhac013. [CrossRef] [PubMed]
22. Chen, Q.; Zhang, J.; Li, G. Dynamic epigenetic modifications in plant sugar signal transduction. *Trends Plant Sci.* **2022**, *27*, 379–391. [CrossRef]
23. Dong, Q.; Hu, B.; Zhang, C. MicroRNAs and their roles in plant development. *Front. Plant Sci.* **2022**, *13*, 824240 . [CrossRef] [PubMed]
24. Deng, Z.; Ma, L.; Zhang, P.; Zhu, H. Small RNAs participate in plant–virus interaction and their application in plant viral defense. *Int. J. Mol. Sci.* **2022**, *23*, 696. [CrossRef] [PubMed]

25. Sun, Y.; Xiong, X.; Wang, Q.; Zhu, L.; Wang, L.; He, Y.; Zeng, H. miR156, miR5488 and miR399 are involved in the regulation of male sterility in PTGMS rice. *Int. J. Mol. Sci.* **2021**, *22*, 2260. [CrossRef]
26. Waheed, S.; Anwar, M.; Saleem, M.A.; Wu, J.; Tayyab, M.; Hu, Z. The critical role of small RNAs in regulating plant innate immunity. *Biomolecules* **2021**, *11*, 184. [CrossRef]
27. Luján-Soto, E.; Dinkova, T.D. Time to wake up: Epigenetic and small-RNA-mediated regulation during seed germination. *Plants* **2021**, *10*, 236. [CrossRef]
28. Zhao, S.; Li, Y. Current understanding of the interplays between host hormones and plant viral infections. *PLoS Pathog.* **2021**, *17*, e1009242. [CrossRef]
29. Aerts, N.; Hickman, R.; Van Dijken, A.J.; Kaufmann, M.; Snoek, B.L.; Pieterse, C.M.; Van Wees, S.C. Architecture and dynamics of the abscisic acid gene regulatory network. *Plant J.* **2024**, *119*, 2538–2563. [CrossRef]
30. Rudolf, J.; Tomovicova, L.; Panzarova, K.; Fajkus, J.; Hejatko, J.; Skalak, J. Epigenetics and plant hormone dynamics: A functional and methodological perspective. *J. Exp. Bot.* **2024**, *75*, 5267–5294. [CrossRef]
31. Yun, P.; Kaya, C.; Shabala, S. Hormonal and epigenetic regulation of root responses to salinity stress. *Crop J.* **2024**, *12*, 1309–1320. [CrossRef]
32. Rajpal, V.R.; Rathore, P.; Mehta, S.; Wadhwa, N.; Yadav, P.; Berry, E.; Raina, S.N. Epigenetic variation: A major player in facilitating plant fitness under changing environmental conditions. *Front. Cell Dev. Biol.* **2022**, *10*, 1020958. [CrossRef] [PubMed]
33. Solanki, M.; Shukla, L.I. Recent advances in auxin biosynthesis and homeostasis. *3 Biotech* **2023**, *13*, 290. [CrossRef]
34. Tiwari, M.; Kumar, R.; Min, D.; Jagadish, S.K. Genetic and molecular mechanisms underlying root architecture and function under heat stress: A hidden story. *Plant Cell Environ.* **2022**, *45*, 771–788. [CrossRef] [PubMed]
35. Shelden, M.C.; Munns, R. Crop root system plasticity for improved yields in saline soils. *Front. Plant Sci.* **2023**, *14*, 1120583. [CrossRef]
36. Pierik, R.; Sasidharan, R.; Voesenek, L.A. Growth control by ethylene: Adjusting phenotypes to the environment. *J. Plant Growth Regul.* **2007**, *26*, 188–200. [CrossRef]
37. Bhoite, R.; Han, Y.; Chaitanya, A.K.; Varshney, R.K.; Sharma, D.L. Genomic approaches to enhance adaptive plasticity to cope with soil constraints amidst climate change in wheat. *Plant Genome* **2024**, *17*, e20358. [CrossRef]
38. Aizaz, M.; Lubna; Jan, R.; Asaf, S.; Bilal, S.; Kim, K.M.; Al-Harrasi, A. Regulatory dynamics of plant hormones and transcription factors under salt stress. *Biology* **2024**, *13*, 673. [CrossRef]
39. Ball, K.; Sadhukhan, S. Epigenetics—The molecular tool in understanding abiotic stress response in plants. In *Biology and Biotechnology of Environment*; Apple Academic Press: Palm Bay, FL, USA, 2023.
40. Sousa, R.O.D.; Jesus, J.F.D.; da Silva, M.N.; Paula-Marinho, S.D.O.; Alcântara Neto, F.D.; Carvalho, H.H.D.; Miranda, R.D.S. Photosynthetic efficiency and water status as determinants for the performance of semiarid-adapted cotton cultivars under drought in greenhouse. *Agronomy* **2025**, *15*, 500. [CrossRef]
41. Kambona, C.M. Memory of Past Drought Stress Exposure Effects Plant Responses in Subsequent Generations in Winter Wheat (*Triticum aestivum* L.). Ph.D. Thesis, Universität Bonn, Bonn, Germany, 2023. Available online: https://hdl.handle.net/20.500.11811/11150 (accessed on 16 March 2025).
42. Zhang, H.; Mu, Y.; Zhang, H.; Yu, C. Maintenance of stem cell activity in plant development and stress responses. *Front. Plant Sci.* **2023**, *14*, 1302046. [CrossRef]
43. Miryeganeh, M. Plants' epigenetic mechanisms and abiotic stress. *Genes* **2021**, *12*, 1106. [CrossRef] [PubMed]
44. Liu, X.; Quan, W.; Bartels, D. Stress memory responses and seed priming correlate with drought tolerance in plants: An overview. *Planta* **2022**, *255*, 45. [CrossRef]
45. Wang, G.L.; Que, F.; Xu, Z.S.; Wang, F.; Xiong, A.S. Exogenous gibberellin altered morphology, anatomic and transcriptional regulatory networks of hormones in carrot root and shoot. *BMC Plant Biol.* **2015**, *15*, 290. [CrossRef] [PubMed]
46. Ahmad, M.; Waraich, E.A.; Skalicky, M.; Hussain, S.; Zulfiqar, U.; Anjum, M.Z.; El Sabagh, A. Adaptation strategies to improve the resistance of oilseed crops to heat stress under a changing climate: An overview. *Front. Plant Sci.* **2021**, *12*, 767150. [CrossRef] [PubMed]
47. Yang, L.; Han, Y.; Li, P.; Li, F.; Ali, S.; Hou, M. Silicon amendment is involved in the induction of plant defense responses to a phloem feeder. *Sci. Rep.* **2017**, *7*, 4232. [CrossRef]
48. Zhao, J.H.; Guo, H.S. RNA silencing: From discovery and elucidation to application and perspectives. *J. Integr. Plant Biol.* **2022**, *64*, 476–498. [CrossRef]
49. Ghosh, M. Climate-smart agriculture, productivity and food security in India. *J. Dev. Policy Pract.* **2019**, *4*, 166–187. [CrossRef]
50. Puy, J.; Carmona, C.P.; Hiiesalu, I.; Öpik, M.; de Bello, F.; Moora, M. Mycorrhizal symbiosis alleviates plant water deficit within and across generations via phenotypic plasticity. *J. Ecol.* **2022**, *110*, 262–276. [CrossRef]

MDPI AG
Grosspeteranlage 5
4052 Basel
Switzerland
Tel.: +41 61 683 77 34

Current Issues in Molecular Biology Editorial Office
E-mail: cimb@mdpi.com
www.mdpi.com/journal/cimb

www.ingramcontent.com/pod-product-compliance
Lightning Source LLC
Chambersburg PA
CBHW061619210926
51850CB00009B/2373
9783725875658